Process Intensification

I0047325

Also of interest

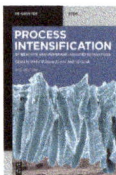

Process Intensification.
by Reactive and Membrane-Assisted Separations
Skiborowski, Górak (Eds.), 2022
ISBN 978-3-11-072045-7, e-ISBN (PDF) 978-3-11-072046-4

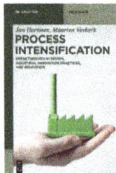

Process Intensification.
Breakthrough in Design, Industrial Innovation
Practices, and Education
Harmsen, Verkerk, 2020
ISBN 978-3-11-065734-0, e-ISBN (PDF) 978-3-11065735-7

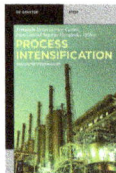

Process Intensification.
Design Methodologies
Gómez-Castro, Segovia-Hernández (Eds.), 2019
ISBN 978-3-11-059607-6, e-ISBN (PDF) 978-3-11-059612-0

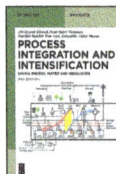

Sustainable Process Integration and Intensification.
Saving Energy, Water and Resources
Klemeš, Varbanov, Wan Alwi, Manan 2018
ISBN 978-3-11-053535-8, e-ISBN (PDF) 978-3-11-053536-5

Process Intensification

by Rotating Packed Beds

Edited by
Mirko Skiborowski, Andrzej Górak

2nd Edition

DE GRUYTER

Editors
Prof. Dr. Mirko Skiborowski
Hamburg University of Technology
Institute of Process Systems Engineering
Am Schwarzenberg-Campus 4 C
21073 Hamburg
Germany
mirko.skiborowski@tuhh.de

Prof. Dr. Andrzej Górak
Łódź University of Technology
Faculty of Process and Environmental Engineering
ul. Wólczańska 175
90-924 Łódź
Poland
andrzej.gorak@p.lodz.pl

ISBN 978-3-11-072490-5
e-ISBN (PDF) 978-3-11-072499-8
e-ISBN (EPUB) 978-3-11-072509-4

Library of Congress Control Number: 2022937420

Bibliographic information published by the Deutsche Nationalbibliothek
The Deutsche Nationalbibliothek lists this publication in the Deutsche Nationalbibliografie;
detailed bibliographic data are available on the Internet at http://dnb.dnb.de.

© 2022 Walter de Gruyter GmbH, Berlin/Boston
Cover image: Foto eines geöffneten Rotors eines RPBs aus dem Technikum / Prof. Andrzej Górak
Typesetting: Integra Software Services Pvt. Ltd.
Printing and binding: CPI books GmbH, Leck.

www.degruyter.com

Preface

Chemical products like basic, commodity, and specialty chemicals, as well as pharmaceuticals, life science, and consumer products, are all of essential importance for every day's life in modern society. However, as one of the most energy-demanding industrial sectors, the chemical processing industries are facing an increasing need to improve energy efficiency and sustainability. The latter also demands a transition of raw materials from those largely based on fossil resources to one that builds on recycling strategies in a circular economy and the use of renewable resources. Yet, production processes still have to be reliable, efficient, and economically viable in rapidly changing markets with global competition. Generally speaking, the necessary development requires an increase in efficiency of existing processes, based on an optimal selection of raw materials, solvents, and energy improvements, as well as innovative technologies and totally new processing pathways.

One important toolset to successfully face these challenges in the chemical process industry is Process Intensification. The first book of our two-book series on Process Intensification, entitled *Reactive and Membrane-Assisted Separations*, provides an introduction and overview over the various technologies, focusing on hybrid and reactive separation processes. This second book of the series provides an elaborate introduction and overview of centrifugally enhanced vapor/gas-liquid separations, focusing on rotating packed beds as innovative technology for (reactive) separations by means of a mass-transfer machine. While the general concept of centrifugally enhanced separations has been proposed almost a century ago and the development of rotating packed beds was first promoted as prime example of the newly termed branch of Process Intensification in the late 1970s, current developments are primarily pursued in the Eastern hemisphere. Depending on the specific applications, this technology has the potential for a significant enhancement of mass transfer and consecutive improvements in reaction rates and selectivities, enabling size reductions up to an order of magnitude.

Beginning with a general introduction on centrifugally enhanced reactive separations, the individual application of rotating packed beds in liquid-liquid, gas-liquid, and vapor-liquid contacting are introduced, covering both academic research and industrial applications. Besides general guidelines for the design and application of rotating packed beds for the different applications, individual chapters are devoted to additive manufacturing of improved packings and multistage configurations. Further focus is placed on advanced modeling methods, covering especially rate-based modeling of multicomponent distillation in rotating packed beds and the simulation of multiphase flow based on computational fluid dynamics.

https://doi.org/10.1515/9783110724998-202

This book is dedicated to our colleague and friend Konrad Gladyszewski, who was an essential part of our team at the Laboratory of Fluid Separations and contributed significantly to the developments and the current state of the research at TU Dortmund and Łódź University of Technology. You and your work have been an important part of the life and the work of all of the authors involved in this book, and you will be missed. Rest in peace.

Contents

List of contributing authors

Daniel Sudhoff
Evonik Operations GmbH
45772 Marl, Germany
Chapter 1

Dennis Wenzel
TU Dortmund University
Laboratory of Fluid Separations
44227 Dortmund, Germany
Chapters 2 and 4

Kai Groß
TU Dortmund University
Laboratory of Fluid Separations
44227 Dortmund, Germany
Chapter 3

Kolja Neumann
TU Dortmund University
Laboratory of Fluid Separations
44227 Dortmund
Chapter 4

Tobias Pyka
TU Dortmund University
Laboratory of Fluid Separations
44227 Dortmund, Germany
Chapter 5

Hina Qammar
TU Dortmund University
Laboratory of Fluid Separations
44227 Dortmund, Germany
Chapter 6

Rouven Loll
TU Dortmund University
Laboratory of Fluid Separations
44227 Dortmund, Germany
Chapter 7

Jens-Uwe Repke
Technische Universität
Process Dynamics and Operations Group
10623 Berlin, Germany
Chapter 8

Markus Illner
Technische Universität
Process Dynamics and Operations Group
10623 Berlin, Germany
Chapter 8

Matthias Hilpert
Technische Universität
Process Dynamics and Operations Group
10623 Berlin, Germany
Chapter 8

Dawid Zawadzki
Łódź University of Technology
Faculty of Process and Environmental
Engineering
90-924 Łódź, Poland
Chapter 9

Maciej Jaskulski
Łódź University of Technology
Faculty of Process and Environmental
Engineering
90-924 Łódź, Poland
Chapter 9

Michał Blatkiewicz
Łódź University of Technology
Faculty of Process and Environmental
Engineering
90-924 Łódź, Poland
Chapter 9

Jörg Koop
TU Dortmund University
Laboratory of Plant and Process Design
44227 Dortmund, Germany
Chapters 5 and 7

https://doi.org/10.1515/9783110724998-204

Daniel Sudhoff

1 Introduction to centrifugally enhanced separations

1.1 Fundamentals

The application of centrifugal forces to separation processes is often referred to as HiGee-technology ("high gravity").This technology employs centrifugal acceleration that exceeds the gravitational acceleration by several magnitudes. HiGee-technology is commonly used for intensified heat and mass transfer between the involved phases and for processing chemicals with special properties (e.g., highly viscous liquids) or designing new equipment (e.g., highly efficient packing). Within this technology, different concepts and devices have been developed such as rotating discs, centrifugal extractors, and rotating packed beds (RPBs). Because the latter are the most important for vapor/gas-liquid separation processes such as distillation, these devices are presented and analyzed in this and the following chapters.

The chapter begins with a historical background on centrifugal processing and the separation principles of RPBs such as the fundamental phenomena and rotor types. Afterwards, a brief overview of the applications of RPBs and its use for reaction, absorption, and distillation is given, and potential future applications are identified based on an analysis of the strength of RPBs. Furthermore, a short overview of the modeling challenges and design rules is presented, which is followed by the three examples of RPB applications for distillation. These examples utilize the special qualities of RPBs for the production of hypochlorous acid for flexible and mobile container systems and high-pressure distillation.

In the following chapter, these topics will be discussed in more detail. A more detailed overview of the different applications of RPBs will be provided in Chapter 2 for liquid-liquid and liquid-solid separations, in Chapters 3 and 4 for deaeration and flue gas cleaning, and in Chapter 6 for distillation. Modeling of separations in RPBs is furthermore addressed in Chapter 8, focusing on rate-based modeling of multicomponent distillation, and Chapter 9 for detailed computational fluid dynamic simulations. The specific design of the equipment and the internals is addressed in Chapter 5, considering different rotor configurations, and Chapter 7, which focusses on different designs of rotors and 3D-printed internals.

Daniel Sudhoff, Evonik Operations GmbH, 45772, Marl, Germany

https://doi.org/10.1515/9783110724998-001

1.1.1 Historical background

Centrifugal acceleration has been used for thousands of years for different purposes, e.g., drying or cleaning dirty water [1]. One of the first reported industrial centrifuges for drying clothes is dated to 1840. The German factory "Maschinenfabrik C. G. Haubold Jr." in Chemnitz reported the drying of nine dozen pairs of socks in less than three minutes using their prototype centrifuge [2]. However, the first reported attempt to industrialize centrifugation for separation processes was performed by Antonin Prandtl [1] in 1864. He invented a milk separator to remove silk from milk quicker than via conventional phase separation in a milk drum. Further improvements in milk separators were made by the development of the well-known "De Lavals separator" [3] (Fig. 1.1) – the first continuously working separator – and "Alfaseparator" [4] (Fig. 1.1) – the first separator with installed blades to increase efficiency.

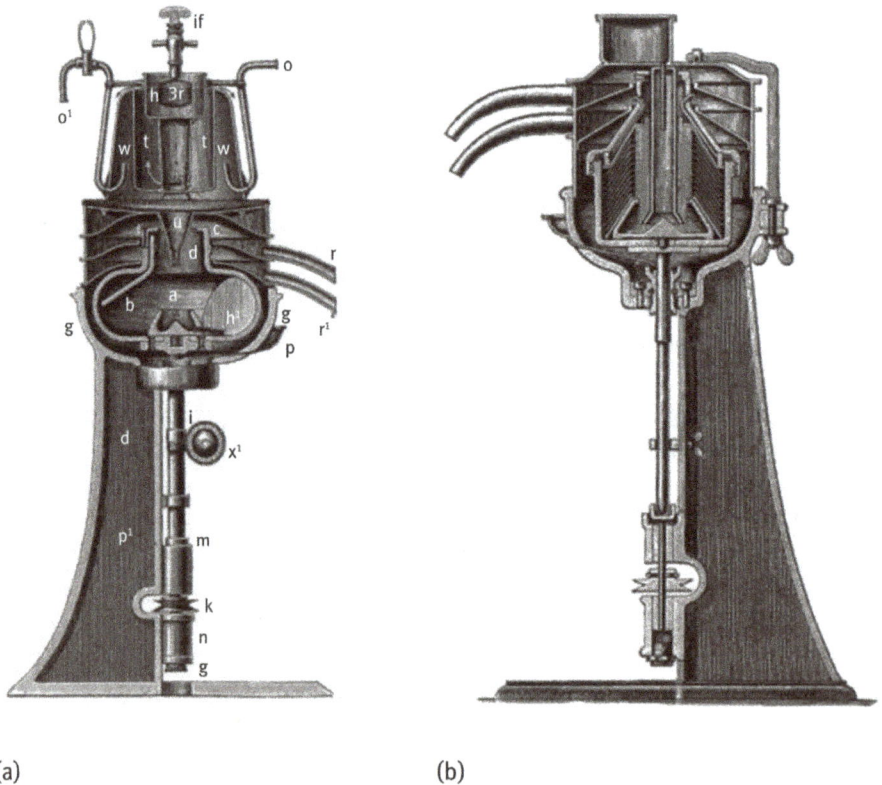

(a)   (b)

Fig. 1.1: De Laval separator (left) and Alfaseparator (right) [4].

The technique to separate two liquids with different specific weights by applying centrifugal force, as in the milk separator, was later applied to liquid-liquid extraction processes [5]. Here, in addition to the separation process, intense mixing of the two liquids had to be enabled. Within this field, an important invention was the continuous centrifugal extractor by Charles Coutor [5], which acted as a multistage mixer settler unit. This machine consisted of a centrifugally rotated drum having several plates that separate sections of the drum with interconnecting pipes. Using the centrifugal field for intense mixing of the phases and more rapid separation, a more efficient and less costly apparatus for extraction processes was patented.

A subsequent development was introduced by George Thayer [6]. Thayer used perforated sheets arranged concentrically. The liquids were forced to flow radially inwards and outwards. In doing so, he realized an extraction column with trays in a centrifugal field. This design was utilized in the so-called "Rotating Zig-Zag Beds" developed for distillation processes [7], which is discussed in more detail in the subsequent sections and which is also discussed in more detail in Chapter 8.

An early application of the centrifugal field for separation processes utilizing liquid and vapor is the washing of gases and vapors. In 1931, Max Aurig [8] patented an apparatus that had corrugated sheets with a cylindrical shape. These sheets rotated while the gas or vapor passed between them. A washing liquid was used to remove particles, dust, or dirt from the gas or vapor stream. The distance between the sheets was decreased compared to conventional washing apparatus by using this new approach, and the liquid with impurities could easily be separated, making the apparatus more effective.

The first attempt to perform mass transfer between a vapor and a liquid in a centrifugal field was conducted by Walter J. Podbielniak [9]. Podbielniak patented an apparatus, the Podbielniak Contactor, that enabled the contact of two fluid phases (liquid-liquid or vapor-liquid) within a small column in a countercurrent manner by exploiting the centrifugal acceleration. In the proposed design, a small column or a pipe was spirally wound around a cone to form a coil. Due to the constantly increasing diameter of the coils, the liquid was centrifugally accelerated. The vapor or lighter liquid flowed in a countercurrent manner within the same pipe, driven by an applied pressure drop. Podbielniak proposed the use of his apparatus for the distillation of very narrow boiling mixtures or very difficult fractionation processes. The advantage of this device was the controlled and short contact time with an intense mass transfer [9].

Initial detailed reports regarding the phase contact between vapor and liquid in a radial countercurrent flow pattern were published by Ramshaw & Mallinson [10]. This concept will be described in detail in the following sections.

1.1.2 Separations principles

For the utilization of centrifugal fields in reaction and separation processes, different types of devices have been developed such as centrifugal extractors [5], spinning disc reactors [11], and RPBs. The different applications of RPBs are discussed in a later section. The concepts and principles of RPBs, the phenomena occurring within the vapor-liquid contacts in RPBs, and the different rotor types are presented in the following section.

Numerous investigations on RPBs have been recently published, primarily in East Asia. Several good reviews focusing on fundamental phenomena, modeling, design alternatives, and current fields of investigation can be found in the literature [12–19].

Concepts

RPBs primarily consist of a rotating packing and a stationary casing (Fig. 1.2). The packing, which can be made from various materials such as stainless steel wire mesh, metal foams, or corrugated sheets [16], is usually shaped as a hollow cylinder and mounted onto a shaft that is rotated at high speed. In the case of vapor-liquid contacting, the liquid is sprayed onto the inner surface of the packing at the center of the rotor. The liquid enters the packing, is accelerated by the centrifugal forces, and flows radially outwards. The liquid exits the packing at the outer end and produces a fog or spray that is collected at the walls of the casing and withdrawn from the device at the bottom. The vapor or gas is introduced into the system at the outer wall of the casing, usually tangential to the packing to create turbulence in the vapor phase. Driven by a pressure difference, the vapor flows radially inwards through the packing in a countercurrent flow pattern to the liquid. The vapor leaves the packing at the center and is withdrawn from the device. To avoid bypassing of the vapor, a seal is necessary between the outer and inner parts of the casing (Fig. 1.2). In addition to these countercurrent flow patterns, co- and crosscurrent flows of liquid-liquid or gas-liquid combinations can be realized in RPBs [13, 17]. These combinations can be advantageous for mixing or reaction processes. For distillation and absorption processes, countercurrent RPBs are employed.

The geometry and accelerating forces inside an RPB differ from conventional columns. However, both devices are comparable; an analogy can be made between them. In Fig. 1.3, a conventional distillation column with feed, a rectifying section, and a stripping section is sketched. To realize this setup using RPBs, a feed has to be introduced into the rotating packing. Because it is difficult to inject liquid at a position along the radial axis, two rotors are typically employed. These rotors can either be installed in separate RPBs [20] or within a single RPB along the same axis [21]. The first option is illustrated in Fig. 1.3, in which the upper RPB serves as the rectifying

Vapour

Packing

Seal

Liquid

Vapour

Liquid

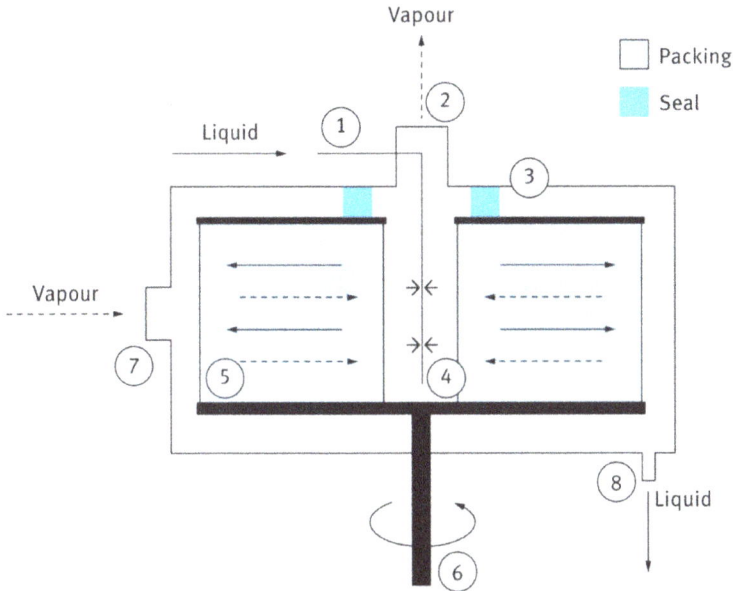

Fig. 1.2: Sketch of a rotating packed bed for high-gravity distillation with a solid packing. (1) liquid inlet; (2) vapor outlet; (3) seals; (4) nozzles; (5) packing; (6) rotating shaft; (7) vapor inlet, and (8) liquid outlet.

section, whereas the lower RPB represents the stripping section. The overall flow direction is vertical in conventional columns and horizontal in RPBs although the axis of the rotors can also be oriented horizontally [22]. Therefore, the height of a packing section in a column corresponds to the radial length of a rotating packing in an RPB and the diameter of the column to the axial length of the rotating packing. For simplification, the radial length of the packing in the RPB determines the separation efficiency, whereas the axial length determines the capacity.

Principles

The fundamental principles of the rotating packing are still not fully understood because noninvasive investigations of the flow pattern within the packing or measurements along the radial direction are very difficult and rarely reported [23]. **The basic processes inside the rotating packing are similar to those inside packed columns. The liquid is distributed over the packing, the vapor passes in a countercurrent flow pattern, and mass transfer occurs at the phases' interface.** The liquid is distributed by spraying it onto the surface of the inner edge of the packing and is accelerated along an inlet region of the packing [24]. Inside the packing, the liquid either flows as a thin film, as droplets, or as rivulets [25]. The

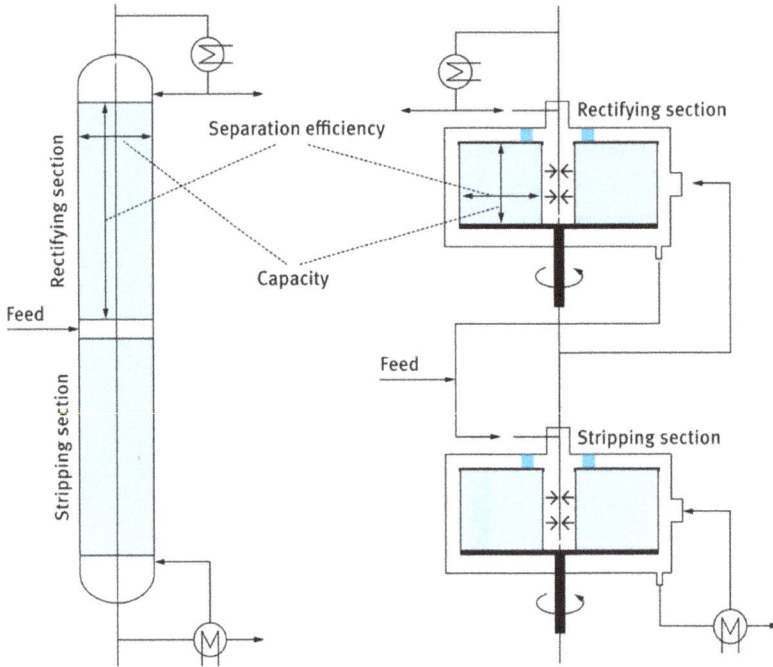

Fig. 1.3: Simplified analogy between a conventional distillation column and two RPBs for a distillation process with a single feed stream. Detailed elaboration can be found in literature [15].

size of the droplets or the thickness of the films is dependent on the type of packing and the rotational speed. However, a definite prediction is not yet possible. The rotating packing that acts like a fan tangentially accelerates the vapor. The turbulence in the vapor phase is not very strong and is often referred to as exhibiting "solid-body behavior" [26]. Packing development, which will be presented later, often aims to increase the turbulence in the vapor phase.

Compared to conventional columns, the mass transfer is intensified. This intensification occurs for two main reasons. First, the liquid-side mass transfer coefficients increase due to the high turbulence present in the liquid phase. Second, the vapor-side mass transfer coefficients do not significantly increase due to the above-mentioned reasons [27, 28]. The liquid distribution, which has a very high surface area for a very small volume, is more important. Thus, very high volumetric mass transfer rates are achieved. The high surface area is produced for two reasons. The first reason is that the liquid film thicknesses or the droplet sizes are very small due to the very high centrifugal forces. Second, due to the high forces, packing types with a very high surface area can be employed (up to 3300 $m^2\ m^{-3}$ [29]), or packing can be manufactured to realize very thin droplets [26]. Thus, the volumetric mass transfer rates can be substantially increased.

The pressure drop over the packing in RPBs is higher than that in conventional distillation columns due to the rotating packing, which acts like a fan [30]. The high centrifugal forces and acceleration of the liquid lead to a very small residence time of the liquid inside the packing (less than 1.8 s [30]) and reduce the liquid holdup (less than 0.3 m^3 m^{-3} [31]). Flooding phenomena in RPBs are not fully understood and differ slightly from those in conventional columns. **The maximum vapor velocities occur at the center of the packing due to its geometry (see below); flooding first occurs at this position. Some additional flooding phenomena, such as separated flow, in which liquid flows outwards at the bottom of the rotor and vapor flows inwards at the top of the rotor, exist in RPBs** [32]. However, the maximum liquid and vapor loads are higher than those in conventional columns, permitting very high capacities for a small packing volume [32, 33]. A comprehensive overview of studies and correlations describing mass transfer and hydrodynamics can be found in the literature [34].

Rotor types

Similar to conventional columns, various types of packing are possible in RPBs. However, the standard packing used in columns cannot be directly used in RPBs due to the radial geometry and higher liquid and vapor loads. The simplest type of packing is a stainless steel wire mesh wound around the center of the rotor [20]. Open foam or sponge packing from metals or ceramics is also commonly used [35]. These packing types are called solid packing. Modern technologies such as 3D-printing offer new possibilities to even enhance geometry and mass transfer in RPBs and will be further elaborated in Chapter 7.

As discussed above, the turbulence in the vapor phase is not very high; therefore, the vapor-side mass transfer coefficients are similar to those in columns. To increase the turbulence in the vapor phase, the concept of a split packing was invented [36]. As shown in Fig. 1.4, rings of packing typically made of metal foam are concentrically and alternately mounted onto two rotors. These two rotors can be operated in a co- or counter-rotating manner. By counter-rotating the split-packing rotors, the direction of the tangential portion of the vapor velocity is inverted at each packing ring, which increases the turbulence in the vapor phase. This type of packing is thought to be well suited for distillation systems, in which the mass transfer resistance lies primarily in the vapor phase [36]. Another important rotor type is the so-called "zig-zag bed" (Fig. 1.4), which can be compared to tray columns [37]. Concentric metal rings are alternately installed on a rotating plate (rotor) and onto the casing (stator). The rotating rings are perforated at the end. The liquid is collected at each rotating ring and flows through the perforated part. Consequently, small droplets (or fog) are created in the spray area. Vapor flows across this area, and mass transfer occurs. These rotating zig-zag beds have a lower pressure drop although they also have a lower separation efficiency. Different types of the latter concept have been developed in recent years [17].

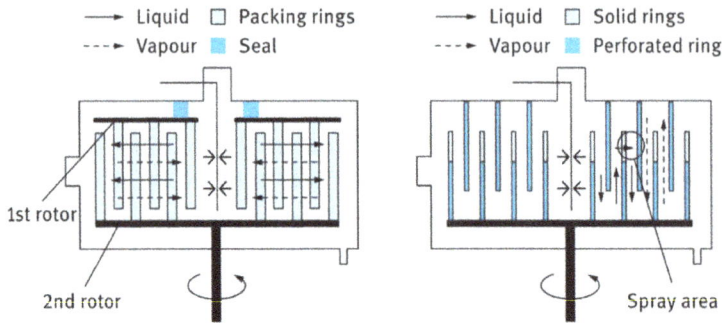

Fig. 1.4: Sketch of a rotating packed bed with split packing (left) and a rotating zig-zag bed (right).

The development of packing and rotor types remains in progress. It can be assumed that packing development along with an increased understanding of the fundamentals inside these packing devices can increase the separation efficiency in RPBs and will provide different packing options as a solution for different separation problems. A detailed discussion about multirotor operation and rotor configuration can be found in Chapter 5.

1.2 Applications

Centrifugal forces have been used for industrial separation processes, such as liquid-liquid extraction, for more than three decades. The extraction of uranium for nuclear power plants and the recovery of penicillin by centrifugal extraction are two well-known examples. **Excluding centrifugation and centrifugal extraction processes, the applications of RPBs can be distinguished between reactive systems and gas-liquid or vapor-liquid separation processes.** The latter can be split into three groups: degassing, absorption or stripping, and distillation. An overview of published applications and investigations of RPBs for these unit operations is provided in the following sections and further chapters of this book, as well as the article of Neumann et al. [15], who also propose a set of heuristics for identifying potential applications.

1.2.1 Liquid phase contacting systems

Reactive systems

The most common application of RPBs as reactors is the production of nanoparticles, on which comprehensive studies and some industrial applications have been reported [38]. The center of this research is located at the Beijing University of Technology, China. Chen & Shao [12] showed that for inorganic nanoparticles, very well-defined sizes and morphologies can be synthesized due to the intense micromixing, high shear forces, and high mass transfer rates in RPBs. Examples of such particles include calcium carbonate (17–36 nm), aluminum hydroxide (1–10 nm), silicon dioxide (30–50 nm), titanium oxide (20–30 nm) [19, 38], and iron oxide [39]. It has also been reported that pharmaceutical nanoparticles can be produced that exhibit various advantageous characteristics, such as higher dissolution rates, a reduced need for complex solubilizers, and finer particle fractions [19]. The excellent micromixing properties along with the short residence time and intense heat transfer in RPBs have proven to be advantageous for very exothermic polymerization processes such as for the butyl rubber synthesis [19]. Another example is the production of polyaniline nanofibers, which can now be produced at higher yields and with more homogeneous morphologies in RPBs compared to standard reactors [40]. In addition, emulsification of liquids is positively affected by the high shear forces and high turbulence in the packing. Reported examples include the emulsification of methanol in diesel fuels [41] and during the formulation of pharmaceuticals. The latter is claimed to be industrially applied in China, thereby reducing electrical power and equipment size requirements [19].

A centrifugally enhanced trickle bed reactor is an additional example. Such a reactor is typically designed similar to RPBs and focuses on intense mixing. For mass transfer-limited reactions, these machines are advantageous. Because the mass transfer rates are intensified by approximately a factor of 40, the reaction rates are significantly increased, and reactor sizes are dramatically reduced [42–44].

Liquid phase separation

The intense micromixing and mass transfer in RPBs are of large interest for liquid phase separations [19]. For these processes, the size of the apparatus should be kept small, whereas the capacity and the mass transfer rate should be as large as possible. All three conditions can be met by RPBs [45–47].

Additionally, combination of high shear forces and short residence times of liquids in RPBs can be utilized for liquids with high viscosity and low thermal stability. Details on liquid-liquid and liquid-solid separations are discussed in Chapter 2.

1.2.2 Gas-liquid contacting systems

The important gas or vapor-liquid contacting processes in RPBs include the degassing of liquids, absorption or stripping, and distillation, which are introduced in the following section and chapters of this book.

Degassing of liquids

The high throughputs and turbulences in the liquid phase of RPBs offer good degassing opportunities, which are used for degassing water or other beverages [48, 49]. The insensitivity to movement offers the possibility to deploy RPBs on floating vessels such as offshore platforms. Thus, RPBs have been used for degassing seawater during the exploitation of oil fields [49, 50]. For further details on degassing and deaeration of liquids in RPBs, refer to Chapter 3.

Absorption and stripping

Absorption or stripping processes are the most studied processes in RPBs and are often applied to investigate hydrodynamic properties such as holdup, pressure drop, and interfacial areas.

The largest interest lies in the absorption of carbon dioxide (CO_2); many studies have been published in the literature. The benefits of RPBs are the large capacity and the intense mass transfer for a small volume. Due to the short residence time and intense mass transfer, higher flue gas temperatures are acceptable [51, 52]. Various solvents have been investigated including caustic soda or typical amines. Selected publications and details on CO_2 absorption in RPBs, the types of RPBs used, and the dimensions are discussed in Chapter 4.

Generally, results show that higher efficiencies can be achieved compared to columns. A very ambitious investigation was conducted by Zhang et al. [53] regarding the absorption of CO_2 using ionic liquids. The mass transfer rates, which are relatively low in conventional columns despite the high absorption capacity of the ionic liquid, were found to be an order of magnitude higher in the RPB. In addition to the higher efficiency during absorption, higher viscosity solvents can also be applied [54]. Chen et al. [55] showed that during the deoxygenation of viscous liquids (up to 80 mPa s), the mass transfer is only slightly reduced with increasing viscosity, and Sudhoff [56] showed that absorption is possible at viscosities of at least 480 mPa s. Further examples include the absorption of volatile organic compounds [34, 57–59], the selective absorption of hydrogen sulfide [60], and absorption using highly corrosive solvents [61–63].

A prominent industrial application of RPBs is the reactive stripping operation for the production of hypochlorous acid [64]. This example will be discussed in more detail in the "Detailed examples" section of this chapter.

Distillation

In 1983, Ramshaw [20, 65] reported on the 'HiGee' distillation concept, which is also known as the first process intensification concept, and also on investigations performed at (Imperial Chemical Industries). They performed intensive pilot- and full-scale studies using separate RPBs for the stripping and rectifying section. Rotors with outer diameters of up to 80 cm and axial lengths of up to 30 cm comprising a wire mesh packing were applied. During their test runs, HETP (height equivalent to theoretical plate) values of 1 to 2 cm at a mean centrifugal acceleration of 1000 times that of gravity were achieved [66]. Although intensive studies have been performed, the detailed results are not accessible in the open literature, and the project has not been further pursued.

More than 10 years later, Kelleher & Fair [35] investigated distillation in RPBs. Their slightly smaller RPB consisted of a single rotor with metal sponge packing although they continued to use a large axial length of 15 cm to allow for large capacities (maximum of 2.5 kg s^{-1}). The RPB was manufactured by Glitsch Inc. and was operated at total reflux. The first experimental results in terms of the number of transfer units (NTU) were presented in the literature. Different experimental and theoretical studies, including modeling approaches, followed in the last decade, utilizing different types of rotors, packing, and chemical systems. Details on advances in and detail about distillation in RPBs are discussed in Chapters 5 and 6.

In addition to solid packing rotor types, some other improvements have been investigated. To improve the vapor-side mass transfer, Chandra et al. [67] used concentric packing rings [split packing] that can be co- or counter-rotating and are meant to improve the turbulence in the vapor phase and the vapor-side mass transfer. As already presented above, another important development has been the tray-like rotor design by Wang et al. [37] which is called the rotating zig-zag bed. This concept has been modified to include multiple rotors in one shaft and a quasi-crossflow design for reducing the pressure drop and electrical power demand [68, 69]. In addition, a combination of this zig-zag design with solid packing has also been investigated [70]. Reactive distillation examples have been analyzed using simple modeling techniques [71]. A more extensive overview on rotor designs and applications is provided in the review article of Cortes Garcia et al. [13].

1.2.3 Potential future applications

Despite the studies on reaction, absorption, and distillation processes in RPBs, outside Asia, only few industrial applications are known, which are summarized in the review article of Neumann et al. [15]. There are several reasons why RPBs are less commonly applied in industry and why conventional columns have been preferred even if their use is more expensive:
– RPBs comprise rotational parts that may require additional maintenance.
– The fundamentals of RPBs are not fully understood; thus, predicting the hydrodynamics or mass transfer is difficult.
– The design methods for RPBs are few, and the accuracy is below those of conventional columns.
– Experience from the long-term usage of RPBs is lacking.

Due to the above-mentioned reasons, RPBs will always compete with conventional columns for standard applications. Further fundamental research must be conducted. **However, the exceptional characteristics of RPBs, such as their additional degree of freedom, rotational speed, and compact design, can already be utilized for special separation tasks. These characteristics offer the opportunity to enlarge the operating window of conventional columns** (see Fig. 1.5). Beginning with the conventional operating window for distillation, several advantages and exceptional characteristics of RPBs are listed (dark gray ring). These advantages lead to several applications (light gray ring), for which conventional distillation columns can either not be applied or may require an enormous amount of effort and cost. These fields of application are briefly discussed in the following section.

Conventional distillation columns are very sensitive to movement. If the column is moved or slightly leaning, both maldistribution of the liquid will occur, and the separation efficiency will decrease. Due to the very strong centrifugal forces in the radial direction, which are more than two orders of magnitude higher than the gravitational forces, movement does not have a strong impact on the performance of RPBs. **In addition to their very compact design, RPBs are good alternatives for mobile applications, such as on floating vessels.** A good example is the floating methanol process on offshore vessels [49, 72].

By adjusting the rotational speed of an RPB, the separation efficiency can be rapidly changed during operation, which is why the operating window for RPBs during operation can be much wider than that for distillation columns. This difference is advantageous for processes in which the feed composition fluctuates [34] or the product purity must be adjusted. This advantage can also be exploited for batch or multipurpose plants.

RPBs can also be used for retrofitting existing distillation plants. Due to their very compact design and the possibility of altering the separation efficiency by adjusting the rotational speed, RPBs can be used as a universal modular distillation

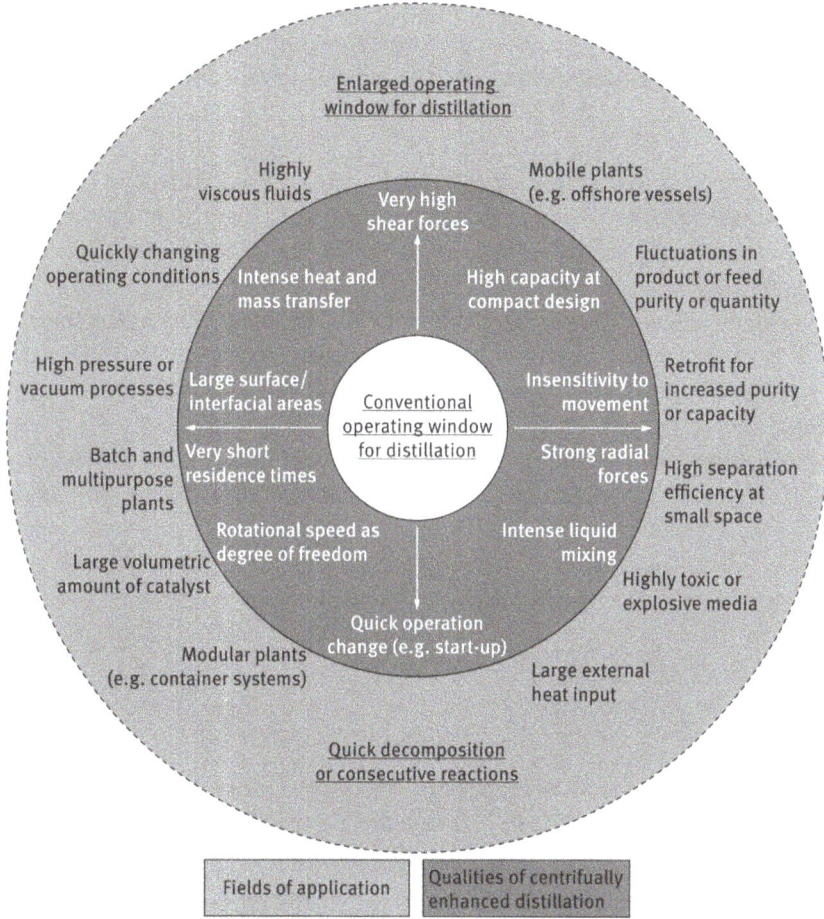

Fig. 1.5: The various qualities (white text on dark gray background) of centrifugally enhanced separation processes to enlarge the conventional operating window for distillation in various applications (black text on light gray background).

device that can be quickly and economically (in terms of space requirements) connected to existing equipment, which also applies for distillation tasks for which a very small space is available.

The liquid in RPBs has a very short residence time. Therefore, toxic, explosive, or thermally unstable liquids can be processed in RPBs more quickly than in conventional columns. Additionally, the compact design reduces the holdup of these substances in the device and minimizes the risk of explosion, decomposition, or exposure [64].

Some reaction systems applied in reactive stripping or reactive distillation have consecutive or decomposition reactions that reduce the selectivity of the product mixture [64]. The short residence time and rapid mass transfer between the phases inside the rotating packing offer the possibility for significantly increasing the selectivity by removing the product from the reactive phase to the nonreactive phase.

The extremely compact design of RPBs also results in a very efficient (small) ratio between the surface area and volume of these devices. If a large heat input is necessary or good insulation or very thick walls are required, RPBs can be more economical than columns. For high-pressure distillation, the maximum pressure that is still economically sensible can be shifted to higher values, making new processes possible.

RPBs can be either horizontally or vertically arranged. They also have a very high capacity with a small footprint. The characteristics make RPBs suitable for modular plants, which may be constructed from different plant segments, such as containers.

The high shear forces and strong radial forces that exist inside the rotating packing of an RPB enable the usage of a larger amount of catalyst, which may be required for heterogeneously catalyzed reactive distillation, and higher viscosities of the processed liquids compared to distillation columns [55]. This difference may enable the use of entirely new educts such as vegetable oils or new process routes.

1.3 Modeling and design

The modeling approaches used to examine the hydrodynamics and separation characteristics in conventional packed or tray columns for distillation or absorption are at a very advanced level; very accurate predictions are possible [73]. The level of development of models for centrifugally enhanced separation processes is much lower. Only a few comprehensive studies regarding the hydrodynamics and mass transfer in RPBs for both the distillation and absorption processes have been published; systematic programs for generating databases, which have been conducted for decades for columns, are still under development.

Although the elementary processes in conventional columns and RPBs are similar, a direct transfer of approaches for columns to centrifugally enhanced processes is not possible. The two main differences between conventional columns and RPBs are the very high centrifugal forces and radial geometry of RPBs, which have several consequences that must be addressed when modeling centrifugally enhanced separation processes. Some considerations that must be taken into account and which are further elaborated in Chapter 3 are as follows:

- The centrifugal acceleration may be up to four orders of magnitude higher than earth's gravity.
- The centrifugal acceleration is not constant (increasing radially).

- The nominal cross-sectional areas for the liquid and gas flows are not constant (increasing radially). The superficial liquid and vapor velocities decrease radially.
- The liquid phase is not necessarily uniformly distributed as a liquid film over the packing; the liquid may also flow as rivulets or droplets through the packing. The form of the liquid distribution not only depends on the packing characteristics and the rotational speed but also may change in the radial direction.
- Flooding of the packing is initiated at the center of the packing, which has the highest liquid and vapor loads.
- The packing tangentially accelerates both the vapor and the liquid phases; therefore, their velocities have radial and tangential components.

If an RPB does not consist of a uniform packing (solid) and instead has counter rotating rings or packing characteristics that are a function of radius, further considerations for the hydrodynamics and mass transfer must be made.

Despite these differences between conventional packed columns and RPBs, the basic approaches for modeling the hydrodynamics and mass transfer of centrifugally enhanced processes are adopted and adjusted from conventional columns, such as in the case of flooding (e.g., Sherwood correlations [74]) or mass transfer coefficients (e.g., Onda correlations [75]). The mass transfer is typically calculated using correlations for the overall volumetric mass transfer coefficients (K_Ga and K_La) based on dimensionless groups and with regression coefficients that are fitted to experimental data. A good overview of the corresponding modeling approaches can be found in the literature [16, 18, 34] and are further described in more detail for multicomponent distillation in Chapter 8. In addition to the hydrodynamics and mass transfer, correlations for predicting the electrical power of the motor [32], the size of the RPB, and the investment and operating costs [18] have also been developed.

1.3.1 Mass transfer evaluation

For evaluating the mass transfer performance of RPBs, concepts similar to those of conventional columns have been applied. Analogously, the "radial distance equivalent to a theoretical plate" (which is referred to as HETP$_{rad}$) is often used for the "height equivalent to a theoretical plate" (HETP) [76]. Due to the varying loads, rotational speeds, and centrifugal forces along a rotor, this value is only of limited use. The HETP$_{rad}$ values are often very small (few centimeters) and are not comparable among different RPB types, rotor sizes, and rotational speeds.

Furthermore, the HTU-NTU (number of transfer units) concept has also been applied to RPBs [77]. TheNTU is the same as for conventional columns although the height of a transfer unit (HTU) refers to the radial distance of the unit. This value does not account for the radial geometry of RPBs. This concept has been further developed to the ATU-NTU concept for RPBs [32]. The height of a transfer unit is replaced by the area of a transfer unit (ATU), which represents a certain ground area of the RPB. This concept can be used for comparing the separation efficiency of RPBs with different packing sizes [35]. However, this concept does not account for the intensity of the centrifugal field.

Details on modeling mass transfer in RPS, especially for multicomponent distillation, are discussed in Chapter 8, while further details on the simulation of multiphase flow are presented in Chapter 9.

1.3.2 Rotor design

A conventional column can be interpreted as a stack of similar packing elements that each contain similar liquid and vapor loads and are within earth's gravitational field. On the contrary, in a rotating packing, the geometry is radial. Consequently, the cross-sectional areas are changing along the radial axis; therefore, the vapor and liquid loads are changing. Additionally, the centrifugal acceleration is not constant and instead increases radially. This complex interplay is illustrated for a normalized radius in Fig. 1.6. High centrifugal accelerations and high liquid and vapor loads are preferable to achieve good separation efficiency. Because these characteristics are opposite, the rotor must be designed to find an optimum functionality.

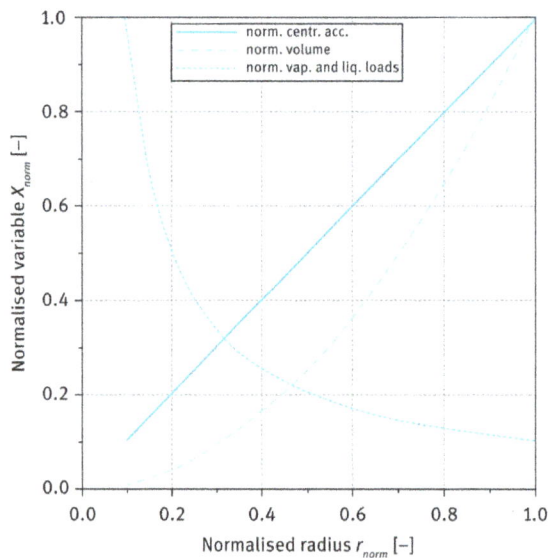

Fig. 1.6: The centrifugal acceleration, the volume of the packing, and the vapor and liquid loads as a function of the radial length of a packing (all values are normalized).

Figure 1.6 also shows that the highest liquid and vapor loads occur at the center of the packing. Therefore, the vapor velocity is the highest at this location. Additionally, the centrifugal acceleration is the smallest at this location, which is why flooding of the rotor also occurs here first. This issue must be addressed during the design process.

Therefore, the main design variables are the inner and outer radius and the axial length of the packing for each rotor. Although the performance of RPBs is dependent on the complex interplay of all variables, the following guidelines for designing a rotor can be made:

- The inner radius of the packing must be sufficiently large (i) to allow enough space for the shaft and the nozzles and (ii) to provide a large cross-sectional area (at a given axial length) at the center of the rotor for high capacities. Moreover, the inner radius must be sufficiently small (iii) to provide a large packing volume and a large interfacial area (at a given outer radius), and (iv) to enable high vapor and liquid loads inside the packing.
- The axial length of the packing must be sufficiently large (i) to provide a large cross-sectional area for high capacities at the center of the rotor and sufficiently small (ii) to realize high vapor and liquid loads along the total radial length, which is necessary for achieving a high separation efficiency.
- The outer radius must be sufficiently large (i) to provide a large packing volume and a large interfacial area and sufficiently small (ii) to minimize the high centrifugal forces on the rotor, shaft, and drives.

In addition to these basic design variables, the rotational speed is an additional operating variable that is unique to RPBs. This variable can be quickly changed during operation to modify the centrifugal acceleration. High centrifugal acceleration leads to intense turbulence and mixing and a large mass transfer. However, high centrifugal forces accelerate the liquid and reduce the residence time inside the packing, which reduces the available time for mass transfer. These contradictory effects lead to a maximum in the separation efficiency (e.g., in terms of the number of theoretical stages) at a specific rotational speed (e.g., at 12.5 s^{-1} [21]), as illustrated in Fig. 1.7. Moreover, the separation efficiency can be adjusted during operation by changing the rotational speed within a given operating window, e.g., in response to changing product requirements or for batch processes. Additionally, high rotational speeds lead to high liquid velocities. Consequently, higher vapor velocities are possible at the center of the packing before flooding is initiated; larger capacities are also possible. Therefore, the rotational speed must be chosen according to the following considerations:

- The rotational speed must be sufficiently high to promote (i) high separation efficiencies and (ii) capacities and sufficiently low (iii) to achieve a sufficient residence time and (iv) to minimize the mechanical forces on the device.

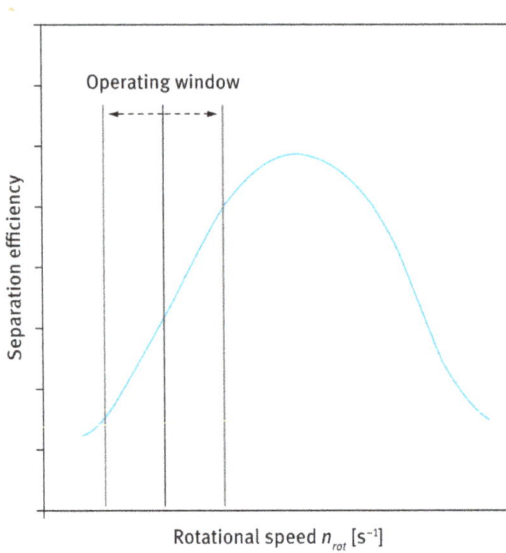

Fig. 1.7: Effect of the rotational speed on the separation efficiency.

Typical values for the inner radius, outer radius, axial length, and rotational speed are 5 to 20 cm, 15 to 80 cm, 2 to 80 cm, and 5 to 50 s^{-1}, respectively. At the outer radius of the packing, the centrifugal acceleration can be 1000 times that of earth's gravity [26].

1.3.3 Design method for RPBs

Modeling the hydrodynamics and mass transfer and predicting the interplay of all relevant design and operating variables of RPBs are very complex tasks. Consequently, very few complete models and comprehensive design methods for the conceptual design of RPBs have been published [18, 78]. The examples of RPBs discussed in the following section include those with a design method for RPBs comprising a detailed model based on generalized correlations taken from the RPB literature and that have been validated. A detailed description of the specific design method and model can be found in Sudhoff et al. [18]. Some particular features of this method are as follows:
- The model discretizes the packing in the radial direction. The packing is divided into discrete rings having the same volume (equiareal discretization). Hence, the radial length of the discrete rings decreases with increasing diameter, which enables a discretized calculation based on volumetric correlations.
- For calculating the hydrodynamics and mass transfer of each discrete ring, generalized correlations from the literature or newly developed correlations suitable for the discretized concept are applied. The experimental data used for the

regression of the parameters and for the validation of the model are obtained from different investigations using different RPBs to model a generalized RPB.

– **The centrifugal acceleration, which increases as a function of the radius, is considered by integrating the centrifugal acceleration over the ground area of a discrete element. This integrated centrifugal acceleration is a mathematical value that represents the "total amount of the centrifugal acceleration" for one discrete ring. This value is used for all correlations because it represents a generalized value.**

– The design method uses the model to calculate all design variables such as the inner and outer diameters, the axial length of the packing, the number of rotors, and the operating variables, including the rotational speed, which is either based on empirical values from the literature (e.g., maximum or minimum diameters) or on mathematical correlations (e.g., vapor loads at flooding).

1.4 Detailed examples

1.4.1 Example 1: production of hypochlorous acid

The process for producing hypochlorous acid (HOCl) is a prominent and the only long-term industrial application of RPBs for gas-liquid contacting systems. This process is a good example to demonstrate how RPBs can provide an economic advantage over conventional equipment. HOCl is produced in a reactive stripping process. This prominent application was developed and implemented by DOW Chemical Company together with ANDRITZ AG [64].

Background

The reaction system is illustrated in Fig. 1.8, which is based on Trent & Tirtowidjojo [79]. Chlorine (Cl_2) and sodium hydroxide (NaOH) are introduced into the contacting device as gas and liquid, respectively. Cl_2 is absorbed into the liquid phase; the very fast reaction with NaOH produces HOCl. Both steps, i.e., the absorption and reaction are limited by the liquid-side mass transfer. In the presence of NaOH, the product decomposes very rapidly to the undesired by-product sodium chlorate. To avoid this decomposition, a rapid desorption of the product into the gas phase is necessary, which is limited by the gas-side mass transfer. In a subsequent step, HOCl is absorbed in fresh water to gain the desired aqueous solution of the product [79].

The conventional equipment used in this absorption, reaction, and desorption system is a spray tower. The spray tower is liquid-side mass transfer limited, which leads to a 20% loss in yield due to the decomposition of HOCl to its by-product [79].

$$Cl_2\,(g) + NaOH\,(l) \xrightarrow[\text{(Absorption and very fast reaction)}]{} HOCl\,(l) \nearrow{\xrightarrow{+\,NaOH\,(l)}} NaClO_3\,(l)\ \text{(Fast decomposition)}$$

$$\searrow HOCl\,(g)\ \ \text{(Quick desorption)}$$

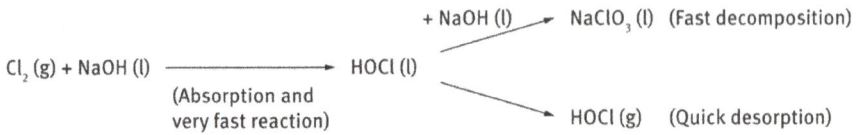

Fig. 1.8: Reaction and absorption system for the production of hypochlorous acid [79].

Additionally, the flow rates for the stripping gas are very high, and the equipment size is very large. These characteristics lead to an infeasible investment and operation process.

The engineers responsible for designing the process needed to determine a solution for this problem; they considered RPBs as alternative equipment. **Although no industrial experience existed, they expected higher yields and lower costs due to the very rapid mass transfer, very short residence time, and small equipment size of RPBs** [64]. A pilot plant was constructed for investigating the important variables, liquid distribution, gas-to-liquid ratio, packing type, and rotor speed [79]. The results were promising; thus, a commercial plant was built. The absorption, reaction, and desorption in the RPB were found to occur within less than a second. Moreover, the very high mass transfer rates reduced the decomposition reaction, providing a product yield exceeding 90%, which exceeds the spray tower by more than 10%, and utilizing smaller amounts of stripping gas (less than half) in smaller equipment (40-fold reduction). Therefore, both the investment and operation costs were reduced, and the economic goals of the process were attained [79].

It has been reported that the aforementioned process, which was implemented at the DOW Chemical Company site, is mechanically reliable, shows no inadmissible vibration, requires little maintenance, allows easy start-up and shut-down procedures, and can be operated for more than 10 years [79, 80]. This example shows that the use of RPBs or general intensified equipment can be implemented in conventional processes to achieve better results and higher profits.

1.4.2 Example 2: modular and flexible container systems

The second example of RPB applications for distillation is modular container systems. The development of modularized equipment has been emphasized, especially for the production of highly specialized, low-capacity, or seasonal products such as agricultural products or pesticides. Moreover, modularized plants have an advantage over permanent large-scale plants [81]. If modularized equipment is standardized, small-scale plants can be quickly assembled for the rapid production of small amounts of pharmaceuticals for test studies or to quickly adjust a process to changing demands such as seasonal changes in agricultural products [82]. Therefore, the application of RPBs for modularized plants is discussed and compared to the other devices below. Additionally, the use of the rotational speed to operate RPBs in a flexible manner is also reviewed.

Background

A primary focus of research lies in the development of modularized container systems that can be flexibly connected. Some recent research projects aiming to develop standardized containers systems include "F3-Factory" [82] and "COPIRIDE" [83]. The main goal is the installation of modularized equipment within standard 20-foot containers. Within these projects, different technologies for the installation of distillation and absorption have been investigated [84]. The investigations have been based on qualitative attributes such as the technical maturity, availability or predictability of different technologies, and evaluations of case studies for distillation and absorption processes. The latter were evaluated in terms of the volumetric capacity, C_{vol} (capacity in relation to the volume of packing) and volumetric efficiency, η_{vol}, of the devices. The efficiency is typically defined as the number of transfer units per unit volume of the packing. Detailed information regarding this technology and a case study on distillation and absorption can be found in the work of Dercks et al. [84], which is further discussed here.

Four different technologies for modular distillation and absorption have been investigated: (1) conventional packed columns, (2) hollow-fiber devices, (3) microchannel devices, and (4) RPBs. All the devices have advantages such as technical maturity (1), compact design (3 and 4), good predictability (1, 3, and 4), and small investment costs (1 and 2) and corresponding drawbacks. The main limitation of conventional columns is the maximum height of the container. For the microchannel and hollow-fiber devices, the small maximum capacity makes parallelization necessary. RPBs have a large capacity with a compact design, making them suitable for modular container plants. However, the low maturity and the lack of standardized apparatuses for the intensified processes have led to the use of conventional columns in container devices [84].

Modularity evaluation

The efficiencies of the studied RPB are recalculated in this work using the model described in Section 1.3 and are analyzed according to the ATU-NTU concept proposed by Kelleher et al. [35]. The test system for this case study, the capacity, and the desired purities are listed in Tab. 1.1 together with the design results for the studied RPB. For the case study, two different RPBs are designed, of which the first is optimized to minimize the required space, whereas the second is optimized for the energy demand. Both results are presented in Fig. 1.9 together with the results

Tab. 1.1: Specifications and results of the case study regarding modular, compact, and efficient apparatuses [84].

Specifications		
Test system	Cyclohexane, n-heptane	
Mass flow feed	$\dot{m}_{feed} = 75$ kg h^{-1}	
Feed composition	$x_{feed,cHex} = 0.5$	
Distillate composition	$x_{dist,cHex} = 0.99$	
Bottom composition	$x_{dist,cHex} = 0.01$	
Number of rotors	$N_{rot} = 2$, middle feed	
Results	**RPB design 1**	**RPB design 2**
Reflux ratio	$v = 3.9$	$v = 2.9$
Rotational speed	$n_{rot} = 33$ Hz	$n_{rot} = 35$ Hz
Inner radius	$R_i = 0.05$ m	$R_i = 0.05$ m
Outer radius	$R_o = 0.38$ m	$R_o = 0.49$ m
Axial length of packing of		
– Lower rotor	$h_{pack} = 0.012$ m	$h_{pack} = 0.007$ m
– Upper rotor	$h_{pack} = 0.010$ m	$h_{pack} = 0.009$ m
Volumetric efficiency	$\eta_{vol} = 4.81 \cdot 10^{-3}$ NTU cm^{-3}	$\eta_{vol} = 6.77 \cdot 10^{-3}$ NTU cm^{-3}
Volumetric capacity	$C_{vol} = 47.67$ L h^{-1}	$C_{vol} = 47.67$ L h^{-1}

from the literature for non-RPB devices. Based on the results, the microchannel and hollow-fiber devices have very high volumetric efficiencies although with only small capacities. Therefore, enormous efforts for numbering-up are necessary. The packed column has smaller efficiencies although higher capacities for processing the mixture in only a few parallel columns or a single column. The two RPBs achieve similar capacities as the column although they have higher volumetric efficiencies. For this particular case study, a single and compact RPB is sufficient. **Hence, RPBs are well suited for modular plants, such as container systems, and show good agreement with the specifications for these processes such as compact design, high capacity, and high volumetric efficiencies.**

Flexibility evaluation

During operation, the performance of a distillation process can be manipulated via several degrees of freedom, mainly operating variables that include the reflux ratio,

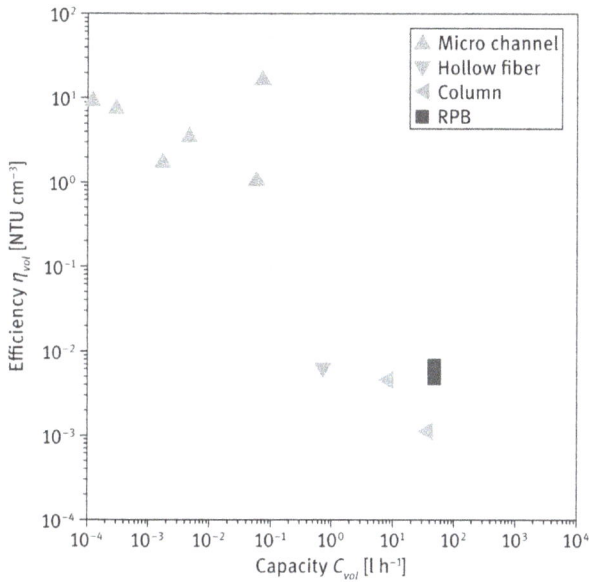

Fig. 1.9: Efficiency analysis of different modularized distillation equipment. Values for non-RPB devices are obtained from the literature [84].

feed streams, or temperatures. For RPBs that are part of a distillation process, the rotational speed is an additional and unique degree of freedom. This characteristic directly affects the centrifugal field and influences the hydrodynamics (e.g., flooding limits and the capacity) and the mass transfer characteristics (e.g., the separation efficiency and the number of theoretical stages) of the RPB. The rotational speed is an operating variable that can be quickly adjusted and is suited for rapid operating condition changes. Consequently, the rotational speed of RPBs contributes to the flexibility of a distillation process and leads to a broader range of operating conditions in modularized equipment. Here, the term flexibility represents the variation in the achieved number of theoretical stages and in capacity. The degree of flexibility is a measure of the range of this variation.

The goal of this study is to estimate if and to what extent the rotational speed can be used to manipulate the capacity and the separation efficiency. For this purpose, an RPB similar to the aforementioned case is designed. The same dimensions and operational conditions are used and kept constant during the experiment. Only the rotational speed is adjusted to obtain different degrees of separation efficiency and capacities. Specifically, the reflux ratio is not changed during this process.

The results for both the capacity and efficiency are presented in Fig. 1.10 (left). For this study, the rotational speed is varied between 10 and 50 s^{-1}. The results show **that by only varying the rotational speed, the capacity varies between 100 and 500 L h^{-1} for the same volumetric efficiency, whereas the volumetric** efficiency varies between $0.17 \cdot 10^{-3}$ NTU cm^{-3} and $0.42 \cdot 10^{-3}$ NTU cm^{-3} at the same capacity corresponding to 15 and 24 theoretical stages, respectively. Therefore, the extent of the variations in efficiency and capacity by adjusting only the rotational speed are remarkably high.

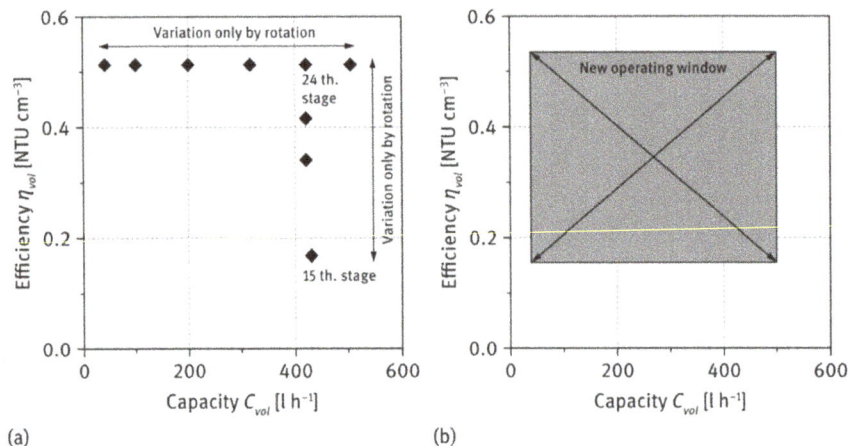

Fig. 1.10: Variations in the efficiency and capacity (left) and the flexible operating window based on rotational speed for an RPB (right).

Based on these results, a new type of operating window can be defined, which is qualitatively illustrated in the diagram in Fig. 1.10 (right). This operating window, which represents the changing capacities and separation efficiencies, can be covered by changing only the rotational speed. All other design variables and operating variables, such as the reflux ratio, remain constant. A conventional distillation column will only have a single operating point for these conditions. This result strongly emphasizes the high operating flexibility of RPBs by exploiting the rotational speed.

This new or extended operating window can be employed for distillation processes with varying feed compositions, product compositions, or capacities. **The rotational speed can be quickly adjusted to guarantee the desired separation within this operating window.** Consequently, RPBs may be applied in (modularized) multipurpose plants or plants with frequently changing operating variables, e.g., for frequent start-ups and shut-downs of batch processes, or in the modular retrofitting of existing plants.

1.4.3 Example 3: high-pressure distillation

The third example of the RPB's potential for distillation is high-pressure distillation. The compact design of RPBs can be advantageous over columns because the main cost driver for high-pressure distillation is the high costs of the thick shells [85].

Background

For industrial distillation processes, the operating pressure is a crucial variable that typically varies between 0.1 and 40 bar [85]. The pressure is often chosen to meet the desired temperature range inside the column. Reducing the pressure, which affects the vapor pressure of the mixture, leads to a decrease in the temperature in the reboiler. In contrast, increasing the pressure leads to an increase in temperature in the condenser. Hence, the advantage of changing the pressure is that appropriate standard heating and cooling fluids can be applied even for very high and low boiling point mixtures. Additionally, the pressure may be chosen to vary the maximum temperature to avoid the decomposition of heat-sensitive components or to positively affect the relative volatilities or azeotropic compositions [85].

High-pressure distillation is typically used for cryogenic distillation to increase the condensation temperature to a sufficiently high level, allowing the use of cooling water or air for cooling instead of expensive refrigeration fluids. Typical examples are the distillation of propylene at approximately 16 [85] to 19 bar [86], the distillation of ethylene at approximately 20 bar [86], and the distillation of methane at 32 bar [85]. When the operating pressure reaches an upper limit of approximately 40 bar, the wall thickness of the column becomes too large, and the distillation columns become uneconomical [85]. In addition to constructive design variables, some physical properties such as densities, viscosities, surface tensions, and relative volatilities are also altered by changing the pressure [85]. However, the influence of these variables is neglected herein.

The application of RPBs for high-pressure distillation may alter the upper limit to higher pressures. The extremely compact design of RPBs leads to comparably small shell areas. Moreover, their cylindrical shape and low heights are favorable for high-pressure vessels. Therefore, it can be expected that RPBs may still be economical at operating pressures that are higher than those of columns.

High-pressure evaluation

As already stated in this chapter, the basic geometry of columns and RPBs differ significantly. Whereas the column has a thin tall structure, the RPB has a squat shape. As illustrated in Fig. 1.11, the RPB has a rather spherical shape and therefore a lower specific shell area compared to columns.

Depending on the number of stages, the volume flow and the separation task the size of RPBs can be significantly smaller than standard columns. If the operating pressure is high, the wall has to be increased, and if the used chemicals are corrosive, more resistant and expensive material of construction has to be used. This is where the RPB will use its advantages.

Conclusively, the pressure has a larger effect on the investment costs of distillation columns than RPBs due to their long and narrow design. For higher pressures, capacities, and separation efficiencies, RPBs tend to be more economical than columns. And, finally, the application of RPBs for high-pressure distillation has the potential to enlarge the distillation operating window to higher operating pressures. Further investigation of the technical feasibility of high-pressure distillation in RPBs must be performed.

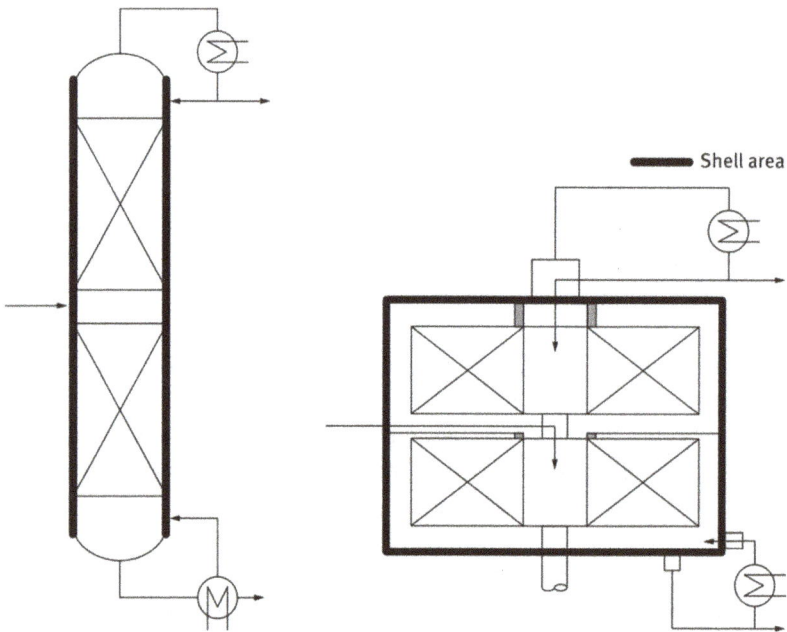

Fig. 1.11: Shell areas of the RPB and columns in comparison.

1.5 Take home messages

- RPBs typically aim for intensified heat and mass transfer between the involved phases although they also employ high centrifugal forces for processing

chemicals with special properties (e.g., highly viscous liquids) or for designing new equipment (e.g., highly efficient packing).
- RPBs consist of one or multiple rotors, in which the liquid usually flows radially outwards and the vapor flows radially inwards.
- Different rotor types have been developed; solid packing, zig-zag packing, and split packing are the most common.
- The axial length of the packing determines the capacity and the diameter the separation efficiency.
- Liquid flows as thin films, droplets, or rivulets through the packing, creating high turbulence and increasing the liquid-side mass transfer coefficients. Vapor flows inwards with less turbulence, which leads to only slight increases in the vapor-side mass transfer coefficients. High interfacial surface areas lead to an intensified volumetric mass transfer on both the vapor and liquid sides.
- In contrast to conventional columns, the liquid and vapor loads and the centrifugal acceleration vary radially, leading to changes in the separation efficiencies.
- The separation efficiency in RPBs can be manipulated via the rotational speed, which represents an additional degree of freedom compared to conventional columns.
- RPBs are applied to reactive systems and to absorption, stripping, and distillation processes. The only published long-term industrial application is the production of hypochlorous acid.
- The special qualities of RPBs enlarge the conventional distillation operating window for fields of application in which conventional columns cannot be economically used.
- The most important qualities of RPBs are the high separation efficiency and capacity with a small footprint, the strong shear and centrifugal forces, the intense heat and mass transfer, the large interfacial areas, the short residence times, and the adjustable rotational speed.
- The most promising fields of application are in mobile and modular plants, for processes requiring a high flexibility in capacity and efficiency and for mixtures with toxic, instable, or explosive properties.
- The two main differences between conventional columns and RPBs in terms of modeling approaches are the very high centrifugal forces and the radial geometry of RPBs.
- The rapid mass transfer between phases in RPBs can be used to remove a product from a reactive phase to a nonreactive one to avoid rapid decomposition, which was successfully implemented in the reactive stripping for the production of hypochlorous acid (example 1).
- Changing the rotational speed of RPBs during operation leads to a high flexibility in the capacity and separation efficiency, which can be used to sketch a new operating window (example 2).

– The compact design and the small shell area of RPBs allow high-pressure distillation processes to be economically conducted at higher pressures compared to conventional columns (exceeding 40 bar).

1.6 Quiz

Question 1. What are the principles of the Podbielniak Contactor? What is it used for?

Question 2. What does the basic design of an RPB (solid packing) for a liquid-vapor contact look like (sketch)? What are the main components? What are the flow paths of liquid and vapor inside the packing?

Question 3. What are the main differences between RPBs and conventional columns (considering the design, flow pattern, and operation)?

Question 4. True or false: in RPBs, liquid flows radially outwards in the packing as thin films.

Question 5. What are the main reasons for mass transfer intensification in RPBs compared to conventional columns?

Question 6. Which parameter are commonly increased in RPBs compared to conventional columns?

– liquid-side mass transfer coefficient
– vapor-side mass transfer coefficient
– interfacial area
– pressure drop
– hold up
– residence times
– capacity
– surface area of packing

Question 7. At which position along the rotor does flooding occur first in RPBs? Why?

Question 8. What are the four most important design and operating variables of RPBs? How do they affect the performance of an RPB?

Question 9. True or false: the maximum rotational speed of an RPB is only limited by the mechanical stability of the device (e.g. shafts and bearings).

Question 10. True or false: conventional packings for columns can easily be used for RPBs.

Question 11. What are the three most often used types of RPBs and the main differences among them?

Question 12. What are the most important characteristics of a good packing for RPBs?

Question 13. What are common values for the size of a rotating packing (e.g., diameter and axial length), the rotational speed, the centrifugal acceleration, and the capacities of RPBs for distillation?

Question 14. True or false: the centrifugal acceleration inside the packing of RPBs is always more than 10 times that of earth's gravity.

Question 15. Why are RPBs suitable for the production of nanoparticles and for polymerization processes? What are the advantages over conventional reactors?

Question 16. Compared to conventional distillation columns, RPBs may be advantageous for what applications? Why?

Question 17. True or false: heterogeneously catalyzed reactive distillation in RPBs typically results in higher yield and selectivity compared to conventional columns.

Question 18. Why is a direct transfer of the mass transfer concepts for distillation columns to RPBs not possible?

Question 19. True or false: the velocities of the vapor and liquid phases have radial and tangential components.

Question 20. True or false: for describing the separation efficiency of RPBs, HETP values and the HTU-NTU concept can be adopted from conventional columns.

Question 21. What are the differences between the HTU-NTU concept and the ATU-NTU concept?

Question 22. What are the advantages of RPBs for the reactive stripping step during the production of hypochlorous acid?

Question 23. What makes RPBs suitable for modular container systems?

Question 24. What is the degree of freedom in RPBs that can be used to sketch a "flexible operating window" for distillation? What are the dimensions of this operating window? How are its limits determined?

Question 25. True or false: RPBs are more economical devices for high-pressure distillation than conventional distillation columns.

Question 26. What is the main reason that RPBs can be more economical for high-pressure distillation than conventional columns?

List of symbols

Latin letters

ATU	area of a transfer unit	m^2
C_{vol}	volumetric capacity	$L\ h^{-1}$
D	diameter	m
h	axial length	m
HETP	height equivalent to a theoretical plate	m
HTU	height of a transfer unit	m
$K_G a$	overall volumetric gas- or vapor-side mass transfer coefficient	s^{-1}
$K_L a$	overall volumetric liquid-side mass transfer coefficient	s^{-1}
$\dot{m}$	mass flow	$kg\ s^{-1}$
n_{rot}	rotational speed	s^{-1}
N_{rot}	number of rotors	–
N_{th}	number of theoretical stages	–
NTU	number of transfer units	–
r	variable radius	m
R	radius	m
x	molar fraction	–

Greek letters

η_{vol}	volumetric efficiency	*NTU* cm^{-3}
v	reflux ratio	–

Subscripts

cHex	cyclohexane
col	column
dist	distillate product
feed	feed
i	inner
o	outer
pack	packing
rad	radially, radius
RPB	rotating packed bed

List of abbreviations

ATU	area of a transfer unit
Cl_2	chlorine

CO_2	carbon dioxide
HETP	height equivalent to a theoretical plate
HOCl	hypochlorous acid
HTU	height of a transfer unit
NaOH	sodium hydroxide
NTU	number of transfer units
RPB	rotating packed bed

References

[1] Prandtl W. Antonin Prandtl Und Die Erfindung der Milchentrahmung Durch Zentrifugieren (Antonin Prandtl and the Invention of Milk Skimming by Centrifugation). 1st edition. München, Germany: Knorr and Hirth; 1938.

[2] Hähnel W. Die erste deutsche Zentrifuge kam aus Chemnitz (The First German Centrifuge came from Chemnitz). Museumskurier des Industriemuseums Chemnitz; 2007; 1–2.

[3] De Laval Separator Company Co. Alpha De Laval Baby Cream Separators. 1st edition. New York: The Library of Congress; 1900.

[4] Meyer J. Butter. In: Meyer Großes Konversations-Lexikon. Meyer J, ed. Leipzig, Germany: Verlag des Bibliographisches Instituts; 1905.

[5] Coutor C. Process and Apparatus for the Continuous Extraction or Treatment of Liquids. U.S. 2036924; 1934.

[6] Thayer G: Centrifugal Countercurrent Contacting Machine. US 2176982; 1939.

[7] Xu ZC, Ji JB, Wang GQ, Li XH, Li YM: Rotating Zigzag Bed Application in Extractive Distillation Process of THF-Methanol-Water System. In: AIChE Annual Meeting: Conference Proceedings; Houston; 2012.

[8] Aurig M. A Method of, and Apparatus for, Washing Gases and Vapours. GB 347303 A; 1931.

[9] Podbielniak WJ. Method of Securing Counter Current Contact of Fluids by Centrifugal Action. US 2044996; 1936.

[10] Ramshaw C, Mallinson RH. Mass Transfer Process. US 4283255; 1981.

[11] van der Schaaf J, Schouten JC. High-gravity and high-shear gas–liquid contactors for the chemical process industry. Curr. Opin. Chem. Eng. 2011;1:84–88.

[12] Chen J, Shao L. Recent advances in nanoparticles production by high gravity technology – From fundamentals to commercialization. J. Chem. Eng. Jpn. 2007;40:896–904.

[13] Cortes Garcia GE, van der Schaaf J, Kiss A. A review on process intensification in HiGee distillation. J. Chem. Technol. Biotechnol. 2017;92:1136–56.

[14] Ghadyanlou F, Azari A, Vatani A. A Review of Modeling Rotating Packed Beds and Improving Their Parameters: Gas-Liquid Contact. Sustainability. 2021;13:8046.

[15] Neumann K, Gladyszewski G, Groß K, Qammar H, Wenzel D, Górak A, Skiborowski M. A guide on the industrial application of rotating packed beds. Chem. Eng. Res. Des. 2018;134:443–62.

[16] Rao DP, Bhowal A, Goswami PS. Process intensification in rotating packed beds (HIGEE): An appraisal. Ind. Eng. Chem. Res. 2004a;43:1150–62.

[17] Sorensen E, Lam L, Sudhoff D. Special Distillation Applications. In: Distillation: Operation and Applications. 1st edition. Górak A, Schoenmakers H, eds. Amsterdam, The Netherlands: Elsevier Ltd; 2014.

[18] Sudhoff D, Leimbrink M, Schleinitz M, Górak A, Lutze P. Modelling, design and flexibility analysis of rotating packed beds for distillation. Chem. Eng. Res. Des. 2015a;94:72–89.

[19] Zhao H, Shao L, Chen J. High-gravity process intensification technology and application. Chem. Eng. J. 2010;156:588–93.

[20] Ramshaw C. 'HIGEE' distillation – An example of process intensification. Chem. Eng.(London). 1983:13–14.

[21] Chu GW, Gao X, Luo Y, Zou HK, Shao L, Chen J. Distillation studies in a two-stage countercurrent rotating packed bed. Sep. Purif. Technol. 2013;102:62–66.

[22] Lin CC, Ho TJ, Liu WT. Distillation in a rotating packed bed. J. Chem. Eng. Jpn. 2002;35:1298–304.

[23] Groß K, Bieberle A, Gladyszewski K, Schubert M, Hampel U, Skiborowski M, Górak A. Analysis of Flow Patterns in High-Gravity Equipment Using Gamma-Ray Computed Tomography. Chem. Ing. Tech. 2019;91:1032–40.

[24] Chu GW, Song YH, Yang HJ, Chen JM, Chen H, Chen J. Micromixing efficiency of a novel rotor-stator reactor. Chem. Eng. J. 2007;128:191–96.

[25] Burns JR, Ramshaw C. Process intensification: Visual study of liquid maldistribution in rotating packed beds. Chem. Eng. Sci. 1996;51:1347–52.

[26] Rao DP, Goswami PS, Bhowal A: A novel rotor design for intensification of gas-phase mass transfer in a HIGEE. In: AIChE Spring Meeting: Conference Proceedings; San Francisco; 2004b.

[27] Chen YS. Correlations of mass transfer coefficients in a rotating packed bed. Ind. Eng. Chem. Res. 2011;50:1778–85.

[28] Chen YS, Lin FY, Lin CC, Tai CYD, Liu HS. Packing characteristics for mass transfer in a rotating packed bed. Ind. Eng. Chem. Res. 2006;45:6846–53.

[29] Reddy KJ, Gupta A, Rao DP, Rama OP. Process intensification in a HIGEE with split packing. Ind. Eng. Chem. Res. 2006;45:4270–77.

[30] Keyvani M, Gardner NC. Operating characteristics of rotating beds. Chem. Eng. Prog. 1989;85:48–52.

[31] Burns JR, Jamil JN, Ramshaw C. Process intensification: Operating characteristics of rotating packed beds – determination of liquid hold-up for a high-voidage structured packing. Chem. Eng. Sci. 2000;55:2401–15.

[32] Singh SP, Wilson JH, Counce RM, Villiersfisher JF, Jennings HL, Lucero AJ, Reed GD, Ashworth RA, Elliott MG. Removal of volatile organic-compounds from groundwater using a rotary air stripper. Ind. Eng. Chem. Res. 1992;31:574–80.

[33] Lockett MJ. Flooding of rotating structured packing and its application to conventional packed columns. Chem. Eng. Res. Des. 1995;73:379–84.

[34] Sudhoff D. Design, Analysis and Investigation of Rotating Packed Beds for Distillation. 1st edition. Munich: Verlag Dr. Hut; 2015b.

[35] Kelleher T, Fair JR. Distillation studies in a high-gravity contactor. Ind. Eng. Chem. Res. 1996;35:4646–55.

[36] Mondal A, Pramanik A, Bhowal A, Datta S. Distillation studies in rotating packed bed with split packing. Che. Eng. Res. Des. 2011;90:453–57.

[37] Wang GQ, Xu OG, Xu ZC, Ji JB. New HIGEE-rotating zigzag bed and its mass transfer performance. Ind. Eng. Chem. Res. 2008;47:8840–46.

[38] Chen J, Wang YH, Guo F, Wang XM, Zheng C. Synthesis of nanoparticles with novel technology: High-gravity reactive precipitation. Ind. Eng. Chem. Res. 2000;39:948–54.

[39] Lin C, Ho J. Structural analysis and catalytic activity of Fe3O4 nanoparticles prepared by a facile co-precipitation method in a rotating packed bed. Ceram. Int. 2014;40:10275–82.

[40] Guo B, Zhao Y, Wu W, Meng H, Zou H, Chen J, Chu G. Research on the preparation technology of polyaniline nanofiber based on high gravity chemical oxidative polymerization. Chem. Eng. Proc. 2013;70:1–8.

[41] Liu YZ, Jiao WZ, Qi GS. Preparation and properties of methanol–diesel oil emulsified fuel under high-gravity environment. Renew. Energy. 2011;36:1463–68.

[42] Dhiman SK, Verma V, Rao DP, Rao MS. Process intensification in a trickle-bed reactor: Experimental studies. AIChE J. 2005;51:3186–92.

[43] Ravindra PV, Rao DP, Rao MS. Liquid flow texture in trickle-bed reactors: An experimental study. Ind. Eng. Chem. Res. 1997;36:5133–45.

[44] Sivalingam G, Radhika M, Rao DP, Rao MS. Process intensification in a model trickle-bed reactor. Ind. Eng. Chem. Res. 2002;41:3139–44.

[45] Li TC, Qi GS. Impinging stream rotating packed bed extractor. Petro-Chem Equip. 2004;133 (14):26–28.

[46] Liu Y, Qi G, Yang L. Study on the mass transfer characteristics in impinging stream rotating packed bed extractor. Chem. Ind. Eng. Prog. 2003;22(10):1108–11.

[47] Modak JB, Bhowal A, Datta S. Extraction of dye from aqueous solution in rotating packed bed. J. Hazard. Mater. 2016;304:337–42.

[48] Park J, Gardner NC. Method for Degassing a Liquid. US 7537644 B2; 2009.

[49] Ramshaw C. Degassing of Liquids. US 4715869; 1987.

[50] Peel J, Howarth CR, Ramshaw C. Process intensification: Higee seawater deaeration. Chem. Eng. Res. Des. 1998;76:585–93.

[51] Joel AS, Wang M, Ramshaw C, Oko E. Process analysis of intensified absorber for postcombustion CO_2 capture through modelling and simulation. Int. J. Green. Gas Control. 2014;21:91–100.

[52] Nessi E, Papadopoulos AI, Seferlis P. A review of research facilities, pilot and commercial plants for solvent-based post-combustion CO2 capture: Packed bed, phase-change and rotating processes. Int. J. Green. Gas Control. 2021;111:103474.

[53] Zhang LL, Wang JX, Liu Z, Lu Y, Chu GW, Wang W, Chen J. Efficient capture of carbon dioxide with novel mass-transfer intensification device using ionic liquids. AIChE J. 2013;59:2957–65.

[54] Yang S, Lin CC, Tseng I, Liu WT, Yu H. Method for Removing Volatile Components from a High Viscosity Liquid by Using Rotation Pack Bed. U.S. 6,884,401 B2; 2005.

[55] Chen YS, Lin CC, Liu HS. Mass transfer in a rotating packed bed with viscous Newtonian and Non-Newtonian fluids. Ind. Eng. Chem. Res. 2005;44:1043–51.

[56] Chen YS, Liu HS. Absorption of VOCs in a rotating packed bed. Ind. Eng. Chem. Res. 2002;41:1583–88.

[57] Hsu LJ, Lin CC. Binary VOCs absorption in a rotating packed bed with blade packings. J. Environ. Manag. 2012;98:175–82.

[58] Lin CC, Lin YC, Chien KS. VOCs absorption in rotating packed beds equipped with blade packings. J. Ind. Eng. Chem. 2009b;15:813–18.

[59] Lin CC, Wei TY, Liu WT, Shen KP. Removal of VOCs from gaseous streams in a high-voidage rotating packed bed. J. Chem. Eng. Jpn. 2004;37:1471–77.

[60] Qian Z, Xu LB, Li ZH, Li H, Guo K. Selective absorption of H2S from a gas mixture with CO_2 by aqueous N-Methyldiethanolamine in a rotating packed bed. Ind. Eng. Chem. Res. 2010;49:6196–203.

[61] Chen YH, Chang CY, Su WL, Chen CC, Chiu CY, Yu YH, Chiang PC, Chiang SIM. Modeling ozone contacting process in a rotating packed bed. Ind. Eng. Chem. Res. 2004;43:228–36.

[62] Chiu CY, Chen YH, Huang YH. Removal of naphthalene in Brij 30-containing solution by ozonation using rotating packed bed. J. Haz. Mat. 2007;147:732–37.

[63] Lin CC, Chao CY, Liu MY, Lee YL. Feasibility of ozone absorption by H2O2 solution in rotating packed beds. J. Haz. Mat. 2009a;167:1014–20.

[64] Quarderer GJ, Trent DL, Steward EJ, Tirtowidjojo D, Mehta AJ, Tirtowidjojo CA. Method for Synthesis of Hypohalous Acid. US 6048513; 2000.

[65] Ramshaw C, Arkley K. Process intensification by miniature mass transfer. Process. Eng. 1983;64:29–31.

[66] Short H. New mass-transfer find is a matter of gravity. Chem. Eng.(London). 1983;90:23–29.

[67] Chandra A, Goswami PS, Rao DP. Characteristics of flow in a rotating packed bed (HIGEE) with split packing. Ind. Eng. Chem. Res. 2005;44:4051–60.

[68] Wang GQ, Guo CF, Xu ZC, Li YM, Ji JB. A new crossflow rotating bed, Part 1: Distillation performance. Ind. Eng. Chem. Res. 2014a;53:4030–37.

[69] Wang GQ, Guo CF, Xu ZC, Yu YL, Ji JB. A new crossflow rotating bed, Part 2: Structure optimization. Ind. Eng. Chem. Res. 2014b;53:4038–45.

[70] Luo Y, Chu GW, Zou HK, Xiang Y, Shao L, Chen J. Characteristics of a two-stage countercurrent rotating packed bed for continuous distillation. Chem. Eng. Proc. 2012b;52:55–62.

[71] Krishna G, Min TH, Rangaiah GP. Modeling and analysis of novel reactive HiGee distillation. Comput. Aided Chem. Eng. 2012;31:1201–05.

[72] Sudhoff D, Neumann K, Lutze P. An integrated design method for rotating packed beds for distillation. Comput. Aided Chem. Eng. 2014;33:1303–08.

[73] Kister HZ. Distillation Design. 1st edition. New York, USA: McGraw-Hill; 1992.

[74] Sherwood TK, Shipley GH, Holloway FAL. Flooding velocities in packed columns. Ind. Eng. Chem. 1938;30:765–69.

[75] Onda K, Takeuchi H, Okumoto Y. Mass transfer coefficients between gas and liquid phases in packed columns. J. Chem. Eng. Jpn. 1968;1:56–62.

[76] Li XP, Liu YZ, Li ZQ, Wang X. Continuous distillation experiment with rotating packed bed. Chin. J. Chem. Eng. 2008;16:656–62.

[77] Lin CC, Liu WT. Mass transfer characteristics of a high-voidage rotating packed bed. J. Ind. Eng. Chem. 2007;13:71–78.

[78] Agarwal L, Pavani V, Rao DP, Kaistha N. Process intensification in HiGee absorption and distillation: Design procedure and applications. Ind. Eng. Chem. Res. 2010;49:10046–58.

[79] Trent DL, Tirtowidjojo D. Intensifying the process. Chem. Eng.(London). 2003:30–31.

[80] van den Berg H: Reactive Stripping in a Rotating Packed Bed for the Production of Hypochlorous Acid. In: PIN NL: Conference Proceeding; Delft; 2010.

[81] Bramsiepe C, Schembecker G. 50% Idea: Modularization in process design. Chem. Ing. Tech. 2012;84:581–87.

[82] Buchholz S. Future manufacturing approaches in the chemical and pharmaceutical industry. Chem. Eng. Proc. 2010;49:993–95.

[83] Fraunhofer ICT-IMM.: COPIRIDE. http://www.copiride.eu; (accessed February 13, 2014).

[84] Dercks B, Frahm B, Górak A, Grünewald M, Kenig EY, Lautenschleger A, Ressler S, Schmidt P, Sudhoff D, Zecirovic R. Intensified Absorption and Distillation Devices for Modular Chemical Production Processes. In: 7th European Congress of Chemical Engineering 7. Prague: Process Engineering Publications; 2010.

[85] Olujic Z. Vacuum and High Pressure Distillation. In: Distillation: Equipment and Processes. 1st edition. Górak A, Olujic Z, eds. Amsterdam, The Netherlands: Elsevier Ltd; 2014.

[86] Assaoui M, Benadda B, Otterbein M. Distillation under high pressure: A behavioral study of packings. Chem. Eng. Technol. 2007;30:702–08.

Dennis Wenzel

2 Rotating packed bed in liquid-liquid and liquid-solid separation

2.1 Fundamentals

In this chapter, liquid-liquid and liquid-solid separation processes in rotating packed bed (RPB) will be presented. In this context, no interactions between liquid and gas phases or liquid and vapor phases will be considered. On the one hand, this includes liquid-liquid extraction processes, in which a solute compound is transferred from one liquid phase to another. On the other hand, this also includes liquid-solid adsorption processes, in which a solute compound is transferred from a liquid phase to a solid phase. Because both process types are used for similar applications and similar considerations play a role for the utilization of RPB, they are considered together in this chapter.

2.1.1 Liquid-liquid extraction

Background and importance

Liquid-liquid extraction is being applied widely in the industry. Its most prominent fields of application are in pharmaceutical and biochemical industries, wastewater treatment, and analytical chemistry [1]. It is the favored separation technology for many large-scale applications due to its relative simplicity, speed, and efficiency [1, 2]. In addition, it can be well applied for studying the equilibrium and the kinetics of complex reactive processes [1]. Refer to Chapter 7 of the first book of this series (Reactive and Membrane-assisted Separations) for an in-depth introduction to reactive extraction processes.

The array of conventional extraction equipment comprises mixer-settlers, spray columns, packed columns, and mechanically agitated columns [3]. While considerable efforts have been put into the enhancement of the mass transfer efficiency in these classic apparatuses [3], they are bound to the gravitational force to enable contacting between the phases. Also, many extraction processes have to be operated with low amounts of solute and short residence times in relatively small apparatus volumes to be economically interesting. However, this is typically in conflict with the extraction efficiency for conventional extraction equipment. In contrast, centrifugal contactors operate by applying a centrifugal field to the contacted

Dennis Wenzel, TU Dortmund University

https://doi.org/10.1515/9783110724998-002

phases. These comprise equipment such as annular centrifugal extractors, rotating spray columns, and the RPBs, which will be the focus of this chapter. The potential benefits of the latter ones for liquid-liquid extraction processes are the fine liquid distribution and the profound liquid droplet breakup, leading to an intensified initial phase contacting and high mass transfer rates between these phases.

Separation principles and parameters

From a physical point of view, liquid-liquid extraction describes a method of separating the solute present in a solution by adding another immiscible liquid, to which the solute is transferred preferentially [3, 4]. Refer to Chapter 3 of the first book of this series (reactive and membrane-assisted separations) for an elaborate introduction to the respective thermodynamic models required for describing the respective thermodynamic equilibria.

Liquid-liquid extraction processes typically comprise four streams: the feed, which consists of the key component (solute) and a carrier liquid, the solvent, the extract (solute-rich solvent), and the raffinate (solute-lean carrier). Furthermore, liquid-liquid extraction processes can be characterized by the capacity K and the selectivity S of the utilized solvent [5]:

$$K_i = x_i^E / x_i^R,$$ (2.1)

$$S_{S/C} = \frac{K_S}{K_C} = \frac{x_S^E / x_S^R}{x_C^E / x_C^R}$$ (2.2)

where K_i is the capacity of the solvent with respect to component i, x_i^E and x_i^R are the molar fractions of component i in the extract and in the raffinate, respectively, $S_{S/C}$ is the selectivity of the solvent with respect to solute and carrier, x_S^E and x_S^R are the molar fractions of the solute in the extract and raffinate, respectively, and x_C^E and x_C^R are the molar fractions of the carrier in the extract and raffinate, respectively [6].

Another key parameter for the assessment of liquid-liquid extraction is the volumetric mass transfer rate. With an increasing volumetric mass transfer rate, the required volume for a given separation task decreases. In this way, the size of the necessary equipment also decreases, saving initial investment costs and reducing the costs of operation [3].

In order to calculate the volumetric mass transfer rate, the mass transfer equation for a discrete, differential liquid volume element (Fig. 2.1) must be solved, which can be described for an aqueous phase as:

$$Q_a \, dC_a = k_L a \, (C^* - C_a) \, dV$$ (2.3)

where Q_a is the flow rate of the aqueous phase, C_a is the concentration of the solute in the aqueous phase, C^* is the equilibrium concentration of the solute, and dV is

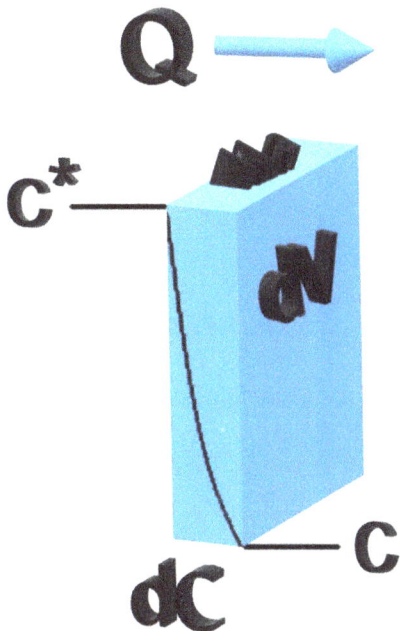

Fig. 2.1: Schematic illustration of a discrete, differential liquid volume and its properties in the mass transfer equation.

the discrete, differential liquid volume. If the aqueous phase constitutes the feed, we obtain the following boundary conditions for the process:

$$\begin{cases} C_{a,\text{in}} = C_F \\ C_{a,\text{out}} = C_R \end{cases} \quad \begin{cases} V_{\text{in}} \\ V_{\text{Out}} \end{cases}.$$

Applying these boundary conditions to eq. (2.3), we obtain:

$$\int_{C_F}^{C_R} \frac{dC}{C^* - C_a} = \frac{k_L a}{Q_a} \int_{V_{\text{out}}}^{V_{\text{in}}} dV \tag{2.4}$$

where C_F and C_R represent the concentration of the solute in the feed and the raffinate, respectively.

In a first approximation, plug flow and full utilization of the packing can be assumed for an RPB. In this case, the term $\int_{V_{\text{out}}}^{V_{\text{in}}} dV$ describes all liquid within the void volume of the packing. It can, therefore, be substituted by the term:

$$V_{\text{void}} = h \cdot \frac{\pi}{2} \cdot (r_o^2 - r_i^2) \cdot \varepsilon \tag{2.5}$$

After solving the integral on the left-hand side and rearranging for $k_L a$, we obtain as expression for the volumetric mass transfer rate:

$$K_L a = \frac{Q_a}{V_{\text{void}}} \cdot \ln\left(\frac{C^* - C_F}{C^* - C_R}\right). \tag{2.6}$$

In addition, many studies use the so-called extraction rate (or removal efficiency) E to characterize the separation performance of an extraction process. This extraction rate is calculated as:

$$E = \frac{C_F - C_R}{C_F}. \tag{2.7}$$

However, since the actual extraction process is limited by the phase distribution equilibrium of the solute, also the equilibrium concentration should be considered for a complete evaluation. This consideration leads to the extraction stage efficiency η, which is calculated as:

$$\eta = \frac{C_F - C_R}{C_F - C^*}. \tag{2.8}$$

2.1.2 Liquid-solid adsorption

Background and importance

Liquid-solid adsorption processes have been widely applied for the removal of trace components, heavy metals, and organic compounds from liquid phases [7]. They are particularly popular in the field of wastewater treatment due to their relative simplicity, their reduced operational costs in comparison to many other options, and the minimum waste disposal [8]. However, adsorption processes also rely heavily on cost- and space-efficient equipment to be economically interesting at a larger scale. Therefore, intensified equipment, such as RPB, has a strong potential in this field. This is due to its ability to process large amounts of liquid in a small equipment volume, by means of higher mass transfer rates than in conventional adsorption equipment.

Separation principles and parameters

In general, liquid-solid adsorption processes rely on the interaction of a compound in solution that is to be removed (adsorptive), with the solid adsorption bed (adsorbent). As soon as it is adsorbed, this compound is considered as adsorbate. Furthermore, it needs to be distinguished between physisorption and chemisorption. During the first one, the adsorptive is bound by physical attraction forces such as dipole, dispersion, or induction forces to the adsorbent [9]. This process is typically reversible by changing the operating conditions (desorption), e.g., the temperature (temperature-swing-adsorption) or the pressure (pressure-swing-adsorption). In contrast, the adsorptive chemically reacts with the packing in a chemisorption process, which is typically irreversible.

With respect to the hydraulics within a packing made from adsorbent material, its porosity, tortuosity, and permeability need to be considered. On the one hand, the porosity describes the void fraction of a packed bed (analogous to eq. (2.5)); on the other hand, the pore size also plays an important role. The tortuosity τ describes the "curvature" of the pores as the ratio between the length of a curved pore C and the direct distance between its ends L:

$$\tau = \frac{C}{L}.$$
(2.9)

Third, the permeability describes the ability of the liquid phase and its compound to pass through a porous packed bed.

For the passage of solution through a RPB, the mass transfer within the liquid phase and between the liquid and the solid phases can be described by different kinetic models. Traditionally, kinetics models consisting of film, surface, and pore diffusion mechanisms are used to describe the external and internal mass transfer [10]. External mass transfer describes the solute compound's transfer through the boundary layer between the bulk solution and the surface of the adsorbent, by means of film diffusion. Internal mass transfer describes the solute compound's transfer through the pores of the adsorbent to an available adsorption spot. Both effects together essentially contribute to the mass transfer rate. In contrast, the actual adsorption of an adsorptive particle onto the adsorbent is typically considered to be infinitely fast and neglected in adsorption kinetics [9].

During the continuous adsorption process, the amount of available adsorption spots in the packed bed constantly decreases in the direction of flow due to the loading with adsorptive (Fig. 2.2). Therefore, the concentration of solute still in solution after passing through the apparatus constantly increases. The time profile for this process can be depicted in a breakthrough curve (Fig. 2.14).

Furthermore, the differential relationship between time and amount of adsorbate can be described by means of the liquid-solid mass transfer rate $k_{L/S}$, the effective interfacial area per unit volume a_{eff}, and the equilibrium amount of adsorbate q_e:

$$\frac{\delta q}{\delta t} = k_{L/S} \cdot a_{eff} \cdot (q_e - q)$$
(2.10)

where q is the amount of adsorbate and q_e is the equilibrium amount of adsorbate, which can be calculated according to:

$$q_e = \frac{Q \cdot (C_0 - C_e)}{W} \tag{2.11}$$

where C_0 is the initial concentration of the adsorptive, C_e is the equilibrium concentration of the adsorptive, and W is the mass of dry adsorbent used.

The relationship between adsorptive load of the solvent and its equilibrium concentration can be described by means of adsorption isotherms. An overview on this topic for a more detailed modeling of adsorption processes in RPB can be found in the research of Li et al. [11]. Finally, for a given time, or a given amount of recycles, the obtained removal efficiency can also be calculated (analogous to eq. (2.7)).

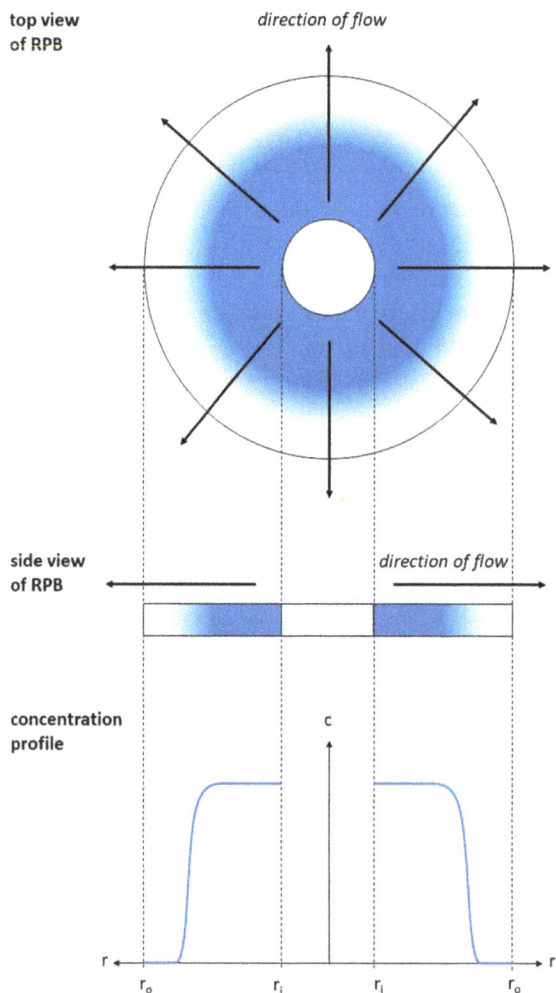

Fig. 2.2: Schematic illustration of the loading of an RPB with adsorptive.

2.2 Applications

While continuous countercurrent operation is common for gas-liquid RPB applications (cf. Chapters 3 and 4), only co-current operation can be applied to liquid-liquid and liquid-solid separation in RPB. This means that the main benefit of using RPB for these applications lies in the efficient phase contacting and the mass transfer between these phases. In contrast, additional equipment is necessary for the separation of the phases in liquid-liquid applications and a discontinuous operation for regeneration of the packed bed (desorption) in liquid-solid applications. Furthermore, multistage operation must be realized with multiple RPB (cf. 2.3 and 2.4), whereas countercurrent operation allows for multiple theoretical stages within a single RPB unit (cf. Chapter 3).

2.2.1 Rotating packed bed liquid-liquid extraction

In the recent years, the most prominent applications for RPB in the field of liquid-liquid extraction processes have been the cleaning of wastewater, the extraction of heavy metals, and the extraction of bio-based chemicals. All these applications focus on the separation of either unwanted or valuable compounds from an aqueous solution. Despite the potential benefits of RPB in liquid-liquid extraction, their application is still limited. A selection of recently published applications is summarized in Tab. 2.1.

Tab. 2.1: List of selected publications on RPB liquid-liquid extraction processes, sorted by year of publication.

Authors	Year	RPB process type	Liquid distribution	Description	Reference
Modak et al.	2016	Co-current; one stage	Distributed flow	Extraction of dye from aqueous solution	[3]
Karmakar et al.	2018	Co-current; one stage	Distributed flow	Extraction of chromium from aqueous solution	[2]
Yang et al.	2018	Co-current; one stage	Impinging stream	Extraction of nitrobenzene from aqueous solution	[4]
Chang et al.	2019	Co-current; one stage	Impinging stream	Extraction of metal ions from aqueous solution	[12]
Karmakar et al.	2020	Co-current; one stage	Distributed flow	Extraction of chromium from aqueous solution	[13]
Wenzel et al.	2020	Counter-current; three stages	Distributed flow	Extraction of γ-valerolactone (GVL) from aqueous solution Detailed example	[6]

Extraction of methyl red dye

Methyl red dye is commonly used as pH indicator and in clinical and microbiological sampling [14]. Colorful dyes are applied widely in the textile, the food, the cosmetics, and many other industries [3]. However, many of these components and their metabolites are toxic, potentially cancerogenic, and harmful to the environment [15, 16]. Xylene, also called xylol, is synonymous for methylbenzene, a benzene ring with two methyl groups in *ortho-*, *meta-*, or *para-*orientation. It is one of the most common hydrocarbon-based solvents produced worldwide on a scale of multiple million metric tons per year [17].

Modak et al. [3] studied the extraction of methyl red dye in an RPB, using xylene as extractant. Methyl red was used in this study as a reference substance for other water-soluble dyes. For their RPB extraction system, and the investigated operating conditions, the authors reported volumetric mass transfer rates between approximately 0.015 and 0.205 s^{-1}. A comparison with values from other extractors is presented in Tab. 2.2 (adapted from [3]). As shown, the values obtained in the RPB are significantly higher than those for other intensified extractors and more than an order of magnitude higher than those for conventional extractors.

Tab. 2.2: List of selected contactors for liquid extraction and the obtained volumetric mass transfer rates. Adapted from [3].

Authors	Year	Extractor	System	K_La [s^{-1}]	Reference
Rotating packed bed extractor					
Modak et al.	2016	RPB	Water-methyl red-Xylene	0.015–0.205	[3]
Other intensified extractors					
Baier et al.	2000	Annular centrifugal extractor	Water-benzyl alcohol-white mineral oil	0.002–0.0127	[18]
Bonam et al.	2009	Rotating spray column	Water-Cr (VI)-Aliquat 336 diluted in kerosene	0.06–0.12	[19]
Conventional extractors					
Seibert and Fair	1988	Spray column	Water-acetone-toluene	0.0005–0.008	[20]
Seibert and Fair	1988	Packed bed column	Water-acetone-toluene	0.0005–0.0055	[20]
Hossein et al.	2006	Mixer-settler	Water-acetone-toluene	0.0015–0.005	[21]

In addition, 95.6% of the target substance could be extracted in the investigated RPB process. In comparison, the volume requirements to achieve the same extraction percentage were estimated to be ten times higher for an annular centrifugal extractor and even up to 500 times higher for a conventional mixer-settler [3]. This means that the necessary equipment volume for extraction processes could be drastically decreased by utilizing RPB.

Modak et al. [3] could further show that the volumetric mass transfer rate more than doubled for an increase in the RPB's rotational speed from 300 rpm to 900 rpm. This can be explained as follows: **with an increasing rotational speed, the distributed liquid droplets become finer and the liquid films thinner, due to the stronger shear under increasing centrifugal force.** Therefore, the mass transfer resistance is reduced, and the available interfacial area for mass transfer is increased. Also, the authors reported that the volumetric mass transfer rate increased for an increasing flow rate of the aqueous phase. On the one hand, the **higher flow rates lead to a better utilization of the packing volume** and, in this way, to a larger interfacial area available for mass transfer [3]. On the other hand, it was shown in RPB mixing studies that **higher liquid distribution velocities – due to higher flow rates through a given liquid distributor – lead to a more vigorous interaction between the distributed liquid phases** [22].

Extraction of hexavalent chromium

Hexavalent chromium (Cr(VI)) can be considered as a model substance for heavy metal contamination of wastewater. It arises in large amounts from the production of chromium ore and in the textile industry. However, it is toxic and cancerogenic and subject to strict regulations. For these reasons, it must be removed or contained wherever possible. Aliquat 336 (methyltrialkyl ammonium chloride) is a quaternary ammonium salt and ionic liquid that is commonly used as phase transfer catalyst and solvent extractant of metals [23]. It is often diluted in hydrocarbons, together with a modifier that prevents the formation of a third, solid phase [2]. It shows a high distribution coefficient towards hexavalent chromium, especially at a low pH.

Karmakar et al. [2] studied the removal of heavy metals from aqueous solution in three different RPB designs. The authors chose hexavalent chromium (Cr(VI)) as the solute and Aliquat 336, diluted in kerosene and 1-decanol, as the extractant.

The authors found that the volumetric mass transfer rate increased with an increasing rotational speed for all three investigated rotor designs. This can be explained by thinner liquid films, smaller liquid droplets, and a more vigorous impingement on the rotor elements. Also, the overall volumetric mass transfer rate increased with an increase of both liquid phases' flow rates. At a rotational speed of 1100 rpm and a total flow rate of 60 L h^{-1}, the authors reported volumetric mass transfer rates between 0.190 m s^{-1} and 0.280 m s^{-1} and stage efficiencies between 96.0% and 99.6%. Under comparable experimental conditions, the estimated $k_L a$ value was 0.015 m s^{-1} in a mixer-settler and only 0.010 m s^{-1} in a fixed bed, with stage efficiencies of 66.8% and

40.0%, respectively. This demonstrates that a substantial intensification of liquid-liquid extraction is possible with RPB.

Extraction of nitrobenzene

Nitrobenzene is a nitrated derivate of benzene. Like methylbenzene (xylene), it is produced on a scale of multiple million metric tons per year. Roughly, 95% of nitrobenzene is used for the production of aniline, an important base chemical for polyurethane foams, rubber products, herbicides, dyes, and pigments [24]. However, nitrobenzene is also an extremely hazardous substance, which is highly toxic, cancerogenic, and dangerous to the environment [25]. Therefore, it is subject to strict regulations and reporting requirements, and its removal from wastewater is of great importance [4].

Yang et al. [4] investigated the extraction of nitrobenzene from aqueous solution in a impinging-stream-RPB (cf. Chapter 2.3.1) using cyclohexane as extractant. Cyclohexane is the fully hydrated derivate of benzene. Like benzene, it also has six carbon atoms arranged in a hexagon shape, but each carbon atom forms a bond with two hydrogen atoms. Besides its use for the production of nylon, it is a popular non-polar organic solvent [26].

In their study, Yang et al. [4] investigated the influence of the rotational speed and the liquid flow rate on the extraction stage efficiency. Interestingly, for both operation parameters, a maximum could be found, although due to two different effects. For the rotational speed, the extraction stage efficiency first increased with an increasing rotational speed, from ~80% at 100 rpm to 99.99% at 400 rpm, but then decreased again. This can be explained by two opposite effects taking place. **On the one hand, the interactions between impinging liquid and rotating packing become more vigorous with an increasing rotational speed. On the other hand, the residence time of the liquid phases in the packing decreases.** A similar efficiency maximum for the rotational speed can be observed in RPB flue gas cleaning (cf. Chapter 4) and RPB distillation processes (cf. Chapter 6).

For the liquid flow rate, the extraction stage efficiency also first increased from ~72% at 20 L h^{-1} to 99.99% at 50 L h^{-1}, before decreasing again at higher liquid flow rates. To understand this effect, it must be considered that the available surface area at the inner rim of the packing is limited. However, this area is necessary to imprint the shear forces of the rotating packing on the impinging liquid. So, while an increasing flow rate has positive effects, at first, as discussed above for the results of Modak et al. [3], **an excessive liquid load on the inner rim of the packing is disadvantageous, as it negatively affects the initial phase contacting.** A similar effect was described for liquid-liquid mixing processes [6, 22].

Extraction of metal ions

Chloride leaching is used in the metallurgical industry to extract nickel from the so-called lateritic nickel ore, which forms the majority of all nickel deposits worldwide. In this process, the nickel is dissolved in chloride leach solution together with further metal ions such as iron, aluminum, copper, zinc, and manganese. However, these metal impurities need to be removed in the downstream to obtain a high-purity product [27].

Chang et al. [12] studied the removal of particular metal ions from chloride leach solution in an impinging-stream RPB. For this task, the extraction of zinc, manganese, and copper from an aqueous solution also containing nickel, cobalt, and chloride was investigated. As solvent, the authors used di-2-ethylhexylphosphoric acid (D2EHPA). D2EHPA is an organophosphorus compound that is typically used together with a diluent agent, e.g., sulfonated kerosene or n-dodecane, for the extraction of metal ions [12, 28].

In order to extract the targeted metal ions only, the authors used the influence of the pH value on the distribution coefficient between the different metals and the chosen extraction system. They found that a pH of 3.5 was optimal for extracting more than 90% of the unwanted zinc, manganese, and copper, without extracting more than 10% of the desired nickel and cobalt. With this pH value, they moved on to investigate the influence of the RPB operation on the extraction rate.

For an increase in the rotational speed from 500 rpm to 900 rpm, the authors found that the extraction rates of copper and manganese increased from around 40% to 87–92%, while the extraction rate of zinc even increased from around 50% to over 99%. At the same time, the co-extraction rate of nickel and cobalt was around 6% for all runs. However, at 1000 rpm, the extraction rate of the unwanted compounds decreased again, which can be related to the shortened contacting time, as elaborated above. In accordance with the extraction rate, also the volumetric mass transfer rate $k_L a$ of copper, zinc, and manganese first increased up to 900 rpm, but then decreased again for a higher rotational speed. Furthermore, an increasing volumetric flow rate was found to have a positive influence on the $k_L a$ value. Exemplary values for the $k_L a$ in Chang et al.'s experiments are shown in Tab. 2.3.

Tab. 2.3: $k_L a$ values for copper, zinc, and manganese in the experiments of Chang et al. [12].

	Copper	Zinc	Manganese
$k_L a$ [s^{-1}], 500 rpm, 30 L h^{-1}	0.017	0.019	0.018
$k_L a$ [s^{-1}], 900 rpm, 30 L h^{-1}	0.060	0.140	0.092
$k_L a$ [s^{-1}], 900 rpm, 50 L h^{-1}	**0.125**	**0.160**	**0.135**
$k_L a$ [s^{-1}], 1000 rpm, 30 L h^{-1}	0.058	0.120	0.090

For the optimum operating conditions, and three theoretical extraction stages, the results shown in Tab. 2.4 were obtained for the feed and the raffinate concentrations of all metal ions, which would correspond to extraction rates E between 90.0% and 99.9%. Afterwards, the co-extracted nickel and cobalt could be scrubbed from the organic phase with hydrochloric acid.

Tab. 2.4: Feed and the raffinate concentrations of all metal ions after three theoretical extraction stages, based on the RPB experimental results of Chang et al. [12].

	Nickel	**Cobalt**	**Copper**	**Zinc**	**Manganese**
C_F [g L^{-1}]	79.08	2.60	1.120	1.040	0.630
C_R [g L^{-1}]	71.14	2.43	0.004	0.001	0.003
E [%]	(10.0)	(6.5)	**99.6**	**99.9**	**99.5**

2.2.2 Rotating packed bed liquid-solid adsorption

Recently, the removal of pollutants by means of packed beds made from different adsorption materials has been in the focus of studies on RPB liquid-solid adsorption processes. An overview of these studies is presented in Tab. 2.5. The investigated operational parameters comprise the RPB rotational speed, the liquid flow rate, and the pollutants' concentrations. Moreover, mass transfer rates, specific surface areas, and removal efficiencies have been determined and compared to other equipment.

Tab. 2.5: List of selected international publications on RPB liquid-solid adsorption processes, sorted by year of publication.

Authors	Year	Adsorbent (packed bed)	Description	Reference
Chang and Lee	2012	Activated carbon	Removal of methomyl from wastewater	[10]
Modak et al.	2016	Rice husk	Adsorption of methylene blue dye from aqueous solution	[7]
Li et al.	2018	Activated carbon	Adsorption of phenol from aqueous solution	[11]
Hashim et al.	2019	Activated carbon	Removal of arsenic from drinking water	[29]
Wu et al.	2019	Clinoptilolite	Removal of ammonium from aqueous solution	[30]
Liu et al.	2020	Copper foam	Chemisorption of potassium dichromate	[31]

Adsorption of methomyl

Methomyl is a compound from the group of carbamates and popularly used as a pesticide due to its insecticidal and nematicidal effects [32]. It is very well soluble in water and moderately soluble in the soil so that it can be expected to infiltrate the groundwater and drinking water resources in areas of extensive use [10]. However, it is also highly toxic and suspected to have endocrine disrupting effects in the human body [33, 34]. Therefore, it does not only pose a direct environmental hazard, but also a hazard to human life, and must be removed from contaminated water.

Chang and Lee [10] combined an RPB with a stirred tank to investigate the removal of methomyl from wastewater. As the adsorbent, the authors used a packed bed made from activated carbon particles. Activated carbon is widely used as an adsorbent for various applications, as also reflected by its multiple appearance in Tab. 2.5. The carbon itself can be produced from virtually any carbonaceous source, whereas the term *activated* refers to a physical or chemical activation step during its production process. Due to this activation, the number of microscopic pores and, thus, the carbon's surface area is strongly increased. Therefore, activated carbon is one of the most popular adsorbents for the elimination of organics from aqueous solutions [10]. The authors showed in their experiments that both for an increasing rotational speed and an increasing liquid flow rate, both the equilibrium concentration and the time to reach equilibrium were strongly decreased [10]. For the rotational speed, this can be explained by higher internal diffusion due to a more intense interaction with the packed bed particles. For the liquid flow rate, improved hydraulic conditions and increased wetting of the packed bed particles lead to more surface area available for internal and external mass transfer. Additional results for the influence of rotational speed and liquid flow rate on the residence time are discussed in Section 2.3.2.

Adsorption of methylene blue dye

Methylene blue (methylthioninium chloride salt) is a popular basic cationic dye. Similar to the methyl red dye, which Modak et al. [3] used in their liquid-liquid extraction experiments, it can be used as a reference substance for other water-soluble dyes that are often toxic to aquatic life, human health, and the entire ecosystem [7].

Modak et al. [7] investigated breakthrough curves for the adsorption of methylene blue dye from aqueous solution. As the adsorbent, the authors used a packed bed of untreated rice husk. Raw rice husk can be considered a low-cost bio-degradable alternative to classic adsorbents. It has been widely investigated over the last decades due to its high porosity, low density, and large external surface area. It accounts for about 20% of the whole rice and contains about 20% silica, is insoluble in water, and possesses a granular structure with high mechanical strength [35]. At the same time, it is available in excess from the industrial rice production.

The authors found that the uptake rate of dye on the packed bed increased with an increasing rotational speed and an increasing liquid flow rate, which was attributed to a higher mass transfer rate and more ideal flow conditions. However, the authors also found that the effective interfacial area available for mass transfer decreased with an increasing radial length of the packed bed. This can be explained by a lower amount of wetted interfacial area per volume at the outer packing radius for a constant liquid flow rate (cf. Chapters 1 and 3).

Consequently, it seems reasonable to use several smaller packed beds in series instead of just one larger packed bed. A respective process concept by Modak et al. is shown in Section 2.3.2.

Based on their experimental results, the authors further calculated the uptake rate of the dye in their RPB and compared them to a conventional fixed bed. On average, they obtained values of 0.34 mg g^{-1} min^{-1}. In comparison, Han et al. [36] reported 0.02 mg g^{-1} min^{-1} for the adsorption of methylene blue onto a rice husk fixed bed. This shows that the uptake rate can be over an order of magnitude higher in an RPB compared to a fixed bed for the same application.

Adsorption of phenol

Phenol, also known as hydroxybenzene, is a derivative of benzene with an additional hydroxyl group. It is a typical pollutant in the wastewater from the coal processing industry [11]. However, it is harmful to the eyes, the skin, and the respiratory tract as well as a serious environmental pollutant [37, 38]. Therefore, the removal of phenol from wastewater is an important task.

Li et al. [11] investigated the adsorption of phenol from aqueous solution. The authors investigated the influence of the RPB rotational speed on the amount of adsorbed phenol in a packed bed made of activated carbon. They found that this amount first increased with an increasing rotational speed, but then decreased again, after passing an optimum value. This indicates that the mass transfer and hydraulic conditions first improved, before the liquid hold-up and the residence time in the packing became unfavorably low for the adsorption to take place. Additionally, some part of already adsorbed phenol might even have been washed-out again. Overall, the authors found that by applying centrifugal force to the (rotating) packed bed, the amount of absorbed phenol could be increased by 40% under otherwise identical conditions [11]. Also, they found that more phenol was adsorbed in experiments with a higher flow rate, but only until a maximum value was reached. This could imply that no further improvement of the hydraulic conditions was possible, when a certain size of liquid elements and liquid hold-up in the packing was exceeded.

Adsorption of arsenic and ammonium

Besides industrial pollutants, also natural pollutants need to be considered in wastewater cleaning. Two natural pollutants, to which RPB liquid-solid adsorption has been applied, are arsenic and ammonium. Arsenic, on the one hand, is one of the most toxic heavy metals to be commonly present in any body of water [29]. It occurs naturally in the soil, but can be released into the groundwater by weathering reactions and microbiological activity [39]. Despite its toxicity, its appearance is so widespread that the WHO estimates that 200 million people worldwide are exposed to drinking water substantially contaminated by arsenic [40]. The consumption of water with an elevated arsenic concentration is linked to nausea, abdominal pain, diarrhea, and multiple forms of cancer [29, 41]. It is apparent that it must be carefully removed from drinking water and that very strict allowance limits have been set by many countries. Hasim et al. [29] investigated the removal of arsenic from water by means of adsorption on an activated carbon RPB. The authors applied the so-called Taguchi method to derive the optimum values of rotational speed, flow rate, and packing density. They found the highest investigated rotational speed, the second-highest liquid flow rate, and the lowest investigated packing density to result in the highest observed removal of arsenic of 94%.

Ammonium, on the other hand, is a natural metabolite and one of the most common forms of nitrogen to appear in water bodies. It is the cationic derivative of ammonia and often found in the form of water-soluble salts, e.g., ammonium nitrate or ammonium chloride. Additional to its natural appearance, its amount in the environment can become excessive due to sewage discharges and heavy use of fertilizers [30]. However, with an increasing accumulation of ammonium in water bodies (eutrophication), the growth of algae and the depletion of oxygen also strongly increases, leading to the harm of other aquatic organisms and the release of climate-damaging methane [42].

For concentrations representative for eutrophic urban ponds, Wu et al. [30] investigated the removal of ammonium from water in a RPB with clinoptilolite as the adsorbent. Clinoptilolite is a type of zeolite, a porous crystalline hydrated aluminosilicate, composed of silicon oxide and aluminum oxide in varying ratios [30]. It is widely available and capable of strongly adsorbing ammonium from the water phase [43]. By means of semi-batch experiments, the authors found that the removal efficiency increases with an increasing rotational speed. This was explained by the increased mass transfer of ammonium from the liquid to the solid phase. At 1500 rpm, and five batch recycles, 90% of the ammonium could be removed from the water phase within less than one hour.

2.3 Modeling, design, and operation

The way RPB can be utilized for liquid-liquid and liquid-solid separation has been investigated by many scientists and in the context of various industries. In this section, the most recent design variants and methods of operation for these separation tasks are presented on the basis of existing and modeled RPB processes.

2.3.1 Rotating packed bed liquid-liquid extraction

Liquid distribution and flow patterns

For the operation of liquid-liquid extraction processes in RPB, both liquid streams are distributed in the middle of the rotor, also called the eye of the rotor, impinge on the rotating packing, and flow through the packing in the direction of the applied centrifugal force to the outer rim. From there, they are ejected into the casing, leaving the RPB through liquid outlets in the casing. For the distribution of the liquid phases, three different designs are the most popular (Fig. 2.3) [44].

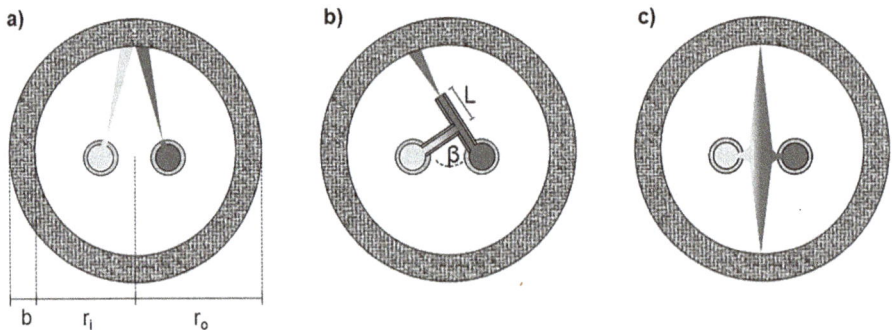

Fig. 2.3: Schematic drawing of the rotating packing and the three most common forms of liquid distribution in an RPB, as seen from the top, where r_o is the outer packing radius, r_i is the inner packing radius, and b is the radial packing depth: (a) Distribution pipes; (b) premixed distributor, with distributor length L and pipe angle β; (c) impinging stream distribution. Reprinted from [44], with permission from Elsevier.

In the first design (Fig. 2.3a), two (or more) separate pipes with small holes at the lower end are used to spray liquid onto the rotating packing. Depending on the exact design, the liquid may be distributed onto a single point of the packing, a narrow area, or a very wide area, overlapping with the distribution of other pipes. In the second design (Fig. 2.3b), two separate distribution pipes are connected to only one pipe with one opening. In this way, the two supplied liquid streams are already pre-mixed within the liquid distributor, before being distributed onto the packing.

The third variant (Fig. 2.3c) is called impinging-stream liquid distribution. In this design, two liquid streams are sprayed against each other in the eye of the rotor before splashing onto the rotating packing [6].

Yang et al. [4] applied an impinging-stream liquid distribution in a horizontal-axis RPB. A schematic of this RPB is shown in Fig. 2.4. The authors reported that benefits over the classical distribution design are stronger dispersion and turbulence. Indeed, slightly better results were reported for liquid-liquid extraction [4] and also for liquid-liquid mixing [44]. Nevertheless, it should also be considered that an impinging stream liquid distribution is also distinctively harder to adjust and less flexible in operation [44].

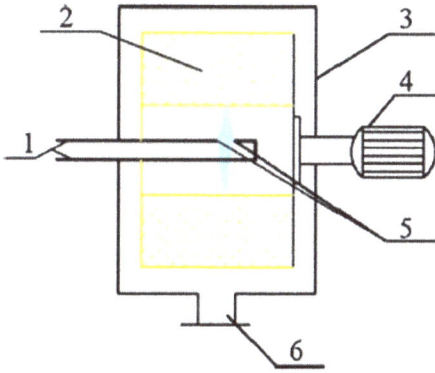

Fig. 2.4: Horizontal-axis RPB with impinging-stream liquid distribution by Yang et al. [4]. 1 – Liquid inlet; 2 – packing; 3 – RPB casing; 4 – electric motor; 5 – impinging-stream liquid distribution; 6 – liquid outlet. Reprinted from [4], with permission from Elsevier.

After being distributed, the liquid phases may flow through the packing in different ways and shapes, depending on the packing design and the operation of the RPB. For the first 7–10 mm of the packing, an impingement zone was suggested [45]. In this zone, an initial vigorous interaction of the distributed liquid with the rotating packing occurs as the momentum of the packing is transferred onto the liquid. Therefore, an intense contacting of liquid phases takes place in this zone. Depending on the liquid flow rate, the liquid distribution design, the ratio of the liquid streams, and the packing design, the liquid hold-up on the inner rim of the packing and inside the impingement zone may strongly vary [6]. Furthermore, Burns et al. [46] found that the liquid hold-up was inversely proportional to the packing radius, with the highest liquid hold-up in the center. However, both the very high and very low liquid hold-up in the impingement zone can negatively affect liquid processing in RPB [22], demonstrating the importance of an informed decision on the operation parameters as well as an appropriate packing design.

After the impingement zone, the liquid phases flow through the inner part of the packing, the so-called bulk zone. Burns and Ramshaw conducted a visual study on this topic with an almost transparent RPB. The authors proposed that liquid flow through the bulk zone could be divided into pore flow, droplet flow, and film flow [47]. Furthermore, Burns and Ramshaw reported that pore (or rivulet) flow dominated at a lower rotational speed, leaving most of the packing starved of liquid, but a more evenly distributed droplet flow set in at higher rotational speed [44]. For the highest rotational speed, film flow was suggested. Based on computed tomography (CT) scans, Yang et al. [48] added that also for different packing types and liquid flow rates, different flow patterns through the packing to its outer rim could be observed. From the outer rim of the packing, drops and strings of liquid are then ejected into the casing [6]. Refer to Chapter 3 for further details on hydraulics as well as CT scans of the liquid distribution inside the packing of an RPB.

Rotor design concepts

As described in chapter 2.1.1, intensifying the phase contacting and increasing the available interfacial area are the key drivers for high mass transfer rates in liquid-liquid extraction. These two parameters not only depend on the mode of operation but also on the RPB rotor and packing design, due to the resulting differences in the liquid flow pattern [2]. In addition to the classical wire mesh or foam packings used as fillings for the rotor, various other concepts for the internal structure of the rotor have been discussed in literature (cf. Chapter 7). For liquid-liquid extraction in RPB, Karmakar et al. [2, 13] studied three different rotor layouts: a rotating spiral bed layout (design A), a rotating zig-zag bed layout (design B), and a rotating (glass bead) bed layout (design C). Schematics of these layouts are shown in Fig. 2.5.

In design A, spiral baffles covering the complete axial height are used. In the conjoined rotor, consisting of an upper and a lower rotor plate with each one spiral, these result in a narrow spiral channel. In this design, the liquids are distributed onto the spiral walls. From there, the two immiscible liquid phases likely flow concurrently on the walls of the spiral as liquid films. Also, it can be assumed that the film thickness decreases with the increasing tangential velocity in the spiral towards its outer circumference. Due to these two effects, the mass transfer resistance is reduced and the extraction rate improved [2].

In contrast, design B comprises concentric ring baffles, which do not cover the complete axial height but leave a small gap. Due to the alternating arrangement of the baffles on the two rotor plates, the conjoined rotor forms a rotating zig-zag bed (cf. Chapters 5 and 7). In this design, the liquid is distributed onto the first baffle. Driven by the centrifugal force, the liquid climbs the baffle in the form of liquid films or rivulets [49] before it is dispersed from the baffle in the form of droplets, flies across the narrow channel, and impinges on the next baffle. There, it again agglomerates into films and rivulets. Thus, the liquid phases repeatedly undergo a dispersion-and-agglomeration-

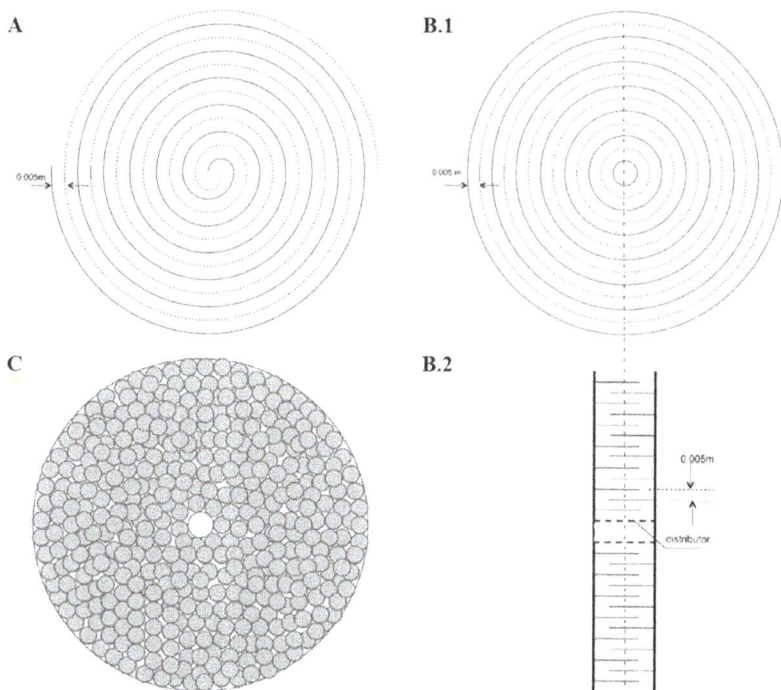

Fig. 2.5: Internal structure of the rotors investigated by Karmakar et al. [2]. A – Front view of rotor A; B1 – front view of rotor B; B2 – side view of rotor B; C – side view of rotor C. Reprinted from [2], with permission from Elsevier.

cycle [50], similar to the phenomenon of split-recombine-mixing [44], which leads to an intense mass transfer and continuous renewal of the interfacial area.

Last, glass beads are used in design C. These beads are about 3 mm in diameter [13]. They are arranged in an irregular pattern between the two rotor plates so that the distributed liquid phases can flow over and past them. For such a design, film flow and pore flow have been reported at lower rotational speed, while film and droplet flow patterns are predominant at higher rotational speed [48].

In comparison to each other, the best results for liquid-liquid extraction were consistently achieved with design A, followed by design B, and then design C. Besides the different hydrodynamics, the different liquid residence time within the rotor packing should be considered the reason [2]. In design A, the liquid is forced to follow the complete length of the spiral, which is 1.90 m in this setup. In design B, the shortest possible path would still amount to roughly 0.22 m, whereas in rotor C, a pathway length of only 0.06 m was reported. As you will see in the following chapters, increasing the liquid's residence time in the rotor packing by increasing its pathway length is also a key parameter of packing design for RPB gas-liquid and distillation processes (cf. Chapters 5 and 7).

Another RPB and rotor design concept that could offer short residence times in small apparatus volumes for liquid-liquid extraction processes is the so-called rotating ring reactor (RRR) [6]. A schematic of such a design is schematically depicted in Fig. 2.6.

(a)   (b)   (c)   (d)

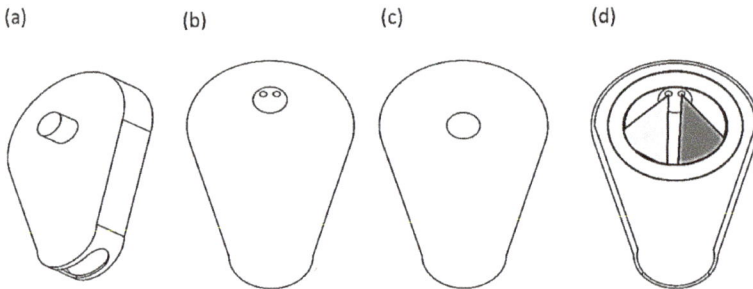

Fig. 2.6: Schematic drawing of the RRR: (a) slanted view from the back; (b) front; (c) back; (d) back view with the back cover removed, nozzles spraying. Reprinted from [52], with permission from Elsevier.

The RRR comprises an annular rotor, which contains a perforated or structured metal ring or a thin ring of structured packing and an elongated casing. The rotor is connected to a motor by the rotor shaft. The rotor shaft is sealed off against the casing and supported by bearings. From the front, liquid distribution pipes with nozzles can be introduced into the casing and the eye of the rotor. Moreover, the current RPB design may lead to an unnecessarily long residence time of liquid in the casing. Therefore, the RRR design comprises a horizontal rotation axis, a small distance between rotor and casing, strongly inclined casing walls, and a large liquid outlet. On the basis of these features, fast liquid processing combined with a low liquid hold-up and a very low residence time in the apparatus can be expected, in combination with a reduced equipment volume, reduced energy consumption, and reduced investment costs. Moreover, multiple RRR modules could be easily arranged in a series for multistage operation, interconnected by pumps, and settling tanks [52].

Process concepts

Modak et al. [3] presented a concept for an one-stage co-current liquid-liquid extraction process in an RPB (Fig. 2.7). The same process was applied by Karmakar et al. [2, 13] to study different rotor setups within the RPB. In this process, the contacting of the phases is conducted continuously, while the separation (settling, not shown) is conducted intermittently.

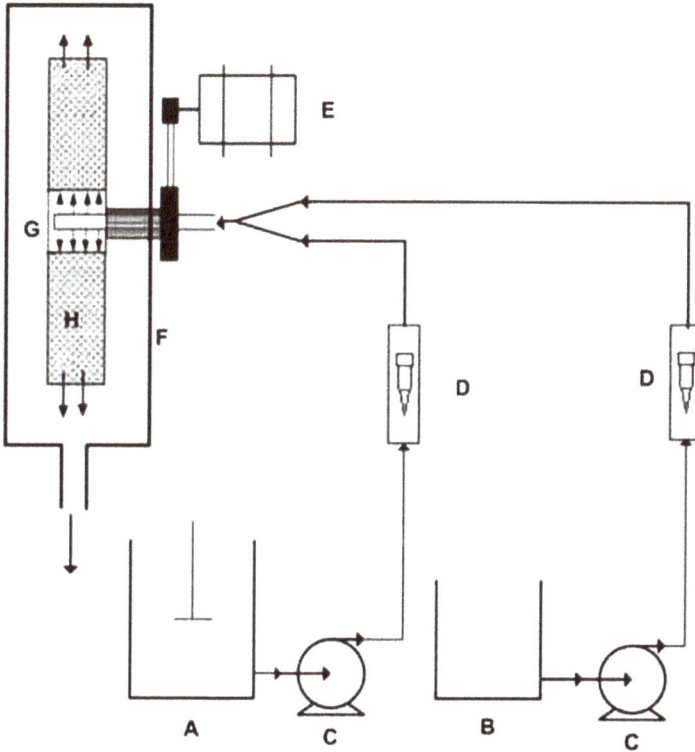

Fig. 2.7: Co-current one-stage liquid-liquid extraction process in a horizontal-axis RPB by Modak et al. [3]. A – Feed tank; B – Solvent tank; C – Pump; D – Rotameter; E – Electric motor; F – RPB casing; G – Liquid distriubtor; H – Packing. Reprinted from [3], with permission from Elsevier.

Yang et al. [4] also applied an on-stage co-current liquid-liquid extraction process in a horizontal-axis RPB. However, the liquid distribution in the eye of the rotor was designed to form two impinging streams. The complete process setup is shown in Fig. 2.8. Again, the contacting of the phases is conducted continuously, while the separation (settling) is conducted intermittently. A similar process concept was used by Chang et al. [12]; however, instead of a settling tank, the authors used a slit separator for de-mixing. From there, the two separate phases were discharged into collecting tanks. A process concept comprising three stages of counter-currently connected RPB will be shown in the detailed example in Section 2.4.

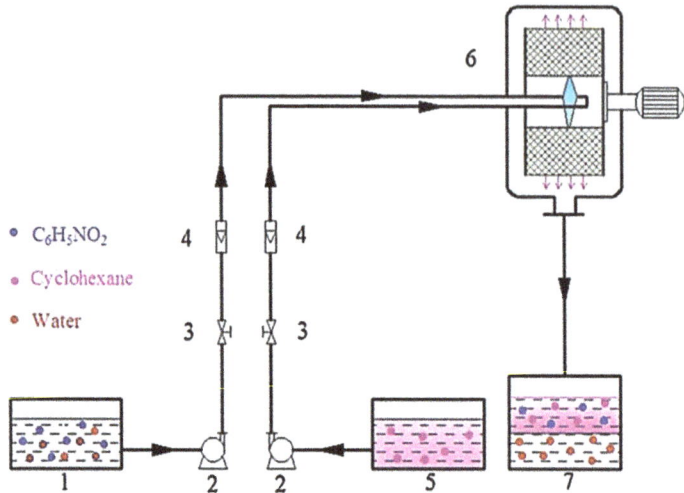

Fig. 2.8: Co-current one-stage liquid-liquid extraction process in a horizontal-axis RPB with impinging-stream liquid distribution by Yang et al. [4]. 1 – Nitrobenzene feed tank; 2 – Pumps; 3 – Valves; 4 – Rotameter; 5 – Cyclohexane feed tank; 6 – RPB; 7 – Settling tank. Reprinted from [4], with permission from Elsevier.

2.3.2 Rotating packed bed liquid-solid adsorption

Rotor design concepts

As highlighted in the applications section, one material particularly popular for the packed bed in RPB liquid-solid adsorption is activated carbon. However, activated carbon is usually supplied in the form of loose adsorbent particles, e.g., small rods or spheres. In order to be useful as an RPB packing, they need to be prepared before operation. First, the adsorbent is sieved and filtered. Afterwards, the adsorbent particles are washed with distilled water to remove dust and debris and then properly dried. Finally, the adsorbent particles are filled into a wire mesh basket to form the packed bed. A scanning electron microscope (SEM) image of loaded activated carbon from RPB experiments by Li et al. [11] is shown in Fig. 2.9.

In their RPB setup (Fig. 2.13), Chang and Lee [10] used a packed bed that was made from activated carbon particles with a median particle size of 0.82 mm and had a relatively low porosity of only 0.47. In comparison, rotating packed beds used for gas-liquid separation typically have a porosity of more than 0.9 (cf. Chapters 3 and 7).

In order to investigate the behavior of this low-voidage packed bed for liquid-solid adsorption, the residence time of liquid in the packing was measured with NaCl as the tracer for varying rotational speed and flow rates. It was shown that the residence time decreased for an increasing rotational speed and increasing flow rates. However, while the residence time decreased by about 60% when the flow

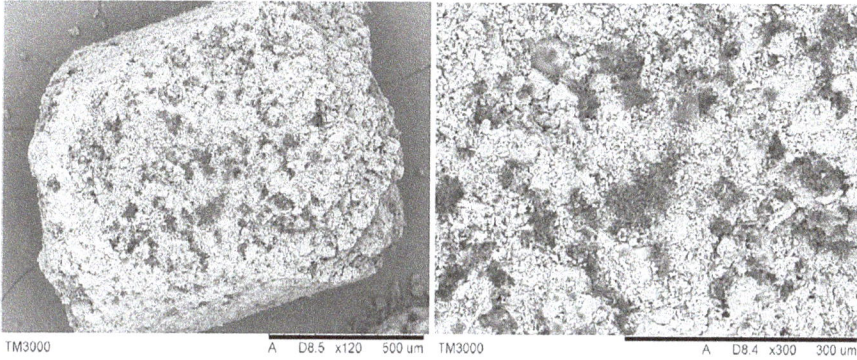

Fig. 2.9: SEM image of loaded activated carbon from RPB experiments by Li et al. [11]. Reprinted from [11], with permission from Elsevier.

rate was increased threefold, it only decreased by about 15% when the rotational speed was increased fivefold. This observation is contradictory to the ones for high-voidage packings [46]. It can be assumed that the low porosity avoided short circuits and increased the tortuosity of the liquid flow [10]. Thus, the influence on the liquid hold-up in the packing was diminished for the rotational speed and intensified for the liquid flow rate. As the hold-up decreased, so did the residence time, but to a much smaller extent.

In addition to activated carbon, various other materials can be used for the packed bed in liquid-solid adsorption. Wu et al. [30] utilized irregular and globular clinoptilolite for this purpose. The clinoptilolite was placed in an annular chamber in the middle of the rotor, which was constrained by rings of stainless-steel wire mesh on the inside and the outside interfaces (Fig. 2.10).

Contrary to the aforementioned adsorbent packed beds, made from activated carbon or clinoptilolite, Modak et al. [7] used a so-called biosorbent packed bed made from biological material. For this, the authors used raw rice husk that was filled into a wire mesh basket. The packed bed voidage was approximately 0.80. A picture of this packed rice husk bed is shown in Fig. 2.11.

As a third, completely different approach, Liu et al. [31] used a metal foam bed for their RPB liquid-solid adsorption process. Instead of a stainless-steel wire mesh basket or chamber filled with loose adsorbent particles, this bed consisted of only one solid but also very porous ring of metal foam (Fig. 2.12). With this packed bed, the authors studied the chemisorption of potassium dichromate from aqueous solution to copper [31]. They investigated foams with 20 ppi (pores per inch) and 40 ppi and found the latter ones, with smaller pores, to be strongly beneficial, especially at higher rotational speed.

In addition, Liu et al. [31] compared the results for both foam configurations to results from a conventional packed bed reactor (PBR) for the same chemisorption process and identical pore densities. The authors found that the mass transfer rate

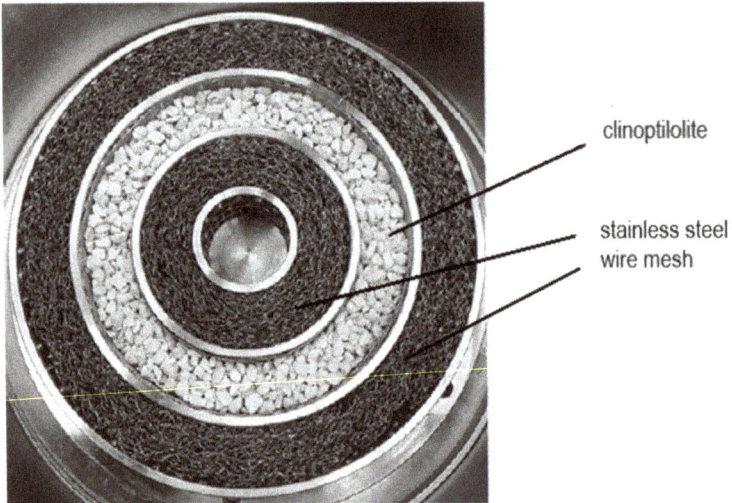

Fig. 2.10: Clinoptilolite packed bed used by Wu et al. [30]. Reprinted from [30], with permission from Elsevier.

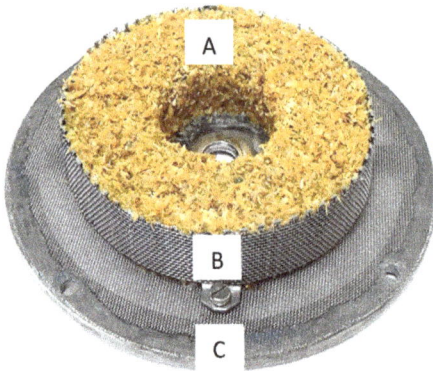

Fig. 2.11: Raw rice husk packed bed used by Modak et al. [7]. A – rice husk adsorbent; B – Wire mesh basket; C – Rotor plate. Reprinted from [7], with permission from Elsevier.

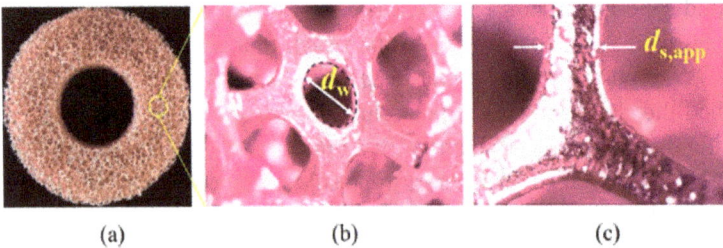

Fig. 2.12: Copper foam bed used by Liu et al. [31]. A – Copper foam ring; B – Detail view of foam pore; C – Detail view of foam strut. Reprinted from [31], with permission from Elsevier.

was up to five times higher for the RPB and ranged from 0.04 s^{-1} to 0.14 s^{-1} compared to 0.008 s^{-1} to 0.045 s^{-1} for the PBR.

Process concepts

Chang and Lee used a laboratory scale setup that combined a stirred tank reactor with a single-stage RPB [10]. A schematic of the process is shown in Fig. 2.13. An analogous process concept was used by Modak et al. [7], but with a horizontal-axis rotor.

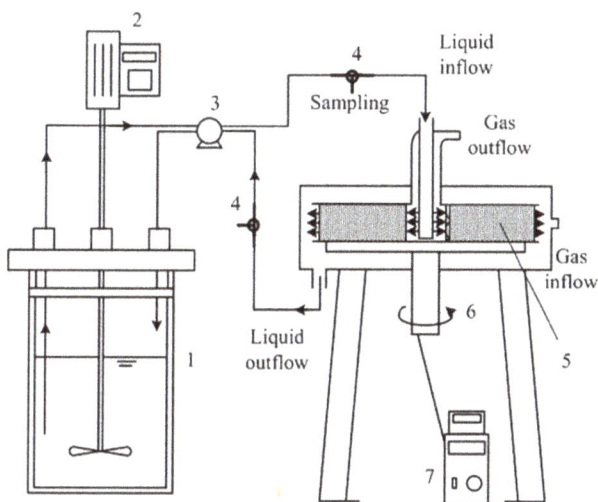

Fig. 2.13: Schematic of the liquid-solid adsorption process investigated by Chang and Lee [10]. 1 – tank; 2 – agitator; 3 – pump; 4 – sampling points; 5 – packed bed; 6 – rotating shaft; 7 – speed controller. Reprinted from [10], with permission from Elsevier.

Modak et al. [7] simulated the breakthrough curves for the adsorption of dye on rice husk and compared the results for one large single-stage RPB to those for a series of multiple smaller RPB. The respective curves are shown in Fig. 2.14. Curves A and B depict the results for one large single-stage RPB, whereas curves A' and B' depict the results for a series of three and six interconnected smaller RPB, respectively. The packing density was the same in all setups. The total volume and the amount of adsorbent in the larger packed beds were the same as the sum of the smaller ones.

As can be obtained from the curves in Fig. 2.14, the effluent concentration from the series of smaller RPB was for both investigated cases smaller than for one larger single-stage RPB. Additionally, the effect was more pronounced for the series of six small RPB (B'). This implies that a multistage setup increases the adsorption process efficiency in comparison with one larger single-stage RPB. It should additionally be noted in this context that also under redundancy and maintenance aspects,

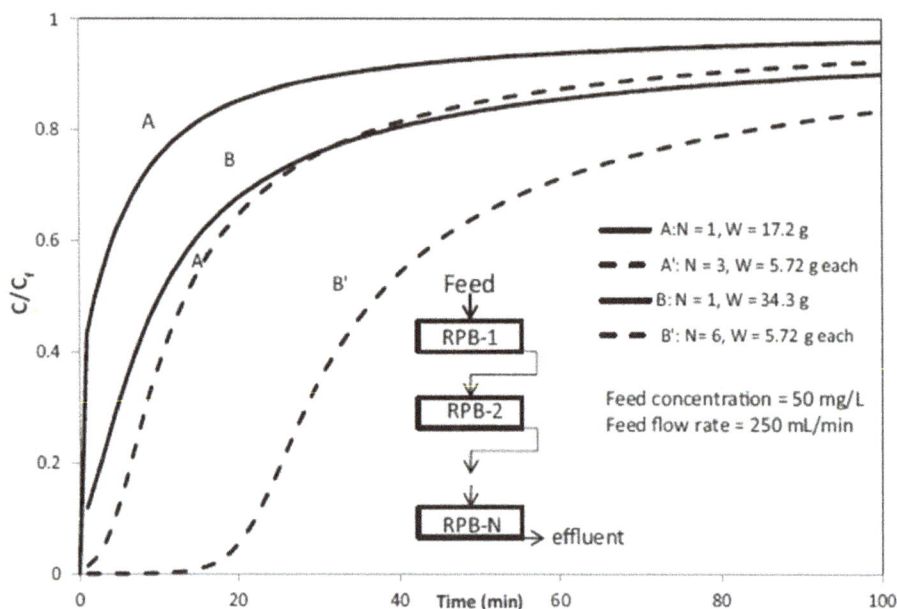

Fig. 2.14: Simulated breakthrough curves for one large single-stage RPB and a series of multiple smaller RPB [7]. Reprinted from [7], with permission from Elsevier.

a series of multiple interconnected RPB is preferrable to a single unit because it can help to reduce shut-down times and production losses.

2.4 Detailed example

Ever since the end of the twentieth century, the limitation of our natural resources, the rapidly increasing CO_2 content in the atmosphere, and the simultaneously increasing global demand for energy have become three of the most important drivers for innovation. Especially the dependence of the transport and the petrochemical sectors on fossil-based fuels and chemicals is one of the biggest political and environmental concerns of the twenty-first century. Alternative solutions could be derived from sustainable, bio-based platform chemicals [53, 54].

One of the currently most promising bio-based platform chemicals is γ-valerolactone (GVL), a "green" and versatile solvent, which could also be used to obtain fuel, polymers, and fine chemicals [55]. GVL is a stable, colorless liquid at room temperature. Its chemical structure is shown in Fig. 2.15, and its liquid-liquid extraction will be in the focus of this detailed example.

GVL can be obtained from water-rich lignocellulosic biomass waste by means of hydrolysis and hydration, with levulenic acid or ethyl luvinate as intermediates

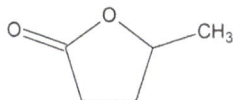

Fig. 2.15: Chemical structure of GVL.

[54, 56]. Consequently, the GVL has to be extracted from a thin aqueous solution, typically by liquid-liquid extraction [56]. Due to its relatively low costs and high selectivity, butyl acetate is currently believed to be one of the most promising solvents for GVL extraction [57]. However, the current extraction methods have not yet met their full potential, leaving room for new, innovative approaches [54]. This is because, in the conventionally utilized liquid-liquid extraction equipment, the economically necessary reduction of equipment volume and amount of solvent is restricted by residence time and mass transfer limitations. In contrast, as shown for the various applications in this chapter, an RPB-based liquid-liquid extraction process requires little space, has a high processing capacity, and is potentially very efficient, with reported extraction efficiencies of close to 100%.

According to Goedeke [58], extraction processes with high energy demand, as it would potentially be the case for an RPB-based process, should be limited to three to four stages to be economically feasible. Therefore, a three-stage process is considered for this example. For feed flow rate, feed composition, and necessary raffinate purity, examplary values reported by Murat Sen et al. [59] are used. Additionally, a pure solvent stream of 20 kmol h^{-1} is assumed, based on considerations regarding the minimum required amount of solvent for the considered three-stage countercurrent extraction process. Based on the very high extraction efficiency reported for RPB in the literature, ideal extraction stages ($\eta = 100\%$) are assumed for this example. Furthermore, ambient pressure and temperature (22 °C) can be assumed for the process.

A complete conceptual design for the considered RPB-based three-stage countercurrent process is shown in Fig. 2.16. The underlying flow rates and the stage construction in the ternary diagram are shown in Tab. 2.6 and Fig. 2.17, respectively.

Tab. 2.6: Molar flow rates and compositions of the concept process presented in this detailed example. The exemplary feed composition was obtained from Murat Sen et al. [59].

	$\dot{n}_{GVL}$[kmol h^{-1}]	$\dot{n}_{H_2O}$[kmol h^{-1}]	$\dot{n}_{ButA}$[kmol h^{-1}]	$\dot{n}_{total}$[kmol h^{-1}]
Feed	31.71	229.12	0.00	260.83
Solvent	0.00	0.00	20.00	20.00
	x_{GVL} [−]	x_{H_2O} [−]	x_{ButA} [−]	Σx_i [−]
Feed	0.122	0.878	0.000	1.000
Solvent	0.000	0.000	1.000	1.000

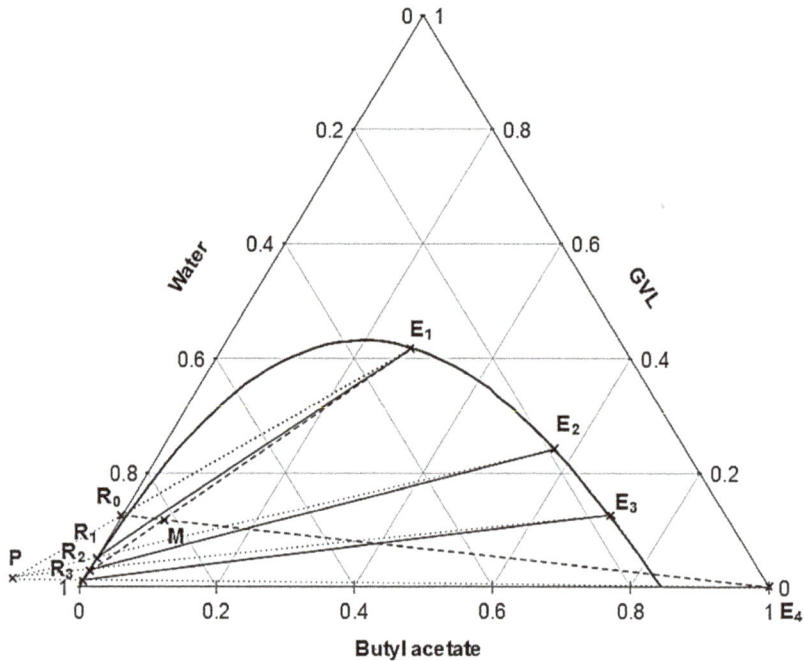

Fig. 2.16: Stage construction of the three-stage countercurrent concept process for the extraction of GVL from water-rich feed with pure butyl acetate as solvent at 22 °C. Raffinate and extract compositions are given as R0 to R3 and E1 to E4, respectively.

The mixing point is denoted as M, and the pole point is denoted as P. Tie-lines are given as full lines, helping lines of the mixing point construction are given as dashed lines, and helping lines of the pole construction are given as dotted lines.

Fresh solvent (E4, pure butyl acetate) is supplied at a rate of 20 kmol h^{-1} from container C4 and pumped to the third RPB unit (RPB3) by pump P9. In RPB3, it is contacted with raffinate R2 from the second RPB unit (RPB2). Both phases leave RPB3 through the liquid outlets (238 kmol h^{-1}) and are pumped to the third centrifuge unit (S3), where they are separated into the final GVL-lean raffinate (R3, 210 kmol h^{-1}) and the first extract (E3, 28 kmol h^{-1}). The final raffinate is stored in container C3, while the extract is pumped to RPB2 by pump P7. Here, the extract is contacted with raffinate R1 from the first RPB unit (RPB1). After leaving the second RPB unit, the liquid stream (253 kmol h^{-1}) is pumped to the second centrifuge unit (S2) by pump P5. There, the liquid stream is separated into the raffinate phase (R2, 218 kmol h^{-1}), which is pumped to RPB3 by pump P6, and the extract phase, which is pumped to RPB R1 by pump P4 (E2, 35 kmol h^{-1}). Here, the extract is contacted with the water-rich feed solution (R0), which is stored in container C1 and pumped to RPB1 by pump P1 (261 kmol h^{-1}). After leaving RPB1, the liquid stream (296 kmol h^{-1}) is pumped to the first centrifuge unit (S1) by pump P2. In centrifuge unit S1, the liquid is separated

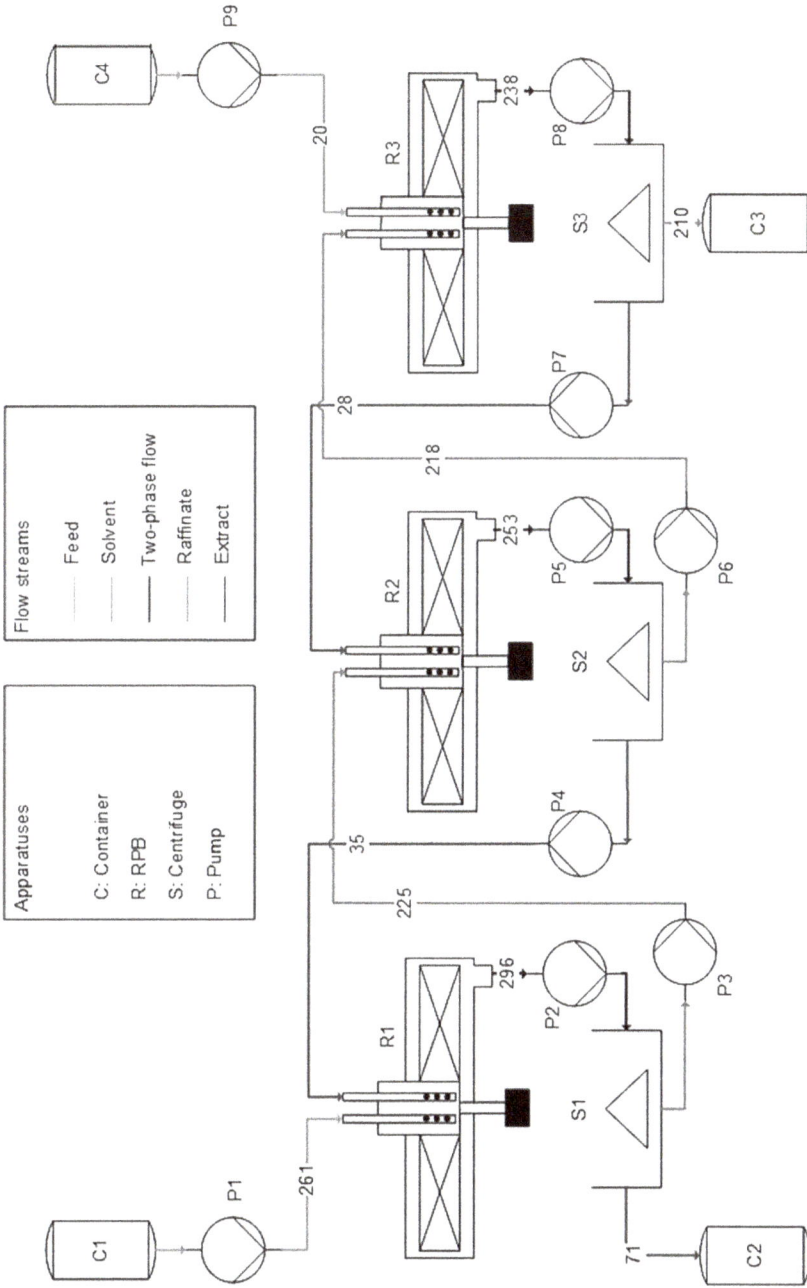

Fig. 2.17: Conceptual design for the three-stage countercurrent RPB extraction process.

into the GVL-rich final extract phase (E1, 71 kmol h^{-1}), which is stored in container C2, and the first raffinate (R1, 225 kmol h^{-1}), which is pumped to the second RPB unit R2 by pump P3. The final extract E1 contains 41.5% GVL, whereas the final raffinate contains 1.2% GVL. With respect to the feed composition, 92.4% of the initial GVL amount could be extracted in the RPB-based concept process with three stages and 20 kmol h^{-1} of pure solvent, resulting in a raffinate purity of 98.8%. In comparison, Murat et al. [59] assumed an extraction column with 20 stages and 90 kmol h^{-1} of solvent stream (70 wt.% butyl acetate, 30 wt.% water) to be necessary to achieve the same raffinate purity in a conventional liquid-liquid extraction process.

In conclusion, the presented detailed example demonstrates a promising extraction process concept by means of three counter-currently connected RPB with associated centrifuge units. In such a process, a final raffinate purity of up to 98.8% could be achieved within three stages due to the very high efficiency of RPB for mixing and extraction tasks, which implies high competitiveness in comparison with the conventional extraction process assumed in the literature.

However, it should also be noted that the research in this field is still in an early stage and that the mass transfer limitations in the system GVL/water/butyl acetate are comparatively low. Therefore, RPB with a high radial packing depth, as considered in this example and presented in the previous sections, might not be necessary to meet the raffinate constraints in this example. Moreover, investment and operation costs of RPB are typically higher than those of conventional equipment. An alternative RPB design, the RRR, was shown in Section 2.3.1. This design could also well be used for the investigated liquid-liquid extraction process and would potentially offer sufficiently high mass transfer rates, reduced equipment volume, lower investment costs, and a lower residence time in each stage.

2.5 Take home messages

– Liquid-liquid extraction and liquid-solid adsorption are popular separation methods for industrial wastewater treatment as well as separation processes in the pharmaceutical and the biochemical industries.
– For liquid-liquid extraction processes, the use of RPB can be advantageous, as they offer a fine liquid distribution and profound liquid droplet breakup, leading to an intensified phase contacting and high mass transfer rates.
– For liquid-solid adsorption processes, RPB have a strong potential due to their ability to process large amounts of liquid in a small equipment volume, by means of higher mass transfer rates than found in conventional adsorption equipment.
– The liquid flow through an RPB packing, and the liquid hold-up inside the packing, largely determine the efficiency of both liquid-liquid extraction and

liquid-solid adsorption processes. Therefore, an informed decision on the operation parameters as well as an appropriate packing design are necessary.

– Ideally, the RPB packing should be designed in a way that allows for a long
residence time of liquid phases in the packing, without leading to an excessive
liquid hold-up.

– For the two most important operational parameters, the rotational speed and the
liquid flow rate, typically an optimum operation point can be found, representing
the best compromise between intensified mass transfer and improved hydraulic
conditions on the one side and a reduced residence time on the other side.

– Once the capacity of an RPB is exceeded, or the process efficiency from a given
RPB does not seem to be sufficient, it is favorable to use a multistage setup
with multiple RPB in series, instead of using one larger single-stage RPB. This
approach is also beneficial under redundancy and maintenance aspects because it can help to reduce shut-down times and production losses.

2.6 Quiz

1. Which are currently the most important fields of application for liquid-liquid
    and liquid-solid RPB processes?
2. What are the two most important types of liquid distribution for liquid-liquid
    extraction? How do they work?
3. What types of liquid flow patterns typically exist in an RPB in liquid-liquid ex-
    traction processes?
4. The extraction rate for a liquid-liquid extraction stage only considers the initial
    and the final concentration of the solute. What else must be considered for the
    calculation of the extraction stage efficiency?
5. Which effects contribute to the influence of the rotational speed on the volu-
    metric mass transfer rate for RPB liquid-liquid extraction?
6. True or false? An increasing rotational speed is always beneficial for RPB liq-
    uid-liquid extraction.
7. Which effects contribute to the influence of the liquid flow rate on the volumet-
    ric mass transfer rate in RPB liquid-liquid extraction?
8. True or false? An increasing volumetric flow rate is always beneficial for RPB
    liquid-liquid extraction.
9. Which types of material have been used in RPB adsorption processes? Which is
    the most popular one?
10. True or false? The loose packing material for adsorption processes can be filled
    directly into the RPB casing.
11. What is the meaning of packing tortuosity? How does it influence RPB adsorp-
    tion processes?

12. True or false? For the same amount of adsorbing packing material, one larger RPB is preferable to a series of smaller RPB.

2.7 Exercise

For this exercise, assume that you are the lead engineer for the revamping of a nitrobenzene production facility. Your boss wants the production capacity to be doubled, but the available space is very limited.

Exercise 1. Why do you need to consider a wastewater treatment for your revised plant? Why could it reasonably be conducted with an RPB by means of liquid-liquid extraction?

You reach out to a vendor of process equipment for a quotation on an RPB and get the offer in Tab. 2.7.

Tab. 2.7: RPB offer you received from the vendor.

	RPB
Inner packing diameter	150 mm
Outer packing diameter	800 mm
Axial packing height	10 mm
Packing voidage	50%
Maximum liquid capacity	200 L h^{-1}
Maximum rotational speed	2000 rpm

The vendor agrees to let you test one sample RPB in your technical hall to see if it fits your requirements. You conduct a series of experiments and obtain the following values in Tab. 2.8.

Tab. 2.8: Data from your experiments.

	RPB
Flow rate	100 L h^{-1}
Rotational speed	1000 rpm
Feed concentration	1200 mg/L
Raffinate concentration	240 mg/L
Equilibrium concentration	120 mg/L

Exercise 2. What is the removal efficiency and the extraction stage efficiency in your experiments?

Exercise 3. What is the $k_L a$ value in your experiments?

Exercise 4. How could the RPB be operated instead to further improve the results? Could there also be a downside to this?

2.8 Solutions

The quiz questions can be answered based on the text. The answers to the exercise, which cannot be directly found within this chapter, but are transfer knowledge are explained in this section.

Exercise 1. Since nitrobenzene is a hazardous, highly toxic, cancerogenic, and environmentally dangerous substance, its removal from wastewater is of great importance (cf. chapter 2.2.1). Therefore, you need to consider a wastewater treatment for your revised plant.

This treatment could reasonably be conducted with an RPB by means of liquid-liquid extraction since RPB offer significant potential benefits for liquid-liquid extraction processes such as a fine liquid distribution and a profound liquid droplet breakup, leading to an intensified initial contacting of phases and high mass transfer rates between these phases (cf. chapter 2.1.1).

Exercise 2. The removal efficiency in your RPB experiments can be calculated according to eq. (2.7) as:

$$E = \frac{C_F - C_R}{C_F} = \frac{1200\,\frac{mg}{L} - 240\,\frac{mg}{L}}{1200\,\frac{mg}{L}} = 0.80 = 80.0\,\% . \tag{2.12}$$

The extraction stage efficiency in your RPB experiments can be calculated according to eq. (2.8) as:

$$\eta = \frac{C_F - C_R}{C_F - C^*} = \frac{1200\,\frac{mg}{L} - 240\,\frac{mg}{L}}{1200\,\frac{mg}{L} - 120\,\frac{mg}{L}} = 0.8\overline{8} = 88.\overline{8}\,\% . \tag{2.13}$$

Exercise 3. The $k_L a$ value in your RPB experiments can be calculated according to eq. (2.5) as:

$$k_L a = \frac{Q_a}{V_{void}} * \ln\left(\frac{C^* - C_F}{C^* - C_R}\right) . \tag{2.14}$$

Here, V_{void} is calculated according to eq. (2.6) as:

$$V_{void} = h * \frac{\pi}{2} * (r_o^2 - r_i^2) * \varepsilon = 10\,\text{mm} * \frac{\pi}{2} * \left(\left(\frac{800\,\text{mm}}{2} \right)^2 - \left(\frac{150\,\text{mm}}{2} \right)^2 \right) * 0.5$$

$$= 1,212,458\,\text{mm}^3 \approx 1.2\,L. \tag{2.15}$$

With this value for V_{void}, we obtain $k_L a$ from eq. (2.14) as:

$$k_L a = \frac{Q_a}{V_{void}} * \ln \left(\frac{C^* - C_F}{C^* - C_R} \right) = \frac{100\,\frac{L}{h}}{1.2\,L} * \ln \left(\frac{120\,\frac{mg}{L} - 1200\,\frac{mg}{L}}{120\,\frac{mg}{L} - 240\,\frac{mg}{L}} \right) = 750\,\text{h}^{-1} = 0.208\bar{3}\,\text{s}^{-1}. \tag{2.16}$$

Exercise 4. When comparing the maximum rotational speed of your RPB and the rotational speed applied in your experiments, it can be seen that the rotational speed could be increased. As discussed for the applications in chapter 2.2.1, an increased rotational speed can lead to higher mass transfer rates due to finer liquid droplets and thinner liquid films. However, a higher rotational speed also leads to a shorter residence time of the liquid phases in the packing, which can be disadvantageous for the extraction process.

Furthermore, based on the values given for the maximum liquid capacity of your RPB and the liquid flow rate applied in your experiments, it can be obtained that the flow rate could also be increased. A higher flow rate can lead to a better utilization of the packing volume, to a larger interfacial area available for mass transfer, and a more vigorous interaction between the distributed liquid phases (cf. chapter 2.2.1). However, a higher flow rate can also lead to a shorter residence time in the packing. Additionally, an excessive liquid load on the inner rim of the packing – due to an excessively high liquid flow rate – can be disadvantageous for the extraction process, as it negatively affects the initial contacting of the phases.

List of symbols

Latin letters

a_{eff}	effective interfacial area per unit volume	$m^2\,m^{-3}$
c	length of a curved pore in RPB packing	m
C^*	equilibrium concentration of the solute	$mol\,m^{-3}$
C_0	initial concentration of the adsorptive	$kg\,m^{-3}$
C_a	concentration of the solute in the aqueous phase	$mol\,m^{-3}$
C_e	equilibrium concentration of the adsorptive	$kg\,m^{-3}$
C_F	concentration of the solute in the feed	$mol\,m^{-3}$
C_R	concentration of the solute in the raffinate	$mol\,m^{-3}$
dV	discrete, differential liquid volume	m^3

H	axial height of RPB packing	m
K	mass transfer rate	m s^{-1}
K_i	capacity of the solvent with respect to component i	–
k_L	mass transfer rate – liquid/liquid	m s^{-1}
$k_{L/S}$	mass transfer rate – liquid/solid	m s^{-1}
$k_{L/S}a$	volumetric mass transfer rate – liquid/solid	s^{-1}
$k_L a$	volumetric mass transfer rate – liquid/liquid	s^{-1}
L	direct distance between a pore's ends in RPB packing	m
$\dot{n}$	molar flow rate	kmol h^{-1}
q	amount of adsorbate	kg
Q	flow rate	$\text{m}^3 \text{ s}^{-1}$
Q_a	flow rate of the aqueous phase	$\text{m}^3 \text{ s}^{-1}$
q_e	equilibrium amount of adsorbate	kg
r_i	inner radius of RPB packing	m
r_o	outer radius of RPB packing	m
$S_{S/C}$	selectivity of the solvent with respect to solute and carrier	–
W	mass of dry adsorbent	kg
x	molar fraction	mol mol^{-1}
x_C^E	molar fraction of the carrier in the extract	mol mol^{-1}
x_C^R	molar fraction of the carrier in the raffinate	mol mol^{-1}
x_i^E	molar fractions of component i in the extract	mol mol^{-1}
x_i^R	molar fractions of component i in the raffinate	mol mol^{-1}
x_S^E	molar fraction of solute in the extract	mol mol^{-1}
x_S^R	molar fraction of solute in the raffinate	mol mol^{-1}

Greek letters

ε	voidage of RPB packing	–
τ	tortuosity of RPB packing	–

Subscripts

0	initial concentration
a	aqueous phase
ButA	butyl acetate
c	carrier
e	equilibrium
eff	effective
F	feed
GVL	γ-valerolactone
H_2O	water
i	index for components
i	inner radius
L	liquid side
o	outer radius
R	raffinate
s	solute

S/C	solute to carrier
total	all components

Superscripts

*	equilibrium
E	extract
R	raffinate

References

[1] El-Nadi YA. Solvent extraction and its applications on ore processing and recovery of metals: Classical approach. Sep. Purif. Rev. 2017;46(3):195–215.

[2] Karmakar S, Bhowal A, Das P. A comparative study of liquid-liquid extraction in different rotating bed contactors. Chem. Eng. Process. Process Intensif. 2018;132:187–93.

[3] Modak JB, Bhowal A, Datta S. extraction of dye from aqueous solution in rotating packed bed. J. Hazard. Mater. 2016;304:337–42.

[4] Yang P-F, Luo S, Zhang D-S, Yang P-Z, Liu Y-Z, Jiao W-Z. Extraction of nitrobenzene from aqueous solution in impinging stream-rotating packed bed. Chem. Eng. Process. Process Intensif. 2018;124:255–60.

[5] Pfennig A, Pilhofer T, Schröter J. Kapitel 10: Flüssig-Flüssig-Extraktion. In: Fluidverfahrenstechnik – Grundlagen, Methodik, Technik, Praxis. Goedecke R, ed. Weinheim, Germany: Wiley-VCH 2011, 907–92.

[6] Wenzel DA. Liquid processing in Rotating Packed Beds: Technische Universität Dortmund; 2020.

[7] Modak JB, Bhowal A, Datta S. Experimental study and mathematical modeling of breakthrough curve in rotating packed bed. Chem. Eng. Process. Process Intensif. 2016;99:19–24.

[8] Ng Y-S, Tan Y-T, Chua ASM, Hashim MA, Sen Gupta B. Removal of nickel from water using rotating packed bed contactor: Parametric studies and mode of operations. J. Water Process Eng. 2020;36:101286.

[9] Bathen D, Breitbach M. Adsorptionstechnik. Berlin, Heidelberg: Springer-Verlag; 2001.

[10] Chang C-F, Lee S-C. Adsorption behavior of pesticide methomyl on activated carbon in a high gravity rotating packed bed reactor. Water Res. 2012;46(9):2869–80.

[11] Li W, Yan J, Yan Z, Song Y, Jiao W, Qi G, et al. Adsorption of phenol by activated carbon in rotating packed bed: Experiment and modeling. Appl. Therm. Eng. 2018;142:760–66.

[12] Chang J, Jia F, Srinivasakannan C, Mumford KA, Yang X. Impure ions removal from multicomponent leach solution of nickel sulfide concentrates by solvent extraction in impinging stream rotating packed bed. Chem. Eng. Process. Process Intensif. 2019;137:54–63.

[13] Karmakar S, Bhowal A, Das P. Process intensification of liquid-liquid extraction in rotating packed bed. Msf. 2020;998:146–50.

[14] Sigma Aldrich. Methyl Red ACS reagent product sheet. [April 03, 2021]; Available from: https://www.sigmaaldrich.com/catalog/product/sial/250198.

[15]   Akgül M. Enhancement of the anionic dye adsorption capacity of clinoptilolite by Fe(3+)-grafting. J. Hazard. Mater. 2014;267:1–8.

[16]   Goel NK, Kumar V, Pahan S, Bhardwaj YK, Sabharwal S. Development of adsorbent from Teflon waste by radiation induced grafting: Equilibrium and kinetic adsorption of dyes. J. Hazard. Mater. 2011;193:17–26.

[17]   Fabri J, Graeser U, Simo TA. Ullmann's Encyclopedia of Industrial Chemistry: Electronic Release. CD. Weinheim: Wiley-VCH; 2003.

[18]   Baier G, Graham MD, Lightfoot EN. Mass transport in a novel two-fluid taylor vortex extractor. AIChE J. 2000;46(12):2395–407.

[19]   Bonam D, Bhattacharyya G, Bhowal A, Datta S. Liquid–liquid extraction in a rotating-spray column: Removal of Cr(VI) by aliquat 336. Ind. Eng. Chem. Res. 2009;48(16):7687–93.

[20]   Seibert AF, Fair JR. Hydrodynamics and mass transfer in spray and packed liquid-liquid extraction columns. Ind. Eng. Chem. Res. 1988;27(3):470–81.

[21]   Hossein A, Ali MM, Reza RS. The effects of a surfactant concentration on themass transfer in a mixer-settler extractor. Iran. J. Chem. Eng. 2006;25:9–15.

[22]   Wenzel D, Gerdes N, Steinbrink M, Ojeda LS, Górak A. Liquid distribution and mixing in rotating packed beds. Ind. Eng. Chem. Res. 2019;58(15):5919–28.

[23]   Litaiem Y, Dhahbi M. Physicochemical properties of an hydrophobic ionic liquid (Aliquat 336) in a polar protic solvent (Formamide) at different temperatures. J Dispers. Sci. Technol. 2015;36(5):641–51.

[24]   Booth G. Nitro Compounds, Aromatic: Ullmann's Encyclopedia of Industrial Chemistry. 6th ed, Weinheim: Wiley-VCH; 2003.

[25]   Xi J-B. Effects of nitrobenzenes on DNA damage in germ cells of rats. Chem. Res. Chin. Univ. 2006;22(1):29–32.

[26]   Philipp B, Stevens P, Philipp/Stevens. Grundzüge der Industriellen Chemie: Ein Einführendes Lehr- Und Lernbuch. Weinheim: VCH; 1987.

[27]   Zhu Z, Yoko P, Cheng CY. Recovery of cobalt and manganese from nickel laterite leach solutions containing chloride by solvent extraction using Cyphos IL 101. Hydrometallurgy. 2017;169:213–18.

[28]   Jafari H, Abdollahi H, Gharabaghi M, Balesini AA. Solvent extraction of zinc from synthetic Zn-Cd-Mn chloride solution using D2EHPA: Optimization and thermodynamic studies. Sep. Purif. Technol. 2018;197:210–19.

[29]   Hashim MA, Kundu A, Mukherjee S, Ng Y-S, Mukhopadhyay S, Redzwan G, et al. Arsenic removal by adsorption on activated carbon in a rotating packed bed. J. Water Process Eng. 2019;30:100591.

[30]   Wu Y, Chang -C-C, Guan C-Y, Chang -C-C, Li J-W, Chang C-Y, et al. Enhanced removal of ammonium from the aqueous solution using a high-gravity rotating packed bed loaded with clinoptilolite. Sep. Purif. Technol. 2019;221:378–84.

[31]   Liu Y, Li Z, Chu G, Shao L, Luo Y, Chen J. Liquid-solid mass transfer in a rotating packed bed reactor with structured foam packing. Chin. J. Chem. Eng. 2020;28(10):2507–12.

[32]   PubChem. Methomyl – Compound Summary: CID 4109. [May 09, 2021]; Available from: https://pubchem.ncbi.nlm.nih.gov/compound/Methomyl.

[33]   Japanese Ministry of Environment. Strategic Programs on Environmental Endocrine Disruptors '98 (SPEED '98). [May 09, 2021]; Available from: http://www.env.go.jp/en/chemi/ed/speed98/sp98t3.html.

[34]   U.S. EPA. Methomyl R.E.D. Facts. [May 09, 2021]; Available from: https://www3.epa.gov/pesticides/chem_search/reg_actions/reregistration/fs_PC-090301_1-Dec-98.pdf.

[35] Chuah TG, Jumasiah A, Azni I, Katayon S, Thomas Choong SY. Rice husk as a potentially low-cost biosorbent for heavy metal and dye removal: An overview. Desalination. 2005;175 (3):305–16.

[36] Han R, Wang Y, Yu W, Zou W, Shi J, Liu H. Biosorption of methylene blue from aqueous solution by rice husk in a fixed-bed column. J. Hazard. Mater. 2007;141(3):713–18.

[37] Hamdaoui O, Naffrechoux E. Modeling of adsorption isotherms of phenol and chlorophenols onto granular activated carbon. Part I. Two-parameter models and equations allowing determination of thermodynamic parameters. J. Hazard. Mater. 2007;147(1-2):381–94.

[38] Merck & Co. The Merck Index: An Encyclopedia of Chemicals, Drugs, and Biologicals. 15th edition, Cambridge: RSC Publ. Royal Soc. of Chemistry; 2013.

[39] Wang S, Gao B, Zimmerman AR, Li Y, Ma L, Harris WG, et al. Removal of arsenic by magnetic biochar prepared from pinewood and natural hematite. Bioresour. Technol. 2015;175:391–95.

[40] Guglielmi G. Arsenic in drinking water threatens up to 60 million in Pakistan. Science. 2017. https://www.science.org/content/article/arsenic-drinking-water-threatens-60-million-paki stan Acessed: 06.06.2022 doi: 10.1126/science.aap7590

[41] Roghani M, Nakhli SAA, Aghajani M, Rostami MH, Borghei SM. Adsorption and oxidation study on arsenite removal from aqueous solutions by polyaniline/polyvinyl alcohol composite. J. Water Process Eng. 2016;14:101–07.

[42] Liikanen A, Martikainen PJ. Effect of ammonium and oxygen on methane and nitrous oxide fluxes across sediment–water interface in a eutrophic lake. Chemosphere. 2003;52 (8):1287–93.

[43] Huo H, Lin H, Dong Y, Cheng H, Wang H, Cao L. Ammonia-nitrogen and phosphates sorption from simulated reclaimed waters by modified clinoptilolite. J. Hazard. Mater. 2012;229-230:292–97.

[44] Wenzel D, Górak A. Review and analysis of micromixing in rotating packed beds. Chem. Eng. J. 2018;345:492–506.

[45] Guo K, Guo F, Feng Y, Chen J, Zheng C, Gardner NC. Synchronous visual and RTD study on liquid flow in rotating packed-bed contactor. Chem Eng Sci. 2000;55(9):1699–706.

[46] Burns JR, Jamil JN, Ramshaw C. Process intensification: Operating characteristics of rotating packed beds – Determination of liquid hold-up for a high-voidage structured packing. Chem Eng Sci. 2000;55(13):2401–15.

[47] Burns JR, Ramshaw C. Process intensification: Visual study of liquid maldistribution in rotating packed beds. Chem. Eng. Sci. 1996;51(8):1347–52.

[48] Yang Y, Xiang Y, Chu G, Zou H, Luo Y, Arowo M, et al. A noninvasive X-ray technique for determination of liquid holdup in a rotating packed bed. Chem Eng Sci. 2015;138:244–55.

[49] Li Y, Lu Y, Liu X, Wang G, Nie Y, Ji J. Mass-transfer characteristics in a rotating zigzag bed as a Higee device. Sep. Purif. Technol. 2017;186:156–65.

[50] Wang GQ, Xu ZC, Yu YL, Ji JB. Performance of a rotating zigzag bed – A new HIGEE. Chem. Eng. Process. Process Intensif. 2008;47(12):2131–39.

[51] Wenzel D, Górak A. Patent: Rotierender Ring Reaktor; 2019.

[52] Wenzel D, Nolte K, Górak A. Reactive mixing in rotating packed beds: On the packing's role and mixing modeling. Chem. Eng. Process. Process Intensif. 2019;143:107596.

[53] Kerkel F, Markiewicz M, Stolte S, Müller E, Kunz W. The green platform molecule gamma-valerolactone – Ecotoxicity, biodegradability, solvent properties, and potential applications. Green Chem. 2021;23(8):2962–76.

[54] Serrano-Ruiz JC, Braden DJ, West RM, Dumesic JA. Conversion of cellulose to hydrocarbon fuels by progressive removal of oxygen. Appl Catal B. 2010;100(1-2):184–89.

[55] Albani D, Li Q, Vilé G, Mitchell S, Almora-Barrios N, Witte PT, et al. Interfacial acidity in ligand-modified ruthenium nanoparticles boosts the hydrogenation of levulinic acid to gamma-valerolactone. Green Chem. 2017;19(10):2361–70.

[56] Alonso DM, Wettstein SG, Dumesic JA. Gamma-valerolactone, a sustainable platform molecule derived from lignocellulosic biomass. Green Chem. 2013;15(3):584.

[57] Braden DJ, Henao CA, Heltzel J, Maravelias CC, Dumesic JA. Production of liquid hydrocarbon fuels by catalytic conversion of biomass-derived levulinic acid. Green Chem. 2011;13(7):1755.

[58] Goedecke R. Fluidverfahrenstechnik – Grundlagen, Methodik, Technik, Praxis: Kapitel 10: Flüssig-Flüssig-Extraktion; pp. 907–92; 2011.

[59] Murat Sen S, Henao CA, Braden DJ, Dumesic JA, Maravelias CT. Catalytic conversion of lignocellulosic biomass to fuels: Process development and technoeconomic evaluation. Chem Eng Sci. 2012;67(1):57–67.

Kai Groß

3 Rotating packed bed hydraulics and liquid-side limited mass transfer

Comparable to other gas- or vapor-liquid contacting equipment (e.g., columns), the design procedure of RPBs consists of hydraulic and mass transfer design requirements. **The superposition of gravitational acceleration with centrifugal acceleration allows to modify existing correlations and transfer knowledge available for conventional gas-liquid contacting equipment (e.g., tray or packed columns) to a high gravity environment.** However, there are still some characteristics in the operation of RPBs, which are not obvious at first sight and require detailed process understanding, which will be pursued in the current chapter.

The hydraulic requirements contain the investigation of valid operating ranges for different gas and liquid flow rates, and pressure drop and power consumption generate important information for sizing and evaluation of energy efficiency of RPBs. Empirical correlations enable a quick assessment of potential RPB configurations. What has been estimated to be hydrodynamically feasible eventually needs further characterization of the mass transfer performance. The determination of HETP-values or mass transfer coefficients is inevitable when designing an RPB.

The current chapter provides a foundation introducing hydrodynamic topics which will be extended in the subsequent chapters for the specific rotor design (cf. Chapter 5), packing design (cf. Chapter 7), and multiphase flow simulations (cf. Chapter 9). Moreover, it connects the introduction provided in Chapter 1 to the specific application-oriented chapters for flue gas cleaning (cf. Chapter 4) and distillation (cf. Chapters 6 and 8).

The focus is set on the hydraulic design of RPBs and the determination of important parameters. Most of the experiments in hydraulic studies are conducted with air and water. The system is very suitable since it relies on nonhazardous, readily available, and cheap substances. The similarity of hydraulic air/water experiments and the determination of the liquid-side limited mass transfer coefficients by deaeration or aeration of water suggest an excursion to the liquid-side limited mass transfer systems which will be done in the later part of this chapter.

The fundamentals section (Section 3.1) addresses specific hydraulic phenomena. The rotational speed as additional degree of freedom and its influence on operating range, pressure drop, liquid hold-up, or power consumption are evaluated. Different technologies developed to investigate the hydraulic behavior of the RPB are introduced and explained. The section concludes on general operating principles and

Kai Groß, TU Dortmund University

https://doi.org/10.1515/9783110724998-003

limitations. Computational fluid dynamics are excluded from the contents of this section and will be addressed in Chapter 9.

Moreover, the fundamentals section addresses the liquid-side limited mass transfer. Starting from two-film theory gradually leading to mass transfer coefficients. The advantages of RPBs and their application in liquid-side limited mass transfer tasks are elucidated. Different zones of mass transfer in the rotating packing and in the enclosing compartment (e.g., static casing) are introduced, and their importance is highlighted.

The modeling and design section (Section 3.2) introduces typical models applied in gas-liquid and vapor-liquid contacting. It includes general considerations necessary when modeling RPB behavior focusing on the changes occurring along the radial packing length. To further foster the gained insights on hydraulic and mass transfer operation, the detailed examples section (Section 3.3) provides an in-depth analysis of the performance of a high surface metal foam. Furthermore, the deaeration of liquids as model system for liquid-side limited mass transfer is presented in more detail.

3.1 Fundamentals

The following section addresses general concepts of the RPB operation. It builds on the introduction provided in Chapter 1, providing a more extensive assessment of the operational characteristics.

3.1.1 Hydrodynamics

Basically, the effects observed in tray and packed columns as described by Sherwood et al. [1], Goedecke et al. [2] or Górak et al. [3–5] can be transferred to the RPB environment. However, it should be noted that:

1. The centrifugal acceleration is an additional degree of freedom influencing performance and cost of the equipment.
2. In a standard rotor geometry with fixed rotor height, the internals need to deal with the changing cross-sectional area from the inner to the outer packing diameter.
3. In radial direction, liquid and vapor or gas experience a change of the centrifugal acceleration.
4. Other non-ideal effects as slippage between fluids and the rotor or backmixing may occur.

The following paragraphs address the different operational aspects related to the hydrodynamics of the RPB.

Pressure drop

The pressure drop observed in RPBs consists of multiple factors. The two main contributions are the frictional pressure loss of the gas flowing through the porous packing and the centrifugal head, a compressor-like behavior, generated by the rotation [6]. With variation in the rotational speed, the pressure drop shows a characteristic behavior (cf. Fig. 3.1). At high rotational speed, a very stable operation is observed. When the rotational speed is decreased, the pressure drop reduces until a minimum is reached. **For moderate liquid loads, the influence of the liquid on the pressure drop can be neglected** [7]. When the minimum in pressure drop is breached by lowering the rotational speed, the liquid starts to have a significant influence on the pressure drop and a steep increase is observed. The pressure drop increases with decreasing rotational speed to a maximum where heavy entrainment of liquid is observed and the machine is inoperable [8].

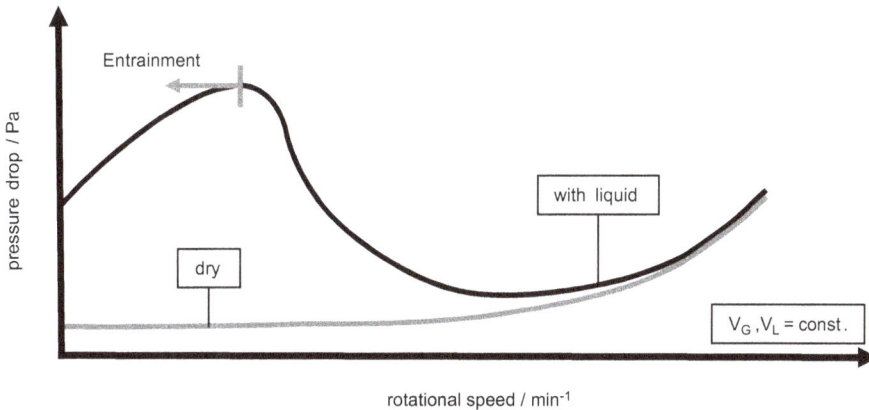

Fig. 3.1: Schematic pressure drop curve for dry and irrigated packed bed varying with the rotational speed.

A large variety of empirical pressure drop correlations are available in the literature. In their review, Zhao et al. [9] provide an overview of different correlations addressing counter-current, cross-flow, or zigzag-flow. Neumann et al. [6] propose a dry pressure drop model where the frictional pressure drop estimation is built on correlations derived for column packings, Mackowiak's so-called "extended channel model" [10]. In their work, two empirical constants are adjusted by a limited number of experiments for metal foam and knitted mesh packing. Recently, Hendry et al. [11] provided experimental results for internal pressure measurements along the radius of the packing. Local measurements are especially of interest since they could facilitate the validation of CFD-simulations.

Operating range

The possibility to adjust the rotational speed and thereby vary the centrifugal acceleration enables a multitude of different operating points. As in columns, the interaction between gas and liquid is crucial. High gas velocities can drag the liquid upwards within the column, and the induced backmixing of the liquid phase reduces the performance of the machine. **The so-called entrainment of liquid against the desired flow direction can be detected visually, evaluating pressure drop data or due to the measurement of the liquid hold-up in the packing.** If the center of the rotor is visually accessible, in the near entrainment region, an accumulation of liquid can be observed before entrainment starts (cf. Fig. 3.2).

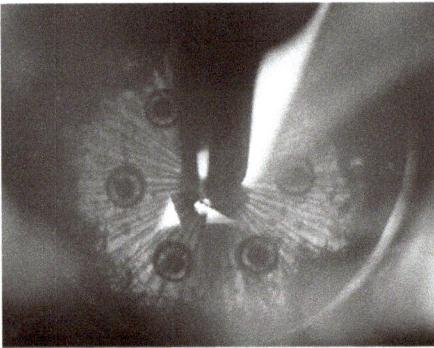

Fig. 3.2: Entrainment start and liquid accumulation in the center of the rotor [7].

As discussed in the previous paragraph, the characteristic pressure drop curve can be aligned with the visual observation. Figure 3.3 illustrates the relations between visual observation and pressure drop. Passing the pressure drop minimum, when reducing the rotational speed, the liquid starts to build up in the center of the rotor. The free cross-sectional area for the gas is reduced by the liquid. Reducing the rotational speed further the liquid starts to build up, a bubble column-like behavior is observed when a phase inversion takes place. The previously continuous gas phase now rises in the form of dispersed gas bubbles in a continuous liquid phase.

To determine the distance from the entrainment region, pressure drop curves have been proven to be a very valuable tool. The visual observation is due to safety reasons not always possible. In high pressure or explosive atmosphere applications, the encapsulation is more important than visual accessibility. Even if sight glasses

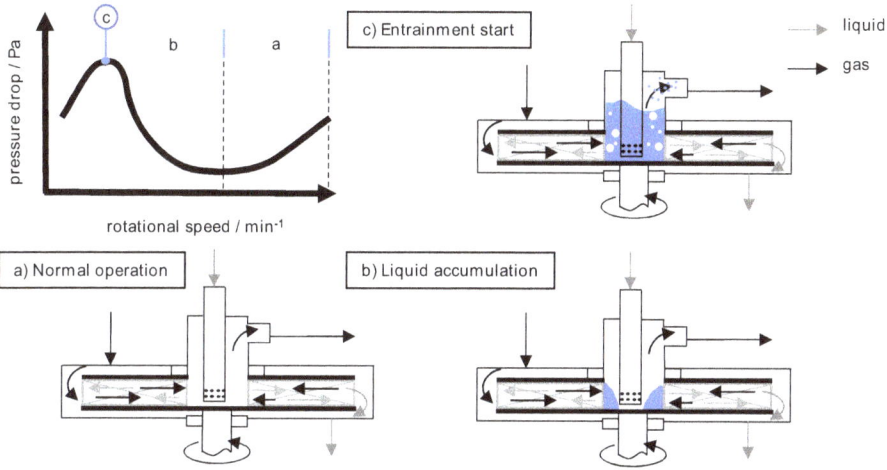

Fig. 3.3: Characteristic pressure drop curve, visual observation, and operating limits [7].

are installed, they might be covered with drops or splashes. Moreover, the pressure drop measurements are easy to implement. There are several definitions of the operating limit in RPBs based on pressure drop analysis [7, 8]. Three commonly used definitions are illustrated in Fig. 3.3. In accordance with the visual observation, Lockett [12] defines the operating limit at the pressure drop maximum. At this position in the pressure drop curve, heavy entrainment is observed. An empirical approach by Singh et al. [13] defines the operating limit where a pressure rise with more than 500 Pa per 100 rpm reduction in rotational speed is observed. However, due to the strong empirical character, the generalized applicability is questionable. The most conservative approach is defined by Rajan et al. [14]. Reducing the rotational speed, they define the observed pressure drop minimum as operating limit, arguing that first entrainment starts when the pressure drop starts to rise again after the minimum is passed. The definition of Rajan et al. is somewhat similar to the definition of loading points in columns, where the first influence of the gas phase on the liquid phase is observed. The advantage of Rajan's defintion is the determination of the point with a high accuracy and reproducibility. At the pressure drop minimum in contrast to the pressure drop maximum, a stable operation is still observed, and large fluctuations due to the dynamic interactions between the fluids are avoided.

As noted in the introduction of the fundamentals section, the cross-sectional area ($A_{C,RPB}$) varies with the radial position (r) within the rotor and is proportional to the rotor diameter. For larger radial packing lengths and small inner packing diameters, the fluid loads at the inner cross-section can be a multiple of the outer cross-section (e.g., in the range of 1 to 10). For larger inner diameters, this effect is less pronounced at equivalent packing length.

$$A_{C,RPB}(r) = 2 \pi \, h_{packing} \, r \tag{3.1}$$

$$LL = u_L = \frac{\dot{V}_L}{A_{C,RPB}} \tag{3.2}$$

$$F_G = u_G \sqrt{\rho_G} = \frac{\dot{V}_G \sqrt{\rho_G}}{A_{C,RPB}} \tag{3.3}$$

When comparing the maximal operating range of different machines or rotors, it is useful to normalize flow rates ($\dot{V}_L$ or $\dot{V}_G$) by the smallest cross-sectional area, usually at the inner radius of the packing. However, one should consider that within one RPB, the load will vary with the ratio of the rotor radii at different positions.

$$LL_1 = LL_2 \, \frac{r_2}{r_1} \tag{3.4}$$

To directly take those changing loads into account, an integral load can be calculated [15], assuming a constant packing height and density (ρ_L or ρ_G) of the fluids.

$$\bar{F}_{G,int} = \frac{\sqrt{\rho_G}}{r_o - r_i} \int_{r_i}^{r_o} u_G(r) \, dr \tag{3.5}$$

Similar to radial-dependent loads, the centrifugal force is calculated. The relative centrifugal force (RCF) is the centrifugal force normalized by earth gravitational acceleration (g).

$$RCF = \frac{\omega^2 r}{g} \tag{3.6}$$

It is reasonable to assume that the knowledge on construction of centrifuges can be directly transferred to the design of RPBs. Centrifuges reaching a circumferential speed of 100 m s^{-1} and diameters up to 3 m are reasonable to construct and commercially available [16]. This can be translated to an RCF of approximately 680. As a rule of thumb, RPBs reaching an RCF in the range of 10 to 1000 are discussed in the literature. The high relative centrifugal force is used to overcome the gravitational limitations which can be seen in column packings. In general, the geometrical surface area in packings is very important to facilitate the mass transfer. High geometrical surface areas increase mass transfer performance, but reduce the allowable loads on the packing [17]. Große [18] has shown that the commonly used foam packings can just be of limited use in columns because pores impose a too-high resistance for the gravitation-driven flow. Small pores were inoperable while for larger pores, the packing showed comparably small operating range at a high liquid hold-up. In contrast, packings with a geometric surface area of 3300 m^2 m^{-3} can be readily used in RPBs with liquid and gas loads even exceeding the allowable loads of columns [19, 20].

Liquid hold-up

For reactive processes as reactive absorption or reactive distillation with liquid phase reactions, the liquid hold-up and the residence time can be important factors. Moreover, a uniform liquid distribution is important for all kinds of gas-liquid contacting processes. The uniform distribution avoids the segregated flow of gas and liquids and enables to use the geometrical packing surface most efficiently. A variety of different methods to analyze the liquid hold-up are available. Starting from visual observation with high speed cameras and strobe lights [21]. First, quantitative methods were developed by Basic et al. [22] and Burns et al. [23] who used electrodes in nonconductive packings to analyze the liquid hold-up. The resolution of conductive methods is limited by the number of electrodes used and how those are positioned in the packing. The application of ring electrodes generates integral results for the liquid hold-up in the volume between the electrode pair. An assumption of flow tortuosity needs to be made to estimate the pathlength of the liquid rivulets. **Nowadays, specially resolved liquid hold-up measurements can be conducted by the application of computer tomographic methods.** X-ray [24] or gamma-ray computed tomography [25] can be used to generate quantitative and local liquid hold-up results. These noninvasive measurements have the great advantage that they do not affect the flow inside the packing. However, for large RPBs made out of stainless steel, the measurement time can be in the range of minutes, and therefore, dynamic phenomena cannot be detected. Groß et al. [25] could show that the radial liquid distribution decreases from the inner to the outer packing radius. Moreover, at low rotational speeds, the liquid builds up at the inner packing radius (cf. Fig. 3.4). When rotational speed was too high, a maldistribution was found induced through the inner packing support ring (cf. Fig. 3.4).

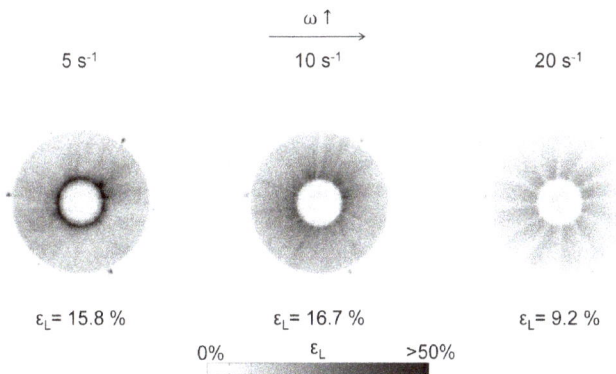

Fig. 3.4: Liquid hold-up entrainment and maldistribution [25].

Power consumption

The power consumption of the motor accelerating the rotor of the RPB is very important. It contributes to the operating costs and adds additional costs to consider compared to columns. Currently, there are two similar correlations from Singh et al. [13] and from Groß [7].

$$P_C = Z_0 + Z_1\, \rho_L r_o^2 \omega^2 \dot{V}_L \qquad (3.7)$$

Both correlations apply a constant (Z_1) that accounts for liquid flow-dependent power consumption. **As a rule of thumb, the ideal kinetic energy of the liquid spinning at the tip speed of the rotor diameter multiplied by a factor in the range of 2–3 can be used to estimate the power consumption.** Additionally, a constant (Z_0) for other contributions, e.g., frictional losses of the bearings, is used. Moreover, Groß could show that a rising liquid hold-up in the casing is tremendously increasing the power consumption [7]. A suitable concept to drain the liquid which is ejected from the rotor into the casing is therefore crucial. A correlation to calculate the power consumption (P_C) of the RPB in W with the resprective constants in Tab. 3.1 can be found in eq. (3.7).

Tab. 3.1: Constants for the power consumption correlations.

Originating literature	Z_0/W	Z_1
Singh et al. [13]	1222	1.1
Groß [7]	744.4	1.43

3.1.2 Mass transfer

The following section addresses the process of aeration and deaeration as a model system for liquid-side limited mass transfer. RPBs are manipulating the contact between different phases of fluids. The thermodynamic fundamentals describing the equilibrium and driving-forces are identical with all other types of contacting equipment. Refer to Chapter 3 of the first book "Reactive and membrane-assisted processes" of this series for an in-depth description of the phase-equilibrium thermodynamics.

Liquid-side limited mass transfer

For the description of gas-liquid contactors, the film theory and the surface-renewal approach are well established [26]. In this chapter, we will follow the two-film approach as it is commonly applied for RPBs. However, both theories and their modifications incorporate empirical factors as film thickness, exposure time, or rate of surface renewal and can be regarded as interchangeable for a wide range of applications [26]. Applying the two-film theory, the mass transfer problem can be divided in a bulk gas phase, a gas-film, a gas-liquid interface, a liquid film, and a bulk liquid phase. Assuming the interface itself has no resistance and the bulk phases of gas and liquid are ideally mixed, the problem can be reduced to individual resistances in the gas film and in the liquid film. Both of the films can contribute to the overall resistance, but in many cases, one of the films is dominant and limiting the flow through the interface [27]. The mass transfer of gases with a low solubility in the liquid phase, such as oxygen, can generally be regarded as liquid-side limited, as the diffusive processes in the gas film are of lesser importance due to the low rate of absorption [28]. Therfore, in the following, only liquid-side limited mass transfer in RPBs will be discussed, while the reader is advised to refer to Chapter 8 for a more detailed description and analysis of mass-transfer modeling for multicomponent systems for distillation.

The experimental system air/water is commonly used for derivation of liquid-side volumetric mass transfer coefficients. The simplicity and the non-hazardousness of the system support its application, dealing with physical ab- or desorption of oxygen. It is used for packed columns [29, 30] as well as for RPBs [31]. Moreover, the system is suitable for large-scale applications [15] as corrosion prevention in boiler or piping systems [32], the extension of shelf life in the food processing industry [33], or in secondary oil recovery in oil fields [28]. A different system used to determine the liquid-side mass transfer coefficient is the absorption of CO_2 [31]. Refer to Chapter 4 for a detailed investigation of gas side limited mass transfer.

Based on liquid-side mass transfer coefficients, the RPB technology has proven to be in reach of one-order higher mass transfer coefficients than columns at equivalent loads [7].

Zones of mass transfer

It is obvious that not only the packing surface might aid to mass transfer but also all other surfaces within the RPB can contribute to the mass transfer as well. The stripping of oxygen in RPBs was extensively investigated by Yu-Shao Chen et al. [34–36]. Based on investigations with adjustable inner and outer packing diameters, a three-zone model was established [35]. The important volumes contributing to the mass transfer where the volume from the center of the rotor to the inner packing radius of the packing (V_i), the packing volume itself (V_p), and the outer cavity zone reaching from the outer packing radius to the inner casing wall (V_c) (cf. Tab. 3.2 and Fig. 3.5.).

Tab. 3.2: Different contributing volumes for the mass transfer according to Chen et al. [35].

Center	Packing	Casing
$V_i = \pi\, R_i^2\, h_P$	$V_p = \pi\, (R_o^2 - R_i^2)\, h_P$	$V_c = \pi\, (R_c^2 - R_o^2)\, h_P$

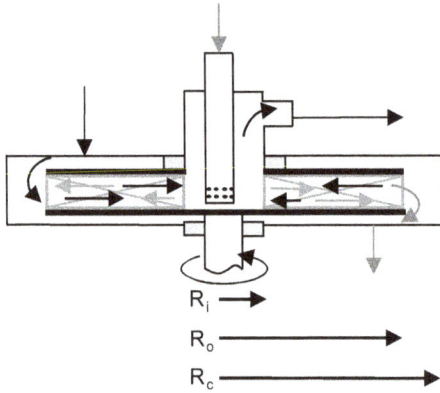

Fig. 3.5: Different zones of mass transfer in an RPB.

All other contributions than the packing are summarized as end effects and need to be considered to avoid an overestimation of the mass transfer coefficient. **The consideration of end effects is especially important for small scale RPBs where the ratio between geometrical surface area of the packing and other potential surfaces contributing to the mass transfer (e.g., casing wall) is considerably lower** [37]. In order to avoid confusion, it has to be mentioned that some authors refer to the inner packing zone that provides the main mass transfer as end-effect zone [24, 38]. Undoubtedly, these highly efficient flow effects in the first centimeters of the packing are very desirable, while the previously mentioned (e.g. casing wall) may increase the experimental bias and should be avoided. More recently, deaeration experiments have been used to investigate the flash degassing of liquids [39].

3.2 Modeling and design

One methodology to start the design of an RPB is to divide the process into hydraulic and mass transfer design steps. The hydraulic design allows to do a rough estimation of the dimension of the machine based on the capacity of the process to be designed. Later, this design is refined with actual mass transfer data from correlation or validation experiments. The capacity of the equipment is defined by the axial height and the inner diameter of the rotor. The mass transfer performance is related to the type of packing and its radial length.

3.2.1 Hydraulic considerations

Starting from the envisioned capacity, the rotor can be dimensioned. The inner diameter of the packing is the decisive factor. In a counter-current process, the operating range is defined by the force balance between frictional force of the gas and the centrifugal force transferred from the rotor onto the liquid. Due to the smallest cross-sectional area at the inner diameter, frictional forces are highest at this position while the centrifugal force is due to the small diameter low. To extend the capacity in the design phase, the inner diameter of the rotor can be increased, while the rotational speed will be reduced to operate at equivalent centrifugal forces. More conveniently, if space is available, the axial packing height can be increased. As mentioned earlier, a variety of correlations for pressure drop estimations is available [6, 9, 11]. Most of the correlations address the dry pressure drop, while for the vast majority of cases the influence of the liquid is negligible, this assumption should be verified for high liquid and gas loads [7]. **While the design of liquid distributors can have a significant influence on the maximal operating capacity [8], the rotational speed offers a large opportunity to work at changing capacities.** However, the maximal capacity anticipated will have an influence on all connecting periphery, the dimensioning of in- and outlets or the mechanical design of the machine itself. It is therefore important to carefully analyze the capacity range needed for the process. Considering the production of high value products with changing capacities, it can be feasible to include a certain flexibility into the design of the machine. A detailed example of a metal foam packing used for hydraulic design studies including pressure drop, operating range, and flexibility can be found in the detailed examples (Section 3.3.1).

3.2.2 Mass transfer considerations

For the deaeration of liquids, the calculation of volumetric mass transfer coefficients was investigated by Yu-Shao Chen et al. [34–36]. The correlations have been validated by experimental results of the authors. Additionally, most of the data from literature [13, 40–44] could be validated with uncertainties, when necessary correlation parameters were not given in the original publication. Further information can be found in [35]. De Beer et al. [39] validated the correlation for co-current flash degassing as well, given that a flash efficiency for the initial flash was calculated a priori. Groß et al. [37] showed that the correlation can predict the co-current deaeration efficiency of a pilot scale packing with a packing length of 0.152 m with an relative error within ±30%. However, it was found that the volumetric mass transfer coefficient reduces radically when the radial packing length is increased to larger diameters, making new packing strategies necessary. A few different options are further outlined in Chapter 7. One of those new concepts was recently published by Hacking et al. [45], proposing a design with several redistribution rings that could generate already promising results compared to conventional single block packings. Geometry, packing type, and centrifugal acceleration can have a significant influence on the separation efficiency. A detailed example for a metal foam packing is further presented in Section 3.3.2.

3.3 Detailed examples

This section will support the creation of a deeper understanding for the topics addressed in the previous sections through the discussion of a detailed example. Based on real numbers, an overview of experimental procedures is given, and the development of an RPB for deaeration purposes is illustrated. More information on the exact procedure and experimental setup can be found in [7].

3.3.1 Hydraulic operating window of a high surface metal foam

For the hydraulic investigation, the determination of operating range and the pressure drop are crucial. The counter-currently operated RPB considered in the current study consists of an inner casing diameter of 0.650 m and enables the investigation of rotors with a maximal outer diameter of 0.500 m (cf. Fig. 3.6).

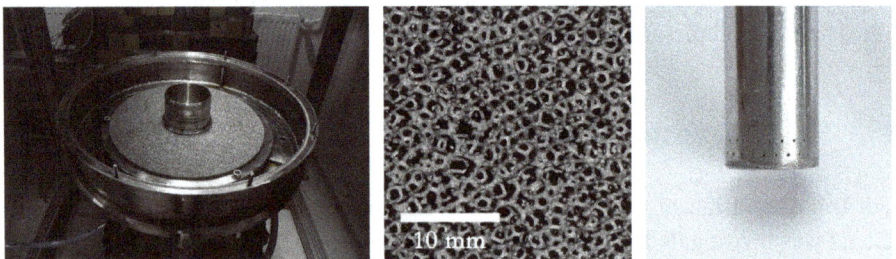

Fig. 3.6: (a) Lower rotor plate with installed metal foam within the RPB casing with opened top cover. (b) Full-jet pipe distributor 360° spraying angle 0.8 mm boreholes 24 holes per row (15°-hole circle division). (c) RECEMAT® NCX 1116 metal foam packing close-up view.

The packing type investigated and maximal possible geometric dimensions are listed in Tab. 3.3. The RECEMAT NCX 1116 metal foam packing provides a geometrical surface area of 1000 m^2 m^{-3} compared to conventional structured packings applied in columns. This is relatively high as commonly packings in the range of 250 to 750 m^2 m^{-3} are used [17].

Tab. 3.3: Packing dimension () maximal allowable packing dimension [7].

Type	RECEMAT® NCX 1116 (metal foam)
a_p/m^2m^{-3}	1000
$\varepsilon/-$	0.92
$d_{packing,i}/m$	0.146
$d_{packing,o}/m$	0.450 (0.500)
$h_{packing}/m$	0.010–0.020 (0.020)

As liquid distributor, a full jet pipe distributor is used. The distributor spray has a limited fraction of small droplet diameters, and as the distributor is installed in a way that the liquid jets directly interact with the packing, the amount of small droplets entrained with the gas is reduced compared to other nozzle systems (e.g., flat fan nozzles [8]). Tap water is used as liquid while the gas flow is provided by a blower or from a compressed air system. The method chosen to investigate the operating range is pressure drop measurement. This includes the advantage that pressure drop data is simultaneously generated with the operating range data.

Pressure drop correlation

In the first step, a pressure drop correlation is set up. Most of the pressure drop correlations are based on dry pressure drop data. As a thumb rule, the pressure drop contribution of the liquid can be neglected for high rotational speeds and a moderate load. The correlation of Neumann et al. [6] was chosen since it can be related to just two empirical constants. The main contributions on the pressure drop are the frictional loss induced by the gas flow and the centrifugal head imposed by the rotation of the packing.

$$\Delta p_{total, dry} = \left(\xi_{CH} + \Delta p_{f, stat} + \Delta p_{f, rotor} + \Delta p_{f, empty} + \Delta p_{rot}\right) \tag{3.8}$$

A small set of dry experiments show that the characteristic shape of the curves can be fitted well with the equations underlying constants. The experimentally determined constants agree well with the constants proposed by Neumann et al. [6] and show a maximal deviation of – 6%. Within an error range of ±30%, the whole set of dry pressure drop data can be reproduced with a sufficient accuracy (cf. Fig. 3.7).

Tab. 3.4: Additional equations for the pressure drop model from Neumann et al. [6] for the dry, unirrigated packing.

With:
Centrifugal head

$$\xi_{CH} = A_{CH} \frac{\rho_G \cdot \omega^2}{2} \left(r_o^2 - r_i^2 \right) \tag{3.9}$$

Static packing frictional pressure drop

$$\Delta p_{f,\,stat} = \Psi_0 \cdot (1-\varphi) \frac{(1-\varepsilon)}{\varepsilon^3} \cdot \frac{\bar{F}_{G,int}^2}{d_p} \cdot \frac{(r_o - r_i)}{K_{Neumann}} \tag{3.10}$$

$$\Psi_0 = \frac{725.6}{\overline{Re}_{G,int}} + 3.203 \tag{3.11}$$

$$\overline{Re}_{G,int} = \frac{d_p}{(1-\varepsilon)v_G} \cdot \frac{\bar{F}_{G,int}}{\sqrt{\rho_G}} \tag{3.12}$$

$$d_p = 6 \cdot \frac{(1-\varepsilon)}{a_{pack}} \tag{3.13}$$

Additional experimental determined contributions: $\tag{3.14}$

$$\Delta p_{f,\,rotor} + \Delta p_{f,\,empty} + \Delta p_{rot} \approx 0$$

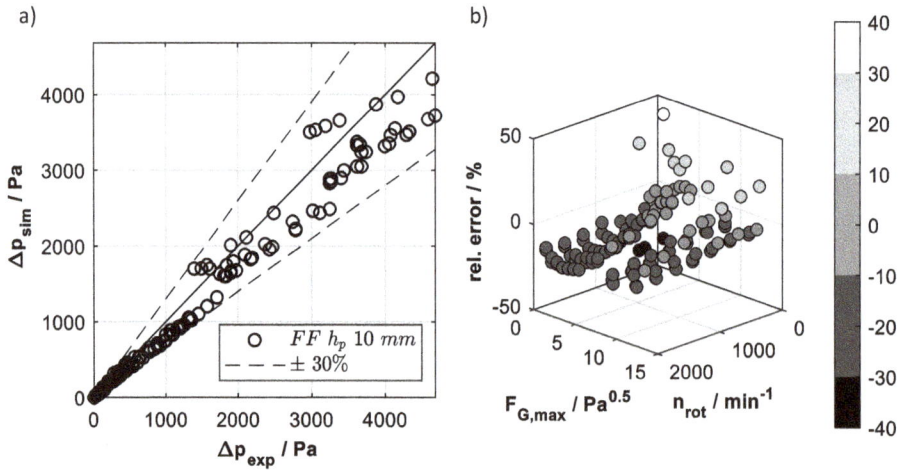

Fig. 3.7: (a) Parity plot for the simulated and the experimental dry pressure drop; (b) overview of the full foam (FF) data with h_p = 10 mm (A_{CH} = 0.88; φ = 0.30) [7].

Additional data was generated to confirm that the liquid is having a minor influence on the pressure drop when having a sufficient distance to the entrainment region. Two different packing heights 10 and 20 mm could confirm the assumption. However, at larger gas loads ($F_{G,m} = 8\ Pa^{0.5}$), an increasing deviation with increasing liquid load could be observed even without vicinity to entrainment region (cf. Fig. 3.8).

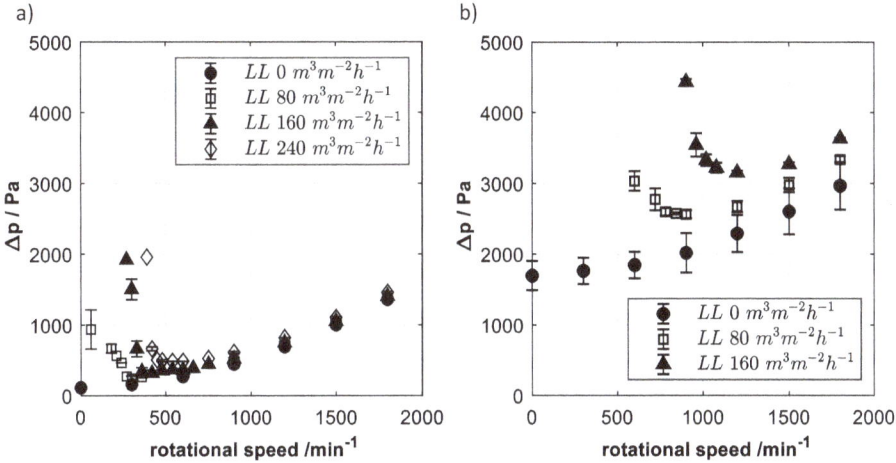

Fig. 3.8: Influence of the liquid load on the pressure drop for the full foam (FF) packing $h_p = 10$ and 20 mm: (a) $F_{G,max}$ 2 $Pa^{0.5}$; (b) $F_{G,max}$ 8 $Pa^{0.5}$. Error bars depict the standard deviation [7].

As discussed earlier, there are several definitions of the operating limit (cf. Section 3.1.1.). In this investigation, the minimum of the pressure drop curve was chosen as it could be reliably measured to generate accurate operating range data. Out of 28 pressure drop curves (cf. Fig. 3.9), an operating map was generated. The determination of the pressure drop minimum was performed automatically in two steps. First, an interpolation curve was calculated based on Piecewise Cubic Hermite Interpolating Polynomial method available in MATLAB R2018a® [46], followed by a

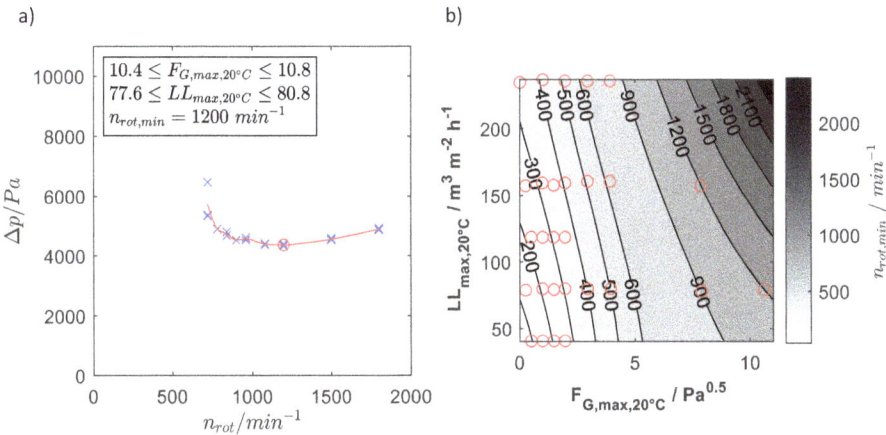

Fig. 3.9: Overview of the surface-fitted data of the metal foam packing: (a) exemplary pressure drop curve from 28 different pressure drop curves to generate the overview surface plot; (b) contour plot of the minimal operational speed ($n_{rot,min}$) (∘ = data points) [7].

simple minimum search for polynomial data. The data was used to fit a three-dimensional surface to generate an operating map displaying the minimal allowable rotational speed for a certain set of gas and liquid loads (cf. Fig. 3.9). It was shown that with increased rotational speeds, the gas and liquid loads could be significantly extended.

3.3.2 Deaeration of liquids

In the previous section, the hydraulic operating characteristics of the RPB were evaluated. For metal foam packing, the model of Neumann et al. [6] was applied to validate the proposed pressure drop model and make it available for future estimations. Moreover, the accurate estimation of the wet pressure drop and corresponding operating limits were investigated. The deaeration of water with nitrogen gas is a process operated at low gas-to-liquid ratios. Whereas, in theory, the co-current operation can already realize very low oxygen concentrations ($<50~\mu g~L^{-1}$) at a volumetric gas-to-liquid ratio of 6 $m^3 h^{-1} m^{-3} h$. The counter-current process provides a very large potential. Already with 1.5 equilibrium stages, a gas-to-liquid ratio of 1–6 $m^3 h^{-1} m^{-3} h$ can reach the envisioned liquid outlet concentration of oxygen (cf. Fig. 3.10).

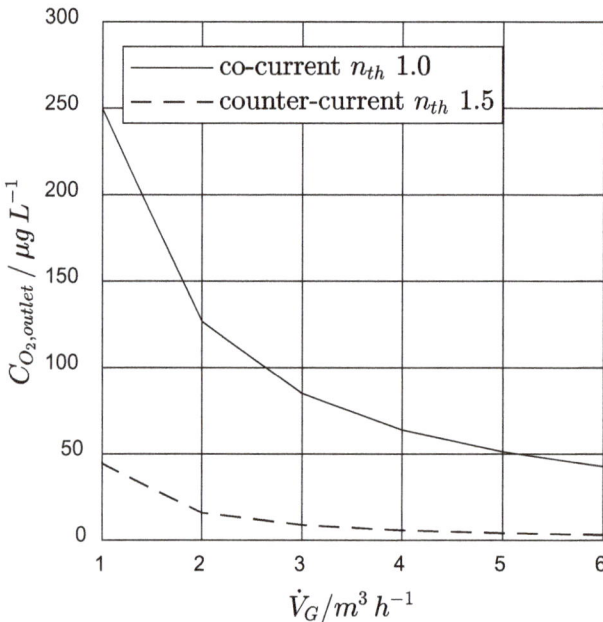

Fig. 3.10: Theoretical liquid outlet oxygen concentration for co- and counter-current operation for saturated liquid $\dot{V}_L = 1~m^3~h^{-1}$, $C_{O_2, inlet} = 8200~\mu g~L^{-1}$ ($p = 1.01325$ bar $T = 25°C$) [7].

While a detailed description of the setup can be found in the thesis of Groß [7], a schematic is presented in Fig. 3.11. The setup consists of the same rotor as in the hydraulics section (cf. 3.3.1). To supply large amounts of water, tap water was used as it naturally contained enough oxygen for deaeration. Bottled nitrogen was used to counter-currently contact the liquid within the packing and casing. The liquid inlet and outlet oxygen concentration were measured with optical oxygen probes.

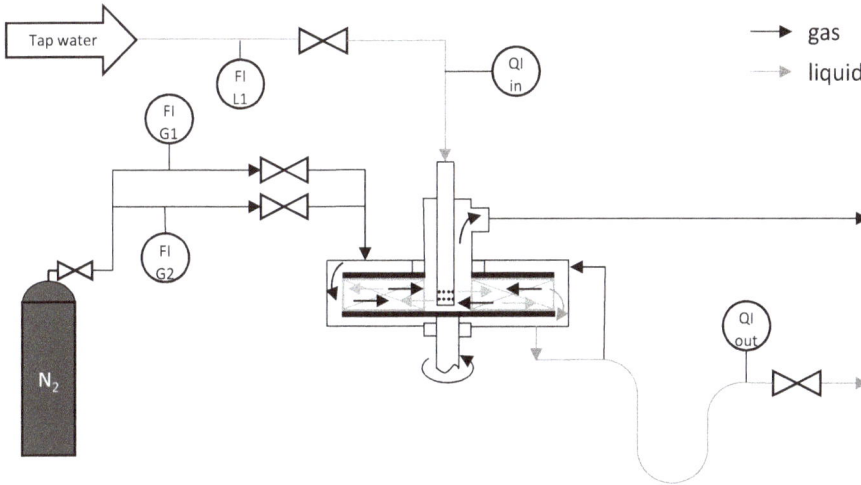

Fig. 3.11: Schematical overview of the deaeration setup [7].

Based on the hydraulic results, a high liquid load (78–210 m^3 m^{-2} h^{-1}) could be selected (cf. Fig. 3.12). The gas velocity was negligible with a maximal $F_{G,max}$ of 0.39 Pa$^{0.5}$. The rotational speed was varied between 300 and 1800 rpm which translates to a minimal RCF (cf. eq. (3.6)) from inner to outer diameter of 8 to 22 and a maximal RCF from 271 to 814, respectively. For a comparable knitted mesh packing and a variety of liquid flow rates, the RPB showed an increasing volumetric mass transfer coefficient ($k_L a$) normalized to the packing volume when the rotational speed was increased to its maximum. At very high rotational speeds, the sensor accuracy limit (<2 µg L^{-1}) was reached in the outlet concentration.

The use of a distributor with a 90° spraying angle compared to a 360° distributor showed no significant difference in the $k_L a$-value. To further analyze the behavior of the process, the gas flow rate was significantly reduced. It could be shown that within the range of 3–6 m^3 h^{-1}, the reduction had no large influence on the $k_L a$-value; however, when reducing the gas flow rate to a minimum, the $k_L a$-value was significantly reduced. When the packing volume was doubled by doubling the axial packing height from 10 to 20 mm, the $k_L a$-value was still found constant at equivalent gas and liquid loads. This confirms the assumption that the scaling of the throughput is possible by extending the axial height of the packing. Compared to a column,

a)

b)

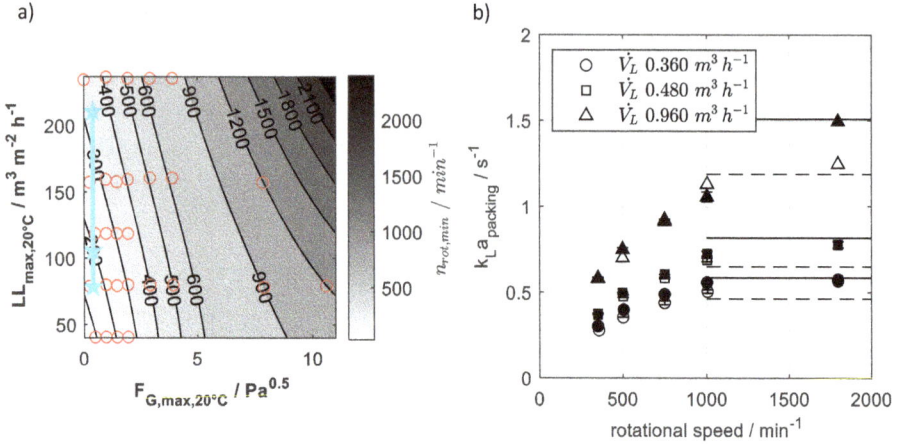

Fig. 3.12: Variation of the k_La-value with increasing rotational speed and different liquid flow rates for the knitted mesh packing. $\dot{V}_G = 6\ m^3\ h^{-1}$; o 360° distributor; • 90° distributor; Solid line (-) denotes the theoretical maximum (<1 µg L^{-1}). Dashed line (–) denotes low outlet concentrations (<6 µg L^{-1}). Error bars depict the standard deviation.

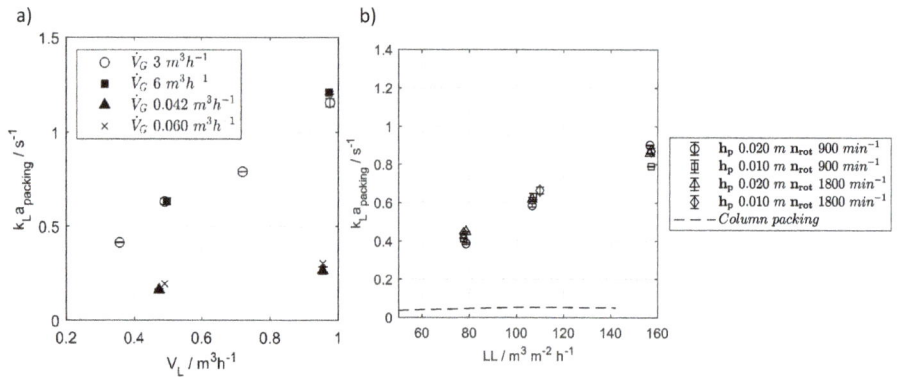

a)

b)

Fig. 3.13: Variation of the k_La-value: (a) with increasing liquid flow rate and different gas flow rates for the metal foam (FF) packing $n_{rot} = 900\ min^{-1}$. Error bars depict the standard deviation (b) with liquid load and rotational speed. For the RPB packing heights (h_p) 0.01; 0.02 m; metal foam packing; $F_G = 0.197\ Pa^{0.5}$ was applied. The column data for one-inch Raschig rings are taken from [29]. Error bars depict the standard deviation [7].

equipped with Raschig ring packings, an approximately one-order of magnitude higher k_La-value could be reached at equivalent liquid loads (cf. Fig. 3.13). The application of a high surface metal foam packing is therefore recommendable. However, a thorough examination of other factors including costs is necessary to wholistically evaluate the feasibility on a commercial scale.

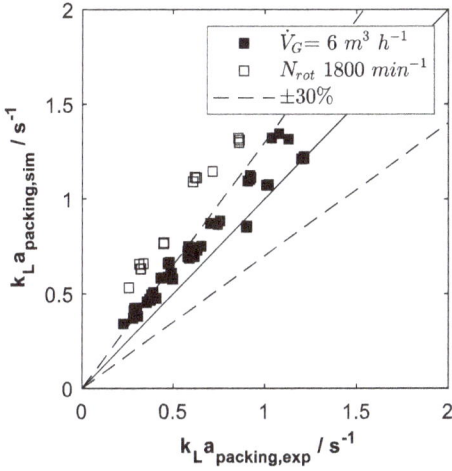

Fig. 3.14: Parity plot of the estimated $k_L a$-value and the experimental $k_L a$-value for full foam (FF) (di/do/hp) 0.146/0.450/[0.01, 0.02] m and knitted mesh packings. 0.146/0.460/0.01. LL_{max} varied between 40 and 213 m^3 m^2 h^{-1} [7].

The applicability of the model from Chen et al. [35] could be confirmed for the data with a uncertainty of ±30% (cf. Fig. 3.14). However, at larger rotational speeds, the correlation exhibits an overestimation of the $k_L a$-value. The deviation could result from the experiments with which the correlation was developed. Those experiments were conducted in a smaller scale RPB, and consequently, at lower centrifugal force. However, it is remarkable that the correlation can be extrapolated to some of the larger radial packing length data of the work shown here.

3.4 Take home message

- The design procedure of RPBs consists of hydraulic and mass transfer design.
- Some of the well-established correlations developed for conventional equipment types can be extended to the high gravity environment.
- It should be carefully checked that the operating characteristics of the RPB are incorporated when correlations are extended.
- The centrifugal acceleration is an additional degree of freedom and variable along the radius. It offers a large opportunity to work at changing capacities.
- The cross-sectional area increases from the inner to the outer packing diameter
- Wet and dry pressure drop are comparable at high centrifugal forces.
- When the rotational speed is too small, a built up of liquid is found at the inner radius of the packing, where the centrifugal acceleration is the smallest, and the velocities of the fluids are high due to a small cross-sectional area.

- The minimal rotational speed or the entrainment region can be investigated visually by examining pressure drop data or due to liquid hold-up measurements. There are several definitions for the minimal rotational speed. However, the pressure drop minimum can be measured most reliably.
- A characteristic pressure drop curve in dependence of the rotational speed exists. High rotational speeds can reduce the pressure drop especially at high liquid loads by the generation of thinner films and smaller droplets.
- Optimal liquid distribution is found between a fan-like maldistribution observed at high rotational speeds and a high liquid accumulation in the center of the rotor when the centrifugal force is too low.
- The power consumption depends on rotational speed, liquid flow rate, and outer rotor diameter. As a rule of thumb, the ideal kinetic energy of the liquid spinning at the tip speed of the rotor diameter multiplied by a factor in the range of 2–3 can be used to estimate the power consumption.
- When mass transfer is evaluated, three important zones can be defined: the center zone of the rotor, the packing, and the casing zone. Each of them contributing differently to the mass transfer. Especially for small rotor types, the additional mass transfer area in the casing can have a significant influence on the mass transfer performance.
- The extended channel model can be used to describe the dry pressure drop behavior of the RPB equipped with a metal foam.
- The model of correlations of Chen et al. [35] described the metal foam packing with a sufficient accuracy of ±30% for most of the data available.

3.5 Quiz

1. Is the application of correlations originally developed for columns in the high gravity environment feasible?
2. What are the two main design tasks when designing an RPB?
3. True or false: The centrifugal acceleration is an additional degree of freedom influencing performance but not the cost of the equipment.
4. True or False: In a standard rotor geometry with fixed rotor height, the internals need to deal with the changing cross-sectional area from the inner to the outer packing diameter.
5. Describe the characteristic difference in the RPB pressure drop curve between wet and dry pressure drop. Which assumption can be made for low loads and high rotational speeds?
6. When the operating limit of the RPB is reached at which position liquid accumulation can be observed?

7. What are the three commonly used definitions for the operating limit in the RPB? Which one of them could be preferred?
8. The liquid hold-up has been measured by many different approaches, which new method was used most recently and what is a general advantage of the technology?
9. Which rule of thumb can be used to estimate the power consumption of an RPB?
10. Describe the different zones of mass transfer within an RPB? What needs to be considered when working on small-scale equipment (e.g., lab RPBs)?
11. What are the two main contributions on pressure drop of an RPB?
12. Regarding operating range, what influence has the rotational speed?
13. What parameter can be used to describe the mass transfer performance of an RPB? Is there a difference between the co- and counter-current operation of the machine?
14. The optimal liquid distribution is found between which extreme cases?

3.6 Exercises

3.6.1 Evaluation of an RPB for air stripping application

Use the content of this chapter to solve the following exercise.

The company you are working in decides to evaluate the high gravity technology for a stripping application. Lately, the company was struggling with the stripping unit which is used to remove volatile organic compounds (VOCs) from a wastewater stream. The changing VOC concentration and changing wastewater flows have led to flooding in an existing stripping column, and purity specifications were not met. Furthermore, the wastewater flows are going to increase in the future due to higher production volumes in different process units. The integration in the existing wastewater plant requires a small and flexible equipment, but before a detailed evaluation with experimental validation in a pilot RPB is justified, the following questions need to be answered.

Exercise 1. The existing column is equipped with one-inch Raschig rings. The 1.5 m diameter column has a packing height of 8 m. Based on volumetric mass transfer coefficients, what range of volume reduction can be expected at comparable gas and liquid loads for a metal foam packing under the assumption that the mass transfer performance is directly proportional to the packing volume?

Exercise 2. In the original process, the gas flow rate was varied to reach the necessary product quality. If the process is running at a liquid load of 150 $m^3\,m^{-2}\,h^{-1}$ and is equipped with a metal foam packing, what gas velocities can be realized at different rotational speeds to meet the purity constraints when ambient air is used as stripping gas. Assuming a wastewater capacity of 25 m³ h⁻¹, how does the higher rotational speed influence the operating costs ($r_o = 0.5$ m, $\rho_L = 1,000$ kg m⁻³)? Assume that the minimal rotational speed used is 300 min⁻¹.

Exercise 3. Estimate two types of RPB rotor dimensions with a smaller and a larger outer diameter. The packing form should be ring-shaped. How do inner and outer diameter influence the design. How does the ratio of inner and outer cross-sectional area change? The packing volume should be around 1 m³ while the radial packing length should be 0.150 m.

3.7 Solutions

Exercise 1. The comparison between volumetric mass transfer coefficients of column and RPB equipped with Raschig rings and metal foam can be found in Fig. 3.15.

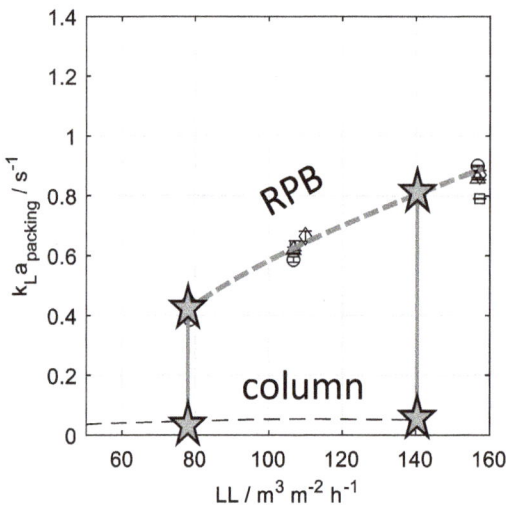

Fig. 3.15: Range of RPB and column $k_L a$ values (cf. Fig. 3.13).

The ratio between $k_L a$-values ranges from

$$\frac{k_L a_{RPB}}{k_L a_{column}} = \frac{0.45}{0.05} = 9 \tag{3.15}$$

to

$$\frac{k_L a_{RPB}}{k_L a_{column}} = \frac{0.82}{0.05} = 16.4 \tag{3.16}$$

The volume of the column packing is

$$V_{\text{column, packing}} = \pi \, D^2_{\overline{4}} \, H = \pi \frac{(1.5 \, \text{m})^2}{4} \, 8 \, \text{m} = 14.13 \, \text{m}^3 \tag{3.17}$$

$$V_{\text{RPB, packing}} \approx \left(\frac{k_L a_{\text{RPB}}}{k_L a_{\text{column}}} \right)^{-1} V_{\text{column, packing}} = 0.86 \, \text{m}^3 \text{ to } 1.57 \, \text{m}^3 \tag{3.18}$$

Exercise 2. The operating diagram for a metal foam packing can be found in Fig. 3.16.

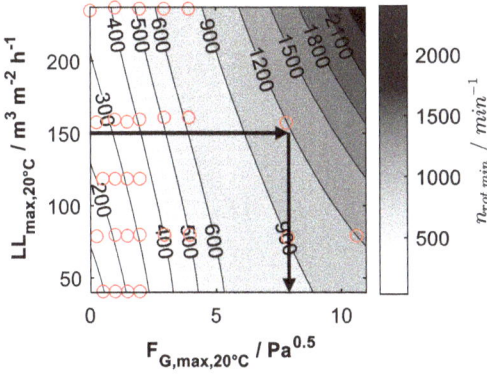

Fig. 3.16: Operating diagram for the specified liquid load of 150 m³ m⁻² h⁻¹ (cf. Fig. 3.8).

For a liquid load of 150 m³ m⁻² h⁻¹, an F_G of up to 8 Pa⁰·⁵ can be realized. To reach this high gas throughputs, the rotational speed has to be increased up to 1200 min⁻¹. The gas velocity can be calculated via the definition of F_G

$$u_G = \frac{F_G}{\sqrt{\rho_G}} = \frac{8 \, \text{Pa}^{0.5}}{\sqrt{1.2 \frac{\text{kg}}{\text{m}^3}}} = 7.3 \, \text{m s}^{-1} \tag{3.19}$$

For the given capacity of 25 m³ h⁻¹ and outer rotor radius of 0.5 m, the power consumption can be calculated by the correlation of Singh et al. [13].

$$P_C = Z_0 + Z_1 \, \rho_L r_o^2 \omega^2 \dot{V}_L = 1220 \, W + 1.1 \, \rho_L r_o^2 \omega^2 \dot{V}_L \tag{3.20}$$

$$P_C = 1220 \, W + 1.91 \, \omega^2 \tag{3.21}$$

$$\Delta P_C = P_{C, \, 1,200 \, \text{rpm}} - P_{C, \, 300 \, \text{rpm}} = 1.91 \left(\frac{1,200}{60} \, 2 \, \pi \right)^2 - 1.91 \left(\frac{300}{60} \, 2 \, \pi \right)^2 = 28.3 \, \text{kW} \tag{3.22}$$

The increased gas flow requires an elevated rotational speed. The higher rotational speed will require additional 28.3 kW which are mainly used to accelerate the liquid in the rotor.

Exercise 3. In this task, multiple solutions are possible; generally, the inner diameter should not be much smaller than 0.150 m to enable accessibility. The smaller rotor can be dimensioned as follows.

$$V_{\text{RPB, Packing}} = 1 \; m^3 = \frac{\pi}{4}\left(d_{\text{out}}^2 - d_{\text{in}}^2\right) h \tag{3.23}$$

When constructing with smallest inner diameters, additional selection on the outer diameter to height ratio needs to be made. A very large packing height will lead to high capacities, while a large outer diameter will result in a longer radial packing length which could facilitate higher separation efficiencies. For a radial packing length of 0.150 m, the outer diameter and the packing height result to:

$$L_{\text{pack}} = 0.150 \, \text{m} = \frac{d_{\text{out}} - d_{\text{in}}}{2} \; \Rightarrow \; d_{\text{out}} = 0.450 \, \text{m} \tag{3.24}$$

$$h = V_{\text{RPB, Packing}} \frac{4}{\pi \left(d_{\text{out}}^2 - d_{\text{in}}^2\right)} \; \Rightarrow h = 7.1 \, \text{m} \tag{3.25}$$

$$\frac{A_{c,\text{in}}}{A_{c,\text{out}}} = \frac{d_{\text{in}}}{d_{\text{out}}} = \frac{1}{3} \tag{3.26}$$

Large packing volumes cannot easily be accommodated in rotors with small inner diameter and fixed axial packing length. The 1 m^3 can more easily be included in a rotor with a larger inner diameter. For an outer diameter of 1 m, the packing height and inner diameter are specified as the following.

$$L_{\text{pack}} = 0.150 \, \text{m} = \frac{d_{\text{out}} - d_{\text{in}}}{2} \; \Rightarrow \; d_{\text{in}} = 0.700 \, \text{m} \tag{3.27}$$

$$h = V_{\text{RPB, Packing}} \frac{4}{\pi \left(d_{\text{out}}^2 - d_{\text{in}}^2\right)} \; \Rightarrow h = 2.5 \, \text{m} \tag{3.28}$$

$$\frac{A_{c,\text{in}}}{A_{c,\text{out}}} = \frac{\pi \, d_{\text{in}} \, h}{\pi \, d_{\text{out}} \, h} = \frac{7}{10} \tag{3.29}$$

The benefit of larger rotors is a better height to diameter ratio and smaller change in cross-sectional area along the radial direction.

List of symbols

Latin letters

a_p	specific geometric packing surface area per unit	m^2 m^{-3}
A_{CH}	RPB configuration dependent parameter to calculate the centrifugal head	–
D	diameter	m
d_p	equivalent spherical diameter,particle diameter	M

F_G	F-factor = $u_g\sqrt{\rho_G}$	$Pa^{0.5}$
g	gravitational acceleration	$m\ s^{-2}$
h	axial length	m
HETP	height equivalent to a theoretical plate	m
$K_{Neumann}$	wall factor	–
$k_G a$	overall volumetric gas- or vapor-side mass transfer coefficient	s^{-1}
$k_L a$	overall volumetric liquid-side mass transfer coefficient	s^{-1}
LL	liquid load	$m^3\ m^{-2}\ h^{-1}$
L_{pack}	Radial packing length	m
n_{rot}	rotational speed	s^{-1}
N_{rot}	number of rotors	–
N_{th}	number of theoretical stages	–
p	pressure	Pa
r	variable radius	m
R	radius	m
Re	Reynolds number	–
u	Velocity	$m\ s^{-1}$
$\dot{V}_G$	gas flow rate	$m^3\ s^{-1}$
$\dot{V}_L$	liquid flow rate	$m^3\ s^{-1}$
v	weakening effect factor	–

Greek letters

ε	porosity	p–
v	kinemativ viscosity	$m^2\ s^{-1}$
ρ	density	$kg\ m^3$
η	dynamic viscosity	$kg\ m^{-1}\ s^{-1}$
η_{vol}	volumetric efficiency	$NTU\ cm^{-3}$
ξ	resistance coefficient	–
v	reflux ratio	–
φ	form factor of dry packing	–
Ψ_0	resistance coeffcient for single-phase flow for classical, noperforated packing elements such as ceramic Raschig rings	–
ω	angular velocity	$rad\ s^{-1}$

Subscripts

CH	centrifugal head
col	column
dry	dry pressure drop in absence of liquid
emtpy	without packing installed
f	frictional
G	gas
i	inner
int	integral
L	liquid
norm	normalized

o	outer
pack	packing
rad	radially, radius
RPB	rotating packed bed
rot	rotation
rotor	related to the rotor
stat	static pressure drop measured without rotation
total	total

List of abbreviations

RCF	relative centrifugal force
RPB	rotating packed bed
RZB	rotating zig-zag bed

References

[1]   Sherwood TK, Shipley GH, Holloway FAL. Flooding velocities in packed columns. Ind. Eng. Chem. 1938;30(7):765–69.

[2]   Goedecke R ed. Fluidverfahrenstechnik: Grundlagen, Methodik, Technik, Praxis. Weinheim: Wiley-VCH; 2006.

[3]   Gorak A, Sorensen E. Distillation: Fundamentals and Principles. 1st edition, s.l.: Elsevier Reference Monographs, 2014.

[4]   Gorak A, Schoenmakers H. Distillation: Operation and Applications. London: Academic Press; 2014.

[5]   Gorak A, Olujic Z. Distillation: Equipment and Processes. Burlington: Elsevier Science; 2014.

[6]   Neumann K, Hunold S, Skiborowski M, Górak A. Dry pressure drop in rotating packed beds – systematic experimental studies. Ind. Eng. Chem. Res. 2017;56(43):12395–405.

[7]   Groß KM. Deaeration in Rotating Packed Beds. Dissertation Dortmund; 2021.

[8]   Neumann K, Hunold S, Groß K, Górak A. Experimental investigations on the upper operating limit in rotating packed beds. Chem. Eng. Process. 2017;121:240–47.

[9]   Zhao B, Tao W, Zhong M, Su Y, Cui G. Process, performance and modeling of CO 2 capture by chemical absorption using high gravity: A review. Renewable Sustainable Energy Rev. 2016;65:44–56.

[10]   Maćkowiak J. Extended channel model for prediction of the pressure drop in single-phase flow in packed columns. Chem. Eng. Res. Des. 2009;87(2):123–34.

[11]   Hendry JR, Lee JGM, Attidekou PS. Pressure drop and flooding in rotating packed beds. Chem. Eng. Process. 2020;151:107908.

[12]   Lockett MJ. Flooding of rotating structured packing and its application to conventional packed columns. Chem. Eng. Res. Des. 1995;73.

[13]   Singh SP, Wilson JH, Counce RM, Villiersfisher JF, Jennings HL, Lucero AJ, et al. Removal of volatile organic-compounds from groundwater using a rotary air stripper. Ind. Eng. Chem. Res. 1992;31(2):574–80.

[14]   Rajan S, Kumar M, Ansari MJ, Rao DP, Kaistha N. Limiting gas liquid flows and mass transfer in a novel rotating packed bed (HiGee). Ind. Eng. Chem. Res. 2011;50(2):986–97.

[15] Neumann K, Gladyszewski K, Groß K, Qammar H, Wenzel D, Górak A, et al. A guide on the industrial application of rotating packed beds. Chem. Eng. Res. Des. 2018;134:443–62.

[16] Mersmann A, Voit H, Zeppenfeld R. Brauchen wir Stoffaustausch-maschinen? Chem. Ing. Tech. 1986;58(2):87–96.

[17] Yildirim Ö, Flechsig S, Brinkmann U, Kenig EY. Bestimmung der Lastgrenzen konventioneller Strukturpackungen und Anstaupackungen mithilfe des Wallis-Plots. Chem. Ing. Tech. 2015;87 (10):1348–56.

[18] Große J. Über Keramische Schwämme Als Kolonneneinbauten Grundlegende Untersuchungen Zu Morphologie, Fluiddynamik Und Stoffübergang Bei der Mehrphasigen Durchströmung Im Gegenstrom. Dissertation Karlsruhe: KIT Scientific Publishing; 2011.

[19] Gładyszewski K, Groß K, Bieberle A, Schubert M, Hild M, Górak A, et al. Evaluation of performance improvements through application of anisotropic foam packings in rotating packed beds. Chem. Eng. Sci. 2021;230:116176.

[20] Rao DP, Bhowal A, Goswami PS. Process intensification in rotating packed beds (HIGEE): An appraisal. Ind. Eng. Chem. Res. 2004;43(4):1150–62.

[21] Burns JR, Ramshaw C. Process intensification: Visual study of liquid maldistribution in rotating packed beds. Chem. Eng. Sci. 1996;51(8):1347–52.

[22] Basic A, Dudukovic MP. Liquid holdup in rotating packed-beds – examination of the film flow assumption. AIChE J. 1995;41(2):301–16.

[23] Burns JR, Jamil JN, Ramshaw C. Process intensification: Operating characteristics of rotating packed beds – Determination of liquid hold-up for a high-voidage structured packing. Chem. Eng. Sci. 2000;55(13):2401–15.

[24] Yang Y, Xiang Y, Chu G, Zou H, Luo Y, Arowo M, et al. A noninvasive X-ray technique for determination of liquid holdup in a rotating packed bed. Chem. Eng. Sci. 2015;138:244–55.

[25] Groß K, Bieberle A, Gladyszewski K, Schubert M, Hampel U, Skiborowski M, et al. Analysis of flow patterns in high-gravity equipment using gamma-ray computed tomography. Chem. Ing. Tech. 2019;91(7):1032–40.

[26] Charpentier J-C. Mass-Transfer Rates in Gas-Liquid Absorbers and Reactors. In: Advances in Chemical Engineering. vol. 11, Elsevier; 1981, 1–133.

[27] Lewis WK, Whitman WG. Principles of gas absorption. Ind. Eng. Chem. 1924;16(12):1215–20.

[28] Peel J. The Mass Transfer and Hydrodynamics of a Gas-Liquid Centrifugal De-oxygenator. Dissertation Newcastle-upon-Tyne; 1995.

[29] McCabe WL, Smith JC, Harriott P. Unit Operations of Chemical Engineering. 5th edition, New York: McGraw-Hill; 1993.

[30] Laso M, de Brito MH, Bomio P, von Stockar U. Liquid-side mass transfer characteristics of a structured packing. Chem. Eng. J. Biochem Eng. J. 1995;58(3):251–58.

[31] Wang Z, Yang T, Liu Z, Wang S, Gao Y, Wu M. Mass transfer in a rotating packed bed: A critical review. Chem. Eng. Process. 2019;139:78–94.

[32] Revie RW, Uhlig HH. Corrosion and Corrosion Control: An Introduction to Corrosion Science and Engineering. Hoboken, New Jersey: Wiley-Interscience, a John Wiley & Sons, Inc., Publication; 2008.

[33] Harbold G, Park J. Single-stage Vacuum Deaeration Technology forAchieving Low Dissolved Gas in Process Water. San Antonio, 2008.

[34] Chen Y-S, Lin -C-C, Liu H-S. Mass transfer in a rotating packed bed with viscous newtonian and non-newtonian fluids. Ind. Eng. Chem. Res. 2005;44(4):1043–51.

[35] Chen YS, Lin CC, Liu HS. Mass transfer in a rotating packed bed with various radii of the bed. Ind. Eng. Chem. Res. 2005;44(20):7868–75.

[36] Chen Y-S, Lin F-Y, Lin -C-C, Tai CY-D, Liu H-S. Packing characteristics for mass transfer in a rotating packed bed. Ind. Eng. Chem. Res. 2006;45(20):6846–53.

[37] Groß K, de Beer M, Dohrn S, Skiborowski M. Scale-up of the radial packing length in rotating packed beds for deaeration processes. Ind. Eng. Chem. Res. 2020.

[38] Luo Y, Chu G-W, Zou H-K, Wang F, Xiang Y, Shao L, et al. Mass transfer studies in a rotating packed bed with novel rotors: Chemisorption of CO 2. Ind. Eng. Chem. Res. 2011;51 (26):9164–72.

[39] de Beer MM, Koolaard AM, Vos H, Bargeman G, Schmuhl R. Intensified and flexible flash degassing in a rotating packed bed. Ind. Eng. Chem. Res. 2018;57(42):14261–72.

[40] Ramshaw C, Mallinson RH. Mass transfer process(U.S. Patent 4,283,255); 1978.

[41] Munjal S, Dudukovic MP, Ramachandran PA. Mass-transfer in rotating packed beds-II. Experimental results and comparison with theory and gravity flow. Chem. Eng. Sci. 1989;44 (10):2257–68.

[42] Keyvani M, Gardner NC. Operating characteristics of rotating beds. Chem. Eng. Prog. 1989;85 (9):48–52.

[43] Kumar M, Rao DP. Studies on a high-gravity gas-liquid contactor. Ind. Eng. Chem. Res. 1990;29(5):917–20.

[44] Chen YH, Chang CY, Su WL, Chen CC, Chiu CY, Yu YH, et al. Modeling ozone contacting process in a rotating packed bed. Ind. Eng. Chem. Res. 2004;43(1):228–36.

[45] Hacking JA, de Beer MM, van der Schaaf J. Gas-liquid mass transfer in a rotating liquid redistributor. Chem. Eng. Process. 2021;108377.

[46] MathWorks. Piecewise Cubic Hermite Interpolating Polynomial (PCHIP): Documentation. [January 17, 2020].

Kolja Neumann, Dennis Wenzel

4 Rotating packed beds in flue gas cleaning

The focus of this chapter is set on flue gas cleaning by means of gas-liquid absorption processes in counter-currently operated rotating packed beds, with a focus on CO_2 absorption. On this topic, respective laboratory-scale, pilot-scale, and industrial-scale absorption processes are presented. Furthermore, one experimental study and one theoretical study will be illustrated further in detail with examples.

4.1 Fundamentals

In the following, a short summary of hydrodynamic and mass transfer phenomena that are important for absorption processes in rotating packed beds (RPBs) is given, including correlations and applications. More general descriptions of mass transfer and hydrodynamics for RPB gas-liquid processes can be found in Chapters 1 and 3.

4.1.1 Fundamentals of liquid and gas phase contacting in RPBs

In RPBs, changes in cross-sectional area and centrifugal force along the packing radius lead to changes in the liquid and gas flow pattern. Hence, along the radius, different zones of mass transfer are present. In counter-current operation, gas flows from the casing through the packing towards the center of the rotor. The tangential velocity of the gas at any radial position in the packing is either larger than the tangential velocity of the rotor or the same [1]. A larger tangential velocity of the gas can exist due to the conservation of angular momentum. In the case of large frictional forces between packing and gas, the gas decelerates to the tangential velocity of the rotor. In RPBs, the cross-sectional area changes proportionally to the radius. Hence, the gas accelerates towards the eye of the rotor. Altogether, **the influence of a standard rotating packing, e.g., knit or wire mesh, on the gas flow pattern in the packing is expected to be of minor importance for intensifying the mass transfer [2]**.

Close to the inner packing radius, a zone of intense mixing is present [3, 4] (cf. also Chapter 3). The **intense mixing occurs due to the slip between the tangential velocity of the entering liquid and the packing** [5, 6]. With increasing radius, dry and wet packing areas are formed as the liquid splits up into rivulets. The degree of

Kolja Neumann, Lanxess
Dennis Wenzel, TU Dortmund University

https://doi.org/10.1515/9783110724998-004

maldistribution depends on the centrifugal force [3] as well as the packing design (cf. Chapter 7). Furthermore, Burns & Ramshaw [3] discussed the presence of three different flow regimes depending on the centrifugal force: rivulets, droplets, and liquid films. A detailed description of the fluid dynamics is provided in Chapters 3 and 9. Burns et al. [7] stated that the liquid holdup decreases with increasing radial position in the packing. For their study, a tracer response technique was used, and the hold-up was measured at different radial positions in a polyvinyl chloride (PVC) foam packing. A sharp decrease of the liquid hold-up with increasing rotational speed was reported [7, 8]. Only close to the upper operating limit, Burns et al. [7] found a dependency of the liquid hold-up on the gas flow rate. They found the volumetric liquid hold-up to be in the range of 0.05–0.3 $m^3\ m^{-3}$. These values are in accordance with the volumetric liquid holdup reported by [4, 9]. Further details on the operating window and liquid hold-up measured by computer tomographic methods can be found in Chapter 3.

At the outer radius of the packing, the liquid is ejected and passes the free space between the rotor and surrounding casing. Sang et al. [10] observed the formation of ligaments (low rotational speeds) and droplets (high rotational speed) at the outer packing radius. A narrow droplet size distribution was measured with average droplet diameters between 1.2 (low rotational speeds) and 0.5 mm (high rotational speed) [10]. With increasing outer packing radius and rotational speed, the droplet size reduces. Simultaneously, the velocity of the droplets in the casing increases due to a larger tangential velocity. The droplets ejected by the rotor impinge on the inner casing wall and form liquid films and back splashing droplets. Due to this flow pattern, a significant contribution of the casing to the overall mass transfer is found for RPBs [11].

4.1.2 CO_2 absorption in RPB

The emission of anthropogenic greenhouse gases is considered to be one of the major contributors to the climate change [12]. CO_2 has the main share of emitted greenhouse gases, with an amount of 76% (49 gt a^{-1} in 2014, see [12]). Approximately half of the emitted CO_2 originates from the energy sector and the chemical industry [12]. The post combustion capture (PCC) by use of reactive (chemical) absorption is a state-of-the-art process to reduce the CO_2 content in waste gas streams [13]. The solvent loaded with CO_2 leaves the absorption column and is regenerated in a stripping column. Mass transfer limitations and large gas flow rates make large conventional equipment volumes necessary, leading to high capital costs [14]. In their comprehensive study, Wang et al. [15] reviewed the potential of different technologies to improve PCC processes, including RPBs. **Due to the achievable size reduction and high capacities, the installation of RPBs in PCC processes is considered to be promising [16]. Besides the equipment selection, the solvent selection is a key factor to improve PCC processes.** The selected solvent determines the thermal efficiency penalty, caused by the regeneration of the solvent, and many publications on this topic have been dealing with the reactive absorption of CO_2 in RPBs. The main findings from these theoretical and experimental studies are summarized in several reviews [15, 17–20]. A discussion of the performance of modified RPB designs, such as crossflow RPBs [21], can be found in the reviews of [17, 18, 20].

Primary amines, secondary amines, tertiary amines, as well as amine blends [22–24, 25] have been investigated in RPBs, mainly on laboratory scale. Tan et al. [26, 27] conducted a comprehensive comparison of different amines and amines mixtures. The main difference between these amines is the reaction mechanism and reaction rate (cf. Chapter 6 of the first book). Primary amines, like monoetha-nolamine (MEA), react quickly with CO_2 forming carbamates but require significant amounts of energy for regenerating the solvent (desorption). Tertiary amines such as methyl-diehtanolamine (MDEA) do not react directly with CO_2, and while the reaction rate is slower than the MEA, the energy demand for regeneration of the loaded solvent is considerably lower, and larger cyclic loadings can be achieved. Amine blends, such as activated MDEA, an aqueous MDEA solution with added piperazine (PZ), are an approach to combine the advantages of the individual solvents [28, 29]. Refer to Chapter 6 of the first book of this series for a more detailed introduction to reactive absorption. Besides amine blends, the use of viscous ionic liquids as solvents was tested for RPBs, as well [30] indicating an increased rate of absorption at the costs of increased corrosivity and viscosity [30].

In many studies, a correlation between an increasing CO_2 removal efficiency and an increasing solvent flow rate is reported. Higher liquid loads lead to an improved packing wetting and, thus, an enhanced CO_2 removal efficiency. In addition, it must be stated that the gas flow is often kept constant, and hence, the G/L ratio decreases, which enhances the CO_2 removal efficiency. Two different effects of the rotational speed on the CO_2 capture are reported in literature. While some studies report an increasing but flattening trend with increasing rotational speed, others report a maximum that is passed. An overview of these results is given in the review of [19]. One explanation for the flattening trend can be found in the fact that at high rotational speeds, the effective interfacial area does not increase much [31]. As the gas flow pattern is not influenced much in a standard RPB mesh packing by the rotational speed, a significant rise in the gas-side mass transfer coefficient is not expected. Therefore, the **low dependency of the effective interfacial area on the rotational speed might be the decisive factor when dealing with fast reactions and high liquid-side mass transfer coefficients. At the same time, the reaction rate is still fast enough to compensate the very short residence time. On the other hand, when the CO_2 removal efficiency is observed to pass a maximum with increasing rotational speed, this can be explained by a slow reaction rate at decreasing residence time.**

Instead of water, solvents such as diethylene glycol have been investigated in order to lower the heat capacity of the solvent and to decrease the energy for regeneration [32, 33]. Several publications are dealing with combinations of diethylene triamine (DETA) and PZ and the combination of MEA and PZ [22, 26, 27, 34, 35]. The authors state larger capture efficiencies for the PZ mixture compared to 30 wt.% MEA solution. This finding is traced back to the faster reaction kinetics, underlining the need for very fast reactions in RPBs, due to the short residence times [35].

Another approach to increase the reaction rate was followed by [23] and Wang et al. [18], who investigated the CO_2 capture in solvents at MEA concentrations ≥ 30 wt.%. The authors followed the idea to compensate the disadvantage of higher MEA concentrations, namely increased corrosivity and viscosity (e.g., reduced rate of diffusion in the liquid), by applying RPB technology. On the one hand, this balancing effect is explained by the small equipment volume of RPBs, requiring a smaller amount of costly corrosion-resistant materials for construction. On the other hand, due to the intensified mass transfer, RPBs can cope with more viscous liquids [23]. In addition, [18] varied the CO_2 loading of MEA and 2-methyl-amino-ethanol and found different optima with respect to removal efficiencies, depending on the solvent concentration and CO_2 loading. This effect was explained by the significant change in liquid viscosity of the solvent with increasing alkanolamine concentration and CO_2 loading [18].

Sheng et al. [36] investigated the performance of DETA with different CO_2 loadings and DETA concentrations in a lab-scale RPB. DETA was found to have a significantly higher CO_2 removal efficiency than MEA. The authors' comparison of the RPB CO_2 capture with data for packed columns, taken from literature, revealed that even with the very short residence time in the RPB of less than 1.5 s, the same CO_2 removal efficiency as in packed columns was achieved due to the intensified mass transfer. Hence, a significant equipment volume reduction was possible with an RPB [36].

In most available studies, the focus is set on the capture efficiency. Stripping and rich solvent loadings are often not considered, impeding the comparison with the state-of-the-art processes. Jassim et al. [23] found the stripping in RPBs to be less efficient compared to packed columns, while [32] reported an improved heat transfer and a size reduction by a factor of 10. Tan et al. [37] conducted experiments using two connected lab-scale RPBs for continuous absorption and stripping studies with solvent mixtures containing PZ, DETA, and the amino acid salt sodium aliphatic diamine sulphonate (NaADS). The authors stated that, besides the capability of amino salts for CO_2 capture, this class of components also reduces the dissolved O_2 concentration and, thus, the corrosivity [37]. The solvent mixture of PZ, DETA, and NaADS is more viscous than a mixture of PZ and DETA, but revealed a similar CO_2 capture efficiency and a slight reduction of energy required for regeneration of the solvent in the presence of NaADS. Xie et al. [38] studied CO_2 removal efficiencies at gas inlet CO_2 concentrations between 2 vol.% and 6 vol.% in potassium sarcosine. Removal efficiencies larger than 80% were found for gas to liquid ratios <150. Besides the development of an optimized solvent for CO_2 absorption in RPBs, several approaches of improved packing designs have been developed which are presented in Chapters 7 and 8. One special example is functionalized mesocellular foam packing, developed by [39], specifically for CO_2 absorption with claimed superior absorption efficiency compared to conventional packings.

4.1.3 Energy transfer in CO_2 absorption process

Coming back to the idea of compensating the short residence times in RPBs by selecting solvents that are more reactive than 30 wt.% MEA solution, the enhanced absorption of CO_2 per packing volume is linked with an increased release of heat. During the reactive absorption of CO_2 in amine solutions, the temperature of the gas and liquid streams changes along the height of a column [40]. This temperature change is primarily influenced by the energy being released during the exothermic absorption reaction and the evaporation of the solvent. The released energy by the reaction in the liquid phase increases the temperature of the solvent. The evaporation of solvent leads to a decrease of the liquid temperature. As a consequence, a temperature bulge forms in the liquid phase [40]. Kvamsdal et al. [40] published a theoretical study about the location and the magnitude of the temperature bulge in packed columns for MEA solutions. Among other operational parameters (e.g., CO_2 concentration in the gas and heat capacity of the solvent), they found the temperature bulge being depend on the liquid to gas ratio (L/G ratio). In Fig. 4.1, the position of the temperature bulge in the packing of a conventional packing is qualitatively plotted for changing L/G ratio. At high L/G ratio ($L/G > 8$ kg kg^{-1}), the temperature bulge is located at the bottom of the column (position of the temperature bulge → 1 in Fig. 4.1). By the reduction of the L/G ratio ($L/G <$ 5 kg kg^{-1}), the temperature bulge location changes and travels towards the top of the column (position of the temperature bulge → 0 in Fig. 4.1). Hence, for low L/G ratio, a significant rise of the gas outlet temperature is observable. The negative effects of the temperature bulge formation on the mass transfer have been subject to several studies, and intercooling of the liquid phase was found to be a promising way to improve the CO_2 removal efficiency in packed columns [41–43]. **For RPBs, the top of a column corresponds to the inner packing radius and the bottom of a column to the outer packing radius. For the same operating conditions and absorbed moles of CO_2, a similar correlation between L/G ratio and the position of the temperature bulge as illustrated in Fig. 4.1 is expected for RPBs.**

RPBs equipped with rotor plates containing coils for heating or cooling, as suggested for spinning disc reactors by [44], can be used in order to reduce the formation of hot spots, to supply heat for endothermic reactions, and to facilitate the intercooling of solvents during reactive absorption processes. Furthermore, RPBs are capable of processing viscous fluids [45, 30]. Hence, RPBs with integrated cooling coils are of interest for viscous solvents like PZ [43, 46] requiring intercooling [47, 48]. Additive manufacturing offers many opportunities to include cooling and heating coils in complex constructional elements and will facilitate the further development of new rotor concepts. Refer to Chapter 7 for further information on the additive manufacturing of innovative RPB packings.

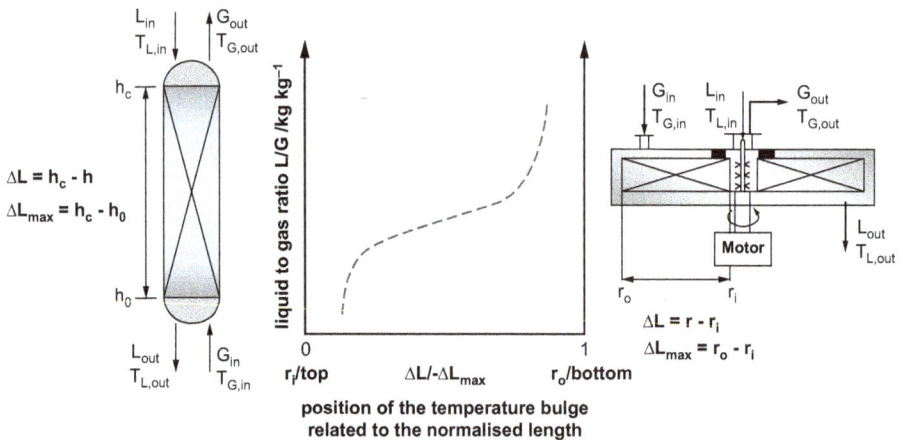

Fig. 4.1: Schematic of a packed column and an RPB used for the absorption of CO_2. The figure in the center illustrates qualitatively the correlation between the position of the temperature bulge in the packing of a column [40] and RPBs, respectively.

4.2 Applications

After Ramshaw and Mallinson [49] filed a patent on their RPB design, [50, 51] reported experimental results and discussed advantages of RPBs such as processing viscous fluids [50], reduced foaming [52], and quick steady-state [51]. As well, several fields of applications are mentioned in the respective publications. For mass transfer equipment made of expensive construction materials, such as high pressure or hazardous inventory [53], the achievable equipment volume reduction was stated as one of the paramount advantages using RPBs [50]. **RPBs have a small footprint and are not influenced by changes in orientation. Hence, the use of RPBs for offshore applications (e.g., on oil platforms) [50] and for modular plants is very promising.** The small footprint is also interesting for retrofitting chemical production sides with space limitations [54]. Furthermore, RPBs are reported to have a larger turndown ratio than conventional packed columns, allowing for a more flexible operation [55]. In the following, the reported pilot and industrial scale gas cleaning applications of RPBs are discussed.

4.2.1 CO$_2$ cleaning processes and carbon capture and storage using RPBs

CO$_2$ must be removed from natural gas prior to its liquefaction, as CO$_2$ may cause corrosion in the pipeline, reduces the calorific value, and CO$_2$ tends to freeze and block the pipes when the natural gas is liquefied [19]. A successfully operated pilot-scale RPB for the removal of CO$_2$ was reported in 1987 by Chevron and Statoil [50]. Around 7 vol.% of CO$_2$ were removed from natural gas using diethanolamine at 69 barg. CO$_2$ capture efficiencies up to 90% and high rich loadings of the solvent were achieved [50]. Unfortunately, no further process details are available. **Besides the reduced equipment volume, [56] emphasized the high operating flexibility (reacting on changing flow rates and compositions) and the quick start-up and shut-down as further advantages of RPBs.** In accordance with [52], the advantage to use mobile RPB units to exploit remote oilfields by treating the natural gas is outlined by [56]. A pilot-scale RPB capturing 40000 t a^{-1} of CO$_2$ was investigated by [56]. Due to the remoteness of gas-fields, mobile units are required, and RPBs can be easily transported without large costs. Around one-third of the world-wide remote natural gas resources are not developed due to the small reserve size impeding the exploitation on a standalone basis [57]. New concepts such as small scale and mobile plants need to be developed in order to develop such remote gas fields and to facilitate the processing of biomass-based feedstocks [58]. The pilot-scale RPB developed by ICI Chemical Group was already mounted on a modular frame weighting 40 tonnes [52]. Nevertheless, this realization of idea has not been reported. A further pilot-scale study dealing with the absorption of CO$_2$ in propylene for natural gas purification was published by [59]. CO$_2$ removal efficiencies in the range from 20% to 65% were reported. Besides temperature and pressure, the rotational speed (290–1450 rpm) was a major operating parameter affecting the CO$_2$ removal efficiency, with a maximum in removal efficiency at around 900 rpm. The idea to use RPBs for the exploitation of remote oil and natural gas fields was picked up by Zhang et al. in their recent review [19]. The authors outlined the increasing demand of natural gas as a substitute for coal and oil during transition phase from fossil-based raw materials towards renewable energy sources for power generation.

Also on pilot-scale, the implementation of RPBs to treat waste gas streams of a steel plant was tested in Taiwan [60]. Instead of amines, an accelerated/high-gravity carbonation (HiGCarb) using basic organic furnace slag was used for capturing CO$_2$ [60]. The carbonation is expected to accelerate due to the enhanced micro-mixing and mass transfer in the centrifugal field. In the pilot-plant, around 165 kg/d of CO$_2$ could be absorbed with an efficiency >90% [60].

4.2.2 RPB NO$_x$ cleaning processes

The term "NO$_x$" refers to all nitrogen oxides, including NO, NO$_2$, N$_2$O$_3$, N$_2$O$_4$, and N$_2$O$_5$. It has been shown that long-term exposure to NO$_x$ increases the risk of respiratory diseases and that NO$_x$ compounds contribute to the formation of acid rain, photochemical smog, and ground level ozone [61]. The largest amount of NO$_x$ emissions originates from flue gases of various sources such as fossil fuel vehicles, fossil fuel combustion plants, and the nitro-chemical industry. Therefore, various technologies have been developed to reduce NO$_x$ emission from flue gas such as absorption processes [62]. However, **conventional absorption columns for NO$_x$ absorption typically require large areas, come with high investment and operation costs, and show only comparatively low removal efficiencies [61], making RPBs an promising alternative.**

Liu et al. investigated the treatment of flue gas consisting mainly of NO$_x$ compounds in a series of two RPBs [61, 63]. The authors used 20 wt.% carbamide as the absorbent, together with an additive to adjust the pH value, and reported a NO$_x$ removal efficiency of more than 96%. Therefore, RPBs could decrease the investment costs by nearly 75% and reduce the operation cost by about 79% compared to conventional absorption columns, according to Liu et al.'s calculations.

Li et al. used nitric acid solution as absorbent to remove NO$_x$ compounds in an RPB [64]. Ozone was additionally introduced to oxidize HNO$_2$ to HNO$_3$ to prevent the decomposition of HNO$_2$ in the liquid phase and to improve the NO$_x$ absorption [61]. For this RPB process, a removal efficiency of NO$_x$ compounds of more than 90% was reported. Zhang et al. used an RPB as gas-liquid reactor to enhance the NO removal efficiency by EDTA solution [65]. For optimum operation conditions, the authors reported a removal efficiency of up to 87%.

Since NO is in many cases the major component in NO$_x$ emissions, but is less soluble in water than, for example, NO$_2$, Sun et al. investigated the pre-oxidation of NO-rich flue gas with ozone, followed by a denitrification step with sodium hydroxide (NaOH) in an RPB [66]. For this process combination, the authors reported a removal efficiency of up to 80%. Additionally, the **NO$_x$ removal efficiency first increased and then decreased with increasing rotational speed of the RPB, demonstrating strong analogy to other unit operations implemented in RPBs.** Refer to Chapter 2 for liquid-liquid and liquid-solid contacting (extraction and adsorption) as well as Chapters 6 and 8 for vapour-liquid conctacting (distillation).

In addition, RPBs have been applied for the combined absorption of NO$_x$ and SO$_2$ as well as the combined absorption of NO$_x$, SO$_2$, and CO$_2$. Pan et al. and Chen et al. summarized many of the published studies on this field of research in their comprehensive reviews [67, 17]. Removal efficiencies between 90% and 99.9% have been reported in combination with significant equipment volume reductions.

4.2.3 RPB H$_2$S and SO$_2$ cleaning processes

Many sulfuric compounds, such as H$_2$S and SO$_2$, are very harmful by inhalation, with increasingly strict emission regulations worldwide. Therefore, flue gas containing these compounds, e.g., from coal or oil-fired power plants, chemical plants, metal smelters, and lime kilns [68], needs to be treated intensively. Furthermore, H$_2$S and SO$_2$ are typical impurities in natural gas. They need to be removed from the gas stream to comply with natural gas network regulations and to enable the production of pure sulfur from sour (i.e., highly sulfuric) natural gas. The efficiency of H$_2$S separation processes is dominated by the choice of a suitable solvent, the process operating conditions, and the chosen equipment [68]. A popular method of removing H$_2$S from a gas stream is the selective removal by reactive absorption. The important feature of this process are the competing chemical reactions of H$_2$S and CO$_2$, limited by chemical equilibrium [69]. One solvent particularly suitable for this application is a tertiary amine, i.e., methyldiethanolamine (MDEA), which reacts with H$_2$S almost instantaneously by proton transfer. In contrast, the reaction of MDEA with CO$_2$ can only occur after the formation of bicarbonate (HCO$_3^-$) in aqueous solution as intermediate, which is a slow reaction [70]. Therefore, it may be assumed that the H$_2$S reaction with MDEA is gas phase-limited, while the CO$_2$ reaction is liquid phase-limited [70], and that the H$_2$S selectivity over CO$_2$ is favored by short contacting times and high liquid side mass transfer [68]. Consequently, **equipment providing high mass transfer coefficients in combination with short contacting times is required in order to improve the H$_2$S selectivity. Those criteria are typically met by RPBs, in which the liquid side mass transfer is significantly larger than in conventional columns [31, 1], and the absorption process takes place in small packing volumes with short contacting times and high gas capacity factors** [71, 72].

The idea to use RPBs for the selective removal of H$_2$S has been first investigated in the 1980s by the company Glitch and Fluor Corp., who reported high loads and selectivities, but no detailed data were given [50]. Quian et al. investigated the selective removal of H$_2$S using a 30 wt.% MDEA solution in laboratory-scale to industrial-scale RPBs [73, 74]. Their work is one of the few published studies covering all steps of classical process development from lab to pilot and finally to industrial scale [68]. For the industrial-scale process, the separation performance of an RPB (packing volume: 0.31 m^3; casing diameter: 1.4 m; casing height: 2.9 m) and a conventional column was compared [74]. For a gas flow rate of 13000 m^3 h^{-1}, the same amount of H$_2$S was absorbed, but the selectivity of the process increased as the CO$_2$ absorption, taking place simultaneously, was reduced from 80% in the column to 9% in the RPB [74]. A volume reduction by a factor of 10 was reported. The gas capacity factor at the inner diameter of the packing of 6.2 Pa$^{0.5}$ underlines the large gas loads that are achievable in RPBs. Furthermore, steady-state operation can be achieved within minutes in RPBs [74], and the rotational speed offers an additional degree of operational freedom.

Similar to other applications, the enlarged hydrodynamic operating window in RPBs is also of major advantage for the absorption of SO_2. Reay et al. reported potential savings in investment costs by 60% if RPBs are used instead of conventional state-of-the-art processes [53]. This investment cost reduction is achieved by smaller equipment volumes, and only two RPBs are required instead of three columns (including an entrainment separator). Zou et al. also published a study covering the steps of classical process development in order to design an RPB for the desulfurization of flue gases [75]. This industrial-scale RPB has an outer rotor diameter of 3.5 m and is designed to treat gas flow rates of 200000 m^3 h^{-1}. An RPB of similar size (outer rotor diameter of 3.6 m) is reported by Pan et al. [17]. This RPB is used to remove 98.8% of SO_2 for gas flow rates of 80000 m^3 h^{-1}. Chu et al. [76] investigated the **removal of SO_2 with ammonia-based solution in a pilot-scale RPB. The authors reported desulfurization efficiencies of 98.5%, which is comparable to results from packed columns. However, RPBs are more compact and require only one-fifth of the space normally occupied by wet spray scrubber or packed columns at the same loading capacities** [76]. The removal of SO_2 with sodium citrate buffer was studied by Jiang et al. [77]. The authors found that the rotational speed of the RPB had the largest positive effect on the separation efficiency. This can be explained with the smaller droplet sizes, more vigorous impingement, and thinner films in the packing, reducing the mass transfer resistance. However, the authors also pointed out that a higher rotating speed will cause higher costs, which must be considered in industrial applications.

Finally, the collection of particulate matter (PM) from flue gas contributes to the advantages of RPBs in SO_2 absorption processes. Since the rotating packing in RPBs generates a strong cyclone-effect [78], and RPBs are not typically susceptible to clocking, a considerable volume reduction in comparison with conventional equipment can be achieved. Therefore, the highest separation efficiency could be obtained with the smallest packing diameter and the highest rotational speed [78]. Furthermore, RPBs combine the working principle of wetted surface and spray collectors in one machine [17].

Zhang et al. [79] investigated the combined SO_2 and PM removal from flue gas [79]. For this study, a pilot-scale RPB was installed in an alkaline sulfite chemical plant, where flue gas streams of up to 20000 m^3 h^{-1} were processed. As the absorbent, an aqueous solution with 0.87 wt.% magnesium hydroxide ($Mg(OH)_2$) was used. The authors reported a removal efficiency for the PM in the range of 89.0% to 96.4%, and a desulfurization efficiency in the range of 81.0% to 82.8% [68]. The reported maximum total pressure drop was below 700 Pa, which is about two times lower than typically for wet dedusting scrubbers with the same capacity [80].

4.3 Modeling and simulation

Developing models and performing simulation studies help to reduce the experimental effort during the process development resulting in savings of time and costs [81]. Refer to Chapters 3 and 6 of the first book of this book series for a detailed description of the thermodynamics and mass-transfer modeling for reactive absorption processes in conventional columns. Different models with different levels of complexity have been proposed to model the CO_2 capture in RPBs. Zhao et al. [20] introduced four categories: mechanistic theory, statistical regression, numerical simulation, and artificial neural networks. The first two model types contain mass transfer and energy balances around a single liquid droplet [82] or for the gas-liquid interphase [83]. Rate-based modeling approaches range from stirred tank in series models [22, 35] to a full discretization of the rotor considering the geometry of the RPB packing geometry [54]. In the recent years, simulations of CO_2 absorption in RPBs involving rate-based approaches using commercial software like gProms®, Aspen Plus®, or Aspen Custom Modeller® gained more attention [27, 84–90, 91].

Rate-based models require correlations to calculate the mass transfer parameters. Correlations published by different authors are often combined in the published studies involving rate-based models in order to identify the combination of correlations that fits best to chosen experimental data. In course of the present chapter, this and further important aspects of this experimental data and correlations are discussed, as many of these studies involve the absorption of CO_2. **For most of the rate-based models, the different zones of mass transfer, such as the inner packing section, packed bed, and casing (cf. Chapter 3) in RPBs are not explicitly considered in modeling.** To overcome this limitation, [92] published a three-zone model, including the three zones mentioned before. The model was validated against experimental data (deviation ± 20%), and a contribution of the casing to the overall mass transfer >50% was found [92]. In recent publications, this approach is now often chosen.

Further simulation approaches dealing with CO_2 absorption are CFD simulations [93] and artificial neural networks [36, 94, 95]. Comprehensive reviews discussing the pro and cons and current state of CFD simulations for RPBs have been published [96]. CFD studies focusing on detailed aspects of RPBs, for example, on the nozzle design for liquid distribution [97] may help improving the understanding of underlying phenomena and result in optimized RPB designs and facilitates a more sound scaling up. Nevertheless, as **many phenomena such as gas-liquid interaction, flow pattern along the radius, and liquid distribution by nozzles have not been fully understood, various simplifications and assumptions are made for most of these CFD studies. Hence, their expressiveness for RPBs in general is limited [19]. As stated by [19], up to now, there are no mature modeling approaches for RPBs available. This especially holds true for commercial process simulation software.**

4.3.1 Correlations for CO_2 absorption

Due to the lack of fully predictive models, mass transfer parameters are required to establish mathematical models. For absorption processes involving fast chemical reactions in the liquid phase, the following mass transfer parameters are of interest: the gas side mass transfer coefficients ($k_{G,i}$) and the volumetric specific effective interfacial area, a_{eff}, also called effective interfacial area. Correlations describing the gas and liquid mass transfer parameters are derived on the basis of three different theories [98]. Mostly, the film theory, developed by [99], is applied for RPBs, but in some cases, the penetration theory and the surface renewal theory are used as well [98]. **If used for modeling approaches, correlations should originate from the same publication in which suitable chemical test systems were applied [100], and a broad range of operating and design parameters was investigated. Furthermore, for RPBs, the experiments must be designed in such a way that a clear distinction between the mass transfer in the casing and the packing can be made. The combination of correlations from different sources, which are often derived with non-standardized test systems, is one of the major drawbacks of publications dealing with rate-based modeling approaches for RPBs.** Hence, the results derived from these simulation studies must be reviewed critically, before deriving conclusions for RPB technology in general. The effects of different operational and design parameters on the mass transfer parameters as well as the main approaches for correlations to calculate $k_{G,i}$ and a_{eff} are discussed in the following and in the comprehensive review of [19].

Gas side mass transfer coefficient

In order to fit coefficients in correlations derived for mass transfer coefficients, experimental investigations are required. The experimental data provide information about the volumetric mass transfer coefficient $k_{j,i} \cdot a_{eff}$, which is the product of the effective interfacial area a_{eff} and the respective mass transfer coefficient $k_{j,i}$. The indices j denotes the phase (either gas or liquid) and i the component in the respective phase. The overall volumetric mass transfer coefficient $K_{j,i} \cdot a_{eff}$ includes the mass transfer resistance in both phases. The basic form of these correlations is given in equation (4.1), which is based on an approach that [101] derived for packed columns.

$$k_{i,j} \cdot a_{eff} = C \cdot Re_j^a Gr_j^b Sc_j^c \tag{4.1}$$

Alternative forms of equation (4.1), where ki,j.aeff is described as a function of We, aP, ε, rO, ri, or Dj, can also be found in literature.

This general equation contains dimensionless numbers (for details, see respective publications) and coefficients that are fitted to the experimental data. The correlations are sometimes extended by the Weber number or terms including design parameters, such as packing properties or rotor dimensions, as well as physical properties, such as surface tension or diffusion coefficients). Saha et al. [102] used experimental data from the literature to train an artificial neural network in order to predict $K_{G,i} \cdot a_{\text{eff}}$, but no correlation was provided.

In Tab. 4.1, the reported dependencies of the (overall) volumetric mass transfer coefficients on the operational parameters are given for selected publications. The investigated RPB types differ in design parameters, such as rotor dimensions or packing type and the chosen operational parameters. In addition, a huge variety of chemical systems and concentrations, as well as unit operations (stripping, distillation, and absorption), was investigated. Finally, in some of the studies, no differentiation between the mass transfer in the casing and the packing was made. As a consequence, the **reported effects of operational parameters on the volumetric mass transfer coefficients are contradictory [103]. Especially for the dependency on the gas flow rate, ambiguous results are reported. Altogether, there is still a lack of correlations being valid for different RPB designs and operating conditions.**

Tab. 4.1: Dependency of the (overall) volumetric liquid and gas side mass transfer coefficients on increasing gas flow $\dot{V}_G$, liquid flow $\dot{V}_L$, and rotational speed n_{rot} reported in the literature. Only gas-liquid systems are considered. Example: [104] reported an increasing $K_G a$ with increasing gas and liquid flow and just a slight increase with increasing rotational speed. Yu et al. [35] reported the same trends for increasing liquid flow and rotational speed, but contrary to [104], they found the $K_G a$ to decrease with increasing gas flow rate. Increase: ↑/Slight increase or flattening trend: ↗/No dependency: →/Decrease: ↓.

	Operational parameter			Reference
	$\dot{V}_G$ ↑	$\dot{V}_L$ ↑	n_{rot} ↑	
$K_G a$	↑	↑	↗	[46, 82, 104–107,[C], 108,[C], 106[C]]
	→	↑	↗	[24[C], 33]
	↓	↑	↗	[35, 38]
	↑		↗	[109]

Index C: Correlation provided; chemical system for gas side mass transfer coefficients: absorption of CO_2, NH_3, and SO_2 in NaOH and amines.

Effective interfacial area

The effective interfacial area is usually determined indirectly, using the absorption of CO_2 in aqueous sodium hydroxide as a standard chemical system [100]. Effective interfacial areas ranging from 200 [6] to 1500 $m^2\,m^{-3}$ [31] have been reported for different RPB types. Again, the differences in design parameters and operational parameters impede a direct comparison. The dependency of a_{eff} on different operating parameters is given in Tab. 4.2. For most of the studies, an increase of a_{eff} with increasing gas flow rate is reported. With increasing gas volume flow, the film thickness decreases and the number of droplets increases [110]. An increasing rotational speed leads to larger values of a_{eff} since smaller droplets and thinner liquid films are formed [110]. Yang et al. [6], determined a_{eff} for a packing width of 0.008 m at a rotational speed of 23.3 s^{-1} to approximately 7000 $m^2\,m^{-3}$. At the same rotational speed and with an increased packing width of 0.14 m, a_{eff} reduced to 500 $m^2\,m^{-3}$. This significant decrease is assigned to the intense mixing in the inner part of the packing, which enhances the mass transfer significantly compared to the bulk zone of the packing [6]. According to these findings, new packings and rotor designs have been developed, which are discussed in Chapters 7 and 8.

Tab. 4.2: Dependency of a_{eff} on increasing gas flow $\dot{V}_G$, liquid flow $\dot{V}_L$, rotational speed n_{rot}, and radial packing width: Δr, specific surface area (packing): a_P, packing porosity ε. Increase: ↑/Slight increase or flattening trend:↗/No dependency: →/Decrease: ↓.

	Operational parameter			Rotor parameters			Reference
	$\dot{V}_G$↑	$\dot{V}_L$↑	n_{Rot}↑	Δr↑	a_P↑	ε↑	
a_{eff}	↑		↗		↑	↓	[5]^C
	↑	→	→				[111]
	↑	↗	↑				[6, 112]
	↑	→	↑		↑		[113],^{C}
				↓	→	↓	[6, 45^C]
		→	↑				[114]

C: Correlation provided; chemical system: absorption of CO_2 in NaOH

4.4 Detailed examples

4.4.1 Absorption of CO_2 in an enzyme-catalyzed aqueous MDEA solvent. Comparison of RPB and a packed column (experimental study]

At the laboratory of fluid separations TU Dortmund, the CO_2 removal efficiency was determined for an aqueous solution containing 30 wt.% methyl diethanolamine (MDEA) and MDEA blended with the enzyme carbonic anhydrase (CA) (MDEA + CA) [115]. The experimentally determined CO_2 capture efficiencies of a pilot-scale packed column and a single-stage counter-currently (SSCC) operated RPB are compared in the following. **Adding a catalyst, in this case CA, provides valuable insights**

about the possibilities to increase the removal efficiency in RPBs by accelerating the reaction speed. For the SSCC-RPB, two different rotor configurations are discussed (see Tab. 4.3).

Tab. 4.3: Rotor configurations tested for CO_2 absorption in RPBs and information about packing used for the packed column.

Equipment	Nozzle liquid distribution	Outer rotor diameter [mm]	Packing	a_P [m²/m³]	ε [m³/m³]
SSCC-1	Flat fan	610	Knitted mesh	857	0.95
SSCC-2	Full jet	610	Knitted mesh	1667	0.91
Column	–	–	Structured	492	Not given

Despite the large shear forces that are present in RPBs, no decrease of the enzyme stability was observed in preliminary experiments. In Fig. 4.2 (left), the absorbed moles of CO_2 (measured in the gas phase) for the solvents MDEA and MDEA + CA are given as a function of the rotational speed at temperatures of 20 and 40 °C, respectively. The values incorporate the total moles of CO_2 that were absorbed in the casing and the packing of the RPB. With MDEA + CA solvent, similar amounts of CO_2 are absorbed at 20 and 40 °C. The low temperature sensitivity is a major advantage using CA as a catalyst since the cyclic loading can be increased, and the energy required for regeneration of the solvent can be further reduced even compared to activated MDEA [28]. Independently of the presence of CA, for MDEA as well as for MDEA + CA doubling, the rotational speed (from 10 to 20 s⁻¹) increased the absorbed moles of CO_2 by 25%. Adding CA to MDEA resulted in an increase of the absorbed moles of CO_2 by a factor of 4 (SSCC-1) to 5 (SSCC-2). This factor is defined as catalytic effect and is of the same order of magnitude as in packed columns [115]. In Fig. 4.2 (left), the influence of the packing on the mass transfer in RPBs is visible. The increase in absorbed moles of CO_2 that was found for the SSCC-2 compared to the SSCC-1 RPB can be traced back to the larger specific surface area and the lower porosity of the packing.

But not only the chosen packing, rotor dimensions and liquid distribution affect the catalytic effect but also the operating conditions. Wojtasik et al. [116] published a study with the same solvent system (30 wt.% MDEA and MDEA + CA] being investigated, but with a different RPB setup (3D-printed foam packing) and higher liquid and gas loads than discussed in this chapter. The authors claim a **catalytic effect in a range from 5 to 18, underlying the potential to further improve RPB designs, especially packings, and to optimize the operating conditions.**

In Fig. 4.2 (right), the moles of CO_2 absorbed per packing volume are illustrated for the SSCC-2 RPB, and the pilot-scale packed column is investigated by [117]. For this comparison, only the moles of CO_2 absorbed in the rotor were considered. The moles of CO_2 were determined by subtracting the CO_2 concentration measured in the

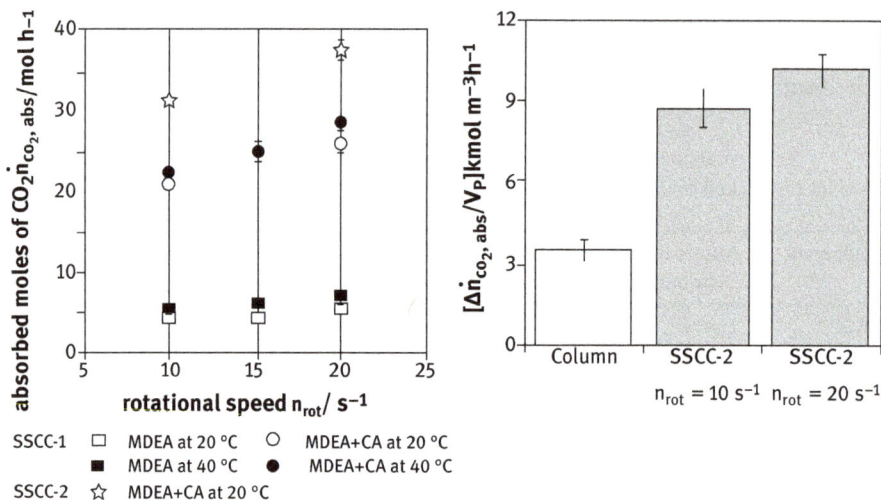

Fig. 4.2: Left: Absorbed moles of CO_2 as a function of the rotational speed and temperatures for the SSCC-1 and SSCC-2 RPB. Operating conditions: volume flow gas = 12.5 m³ h⁻¹; $y_{CO2,in}$ = 15 vol.%; mass flow solvent = 120 kg h⁻¹. Right: Moles of CO2 absorbed per packing volume for a packed column and the SSCC-2 RPB at 10 and 20 s−1. F-factor: 0.9 Pa0.5; liquid load = 26 m3 m−2 h−1.For the RPB, the gas capacity factor and the liquid load are related to the inner packing diameter.

gas outlet from the concentration measured in the casing. Compared to the packed column, the moles of CO_2 absorbed per packing volume are increased by a factor of 3 for the SSCC-2 RPB.

4.4.2 Absorption of CO_2 in MEA: comparison RPB and packed column (theoretical study)

The potential of RPBs to intensify reactive CO_2 capture processes is discussed in the following by means of simulation results comparing a conventional packed column and an RPB. Two rate-based models were used: a packed column (Aspen Plus®) and an RPB model. The RPB model builds on the modeling approach proposed by [54], which was initially developed for distillation in RPBs. The RPB model was adapted for CO_2 absorption using an implementation in Aspen Custom Modeller®. For the RPB model as well as for the packed column model, no reaction in the gas film was assumed, and the reaction solely takes place in the liquid film and liquid bulk phase. The same properties file to calculate property data using the ELECNRTL model [118] is used for both models. Reaction kinetics were taken from literature [119–121]. The rate-based RPB model considers solely the mass transfer in the packing; hence, the applied correlations shall be derived for the rotor excluding the mass transfer in the casing. As discussed in Section 4.31, suitable chemical test systems shall be applied

[100], and a broad range of operating and design parameters shall be investigated for deriving these correlation. None of the published RPB correlations for mass transfer parameters satisfies all criteria. Nevertheless, the publication of [109] meets most of the criteria, except a correlation for a_{eff} is missing. Therefore, own experimental data were used to derive a correlation for a_{eff} basing on the centrifugal acceleration and the packing depth. To validate the model of the packed column, the experimental results [122] published for a pilot-scale packed column were used. The RPB model was validated on the basis of own experimental data and experimental data published by [123]. The deviation between experimental and simulation results was found to be within ±20% (see Fig. 4.3).

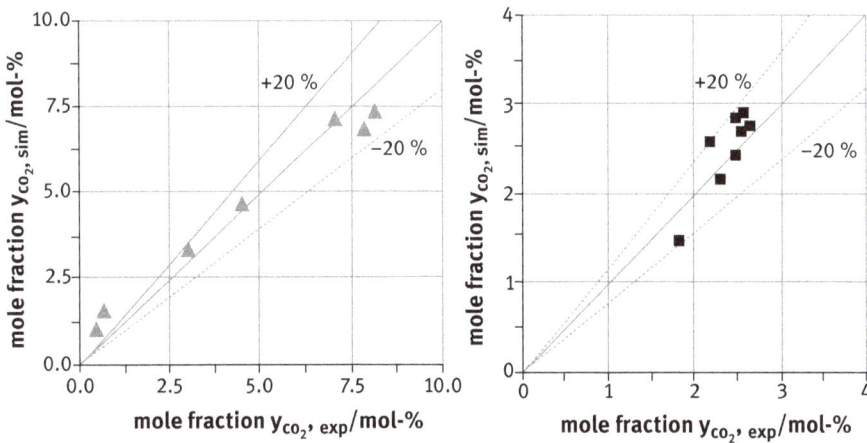

Fig. 4.3: Parity plots for the experimental and calculated CO_2 mole fraction in the column (left) and RPB (right) gas outlet, respectively. Experimental data for the absorption of CO_2 in MEA in conventional column in RPBs were taken from [122] and for RPBs from [123].

For the comparative study, the CO_2 capture efficiency was set to 90%, and the following operating and design parameters were chosen. The outer packing radius of the RPB was fixed to 0.5 m, and the rotational speed was set to $n_{rot} = 15$ s^{-1}. This is the lowest rotational speed (lowest power consumption and pressure drop] at which the objective of 90% CO_2 removal is satisfied with the chosen rotor. For the packed column and the RPB, the same operating conditions were defined. The diameter of the column was determined by limiting the maximal gas capacity factor to $F_G = 2.1$ Pa$^{0.5}$. The height of the column was determined by setting the CO_2 capture efficiency to 90%. Table 4.4 summarizes the main design and operating parameters chosen for the simulation study.

The potentially achievable size reduction by use of RPBs is illustrated in Fig. 4.4 (left) by plotting the factor of volume reduction against the normalized length. The factor of volume reduction is calculated by the ratio of CO_2 absorbed in the RPB and in the packed column. A normalized length = 1 (gas inlet) corresponds to the outer

Tab. 4.4: Specifications of the gas and liquid inlet streams and selected design parameters of the RPB and packed column (L/G: liquid to gas ratio). Gas capacity factor for RPB is calculated at the inner packing diameter.

	RPB	Packed column
Design parameter	$V_P = 0.08$ m^3	$V_P = 0.76$ m^3
	$a_P = 2700$ m^2 m^{-3}	$a_P = 250$ m^2 m^{-3}
Operational parameter	$F_G = 2.1$ Pa$^{0.5}$	$F_G = 2.1$ Pa$^{0.5}$
	$L/G = 4.4$ kg kg^{-1}	$L/G = 4.4$ kg kg^{-1}

packing radius of the RPB and bottom of the column, respectively. Consequently, a normalized length = 0 (gas outlet) corresponds to the inner packing radius of the RPB and top of the column. The factor of volume reduction is high at low normalized lengths as the reaction is fast due to low solvent loadings. At this inner part of the RPB packing, an intense mixing of the phases takes place. Consequently, **large factors of volume reduction are achievable. After passing a maximum, the factor decreases for increasing normalized lengths. The maximum can be traced back to an accelerating reaction speed with increasing solvent temperature. After passing the maximum, the factor of volume reduction reduces since reaction speed slows down with increasing solvent loading and the decreasing gas capacity factor and liquid load in RPBs.** Furthermore, the solvent temperature increases due to the released heat, lowering the solubility of CO_2 in the solvent.

In Fig. 4.4 (right), the dependency of the CO_2 capture efficiency on the lean loading of the solvent is illustrated. With increasing the loading, the reaction speed slows down [29, 124]. For low loadings, larger CO_2 capture efficiencies are achievable in RPBs for larger rotational speeds due to the increased effective interfacial. In accordance with the findings of [125], a **larger effective interfacial area in RPBs is only exploited for sufficiently fast reactions.** Hence, for loadings larger than 0.35, the CO_2 removal efficiency becomes independent of the rotational speed.

Altogether, the simulation results indicate that, **overall, a reduction of the packing volume by a factor of 10 can be achieved by the use of RPBs instead of packed columns.** This result is in accordance with the factors [126] and [56] reported. On the basis of theoretical studies, [126] found factors of volume reduction between 10 and 15 for rotating mass transfer machines compared to conventional static separation equipment. In the RPB, 27.0 kmol h^{-1} m^{-3} of CO_2 are absorbed per packing volume. This value is close to the value of 23.5 kmol h^{-1} m^{-3} that [123] measured during his experimental investigations. Interestingly, in the packed column, similar amounts of CO_2 are absorbed per packing volume (2.7 kmol h^{-1} m^{-3}) using MEA and the MDEA + CA system which is discussed in the previous example (Section 4.4.1). Assuming a pressure drop of 150 Pa m^{-1} for the column [127], a pressure

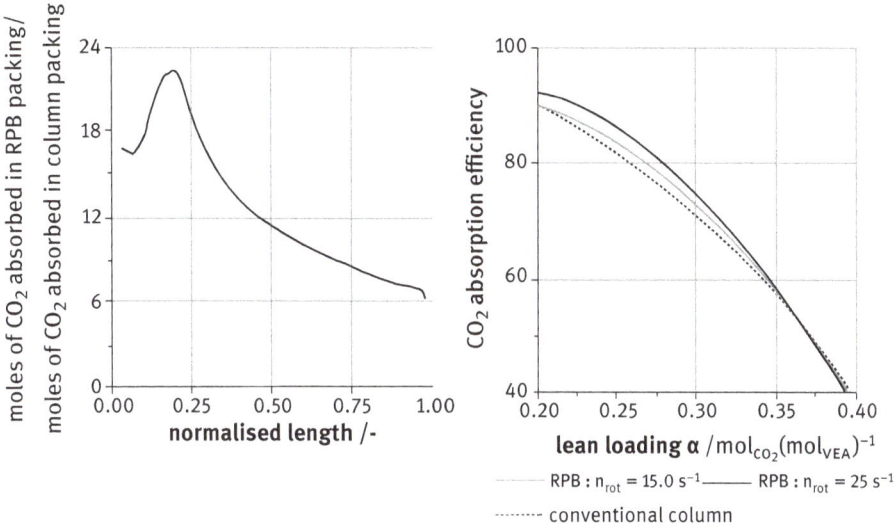

Fig. 4.4: Mole fraction of CO_2 in the gas phase (right) as a function of the normalized length (left). Normalized length of 0 corresponds to the inner packing radius in the RPB and the top of the column, respectively. CO_2 capture efficiency as a function of the lean loading (right).

drop 1418 Pa is calculated for the column. The calculated pressure drop in the RPB is in a similar range with 1424 Pa.

4.5 Take-home messages

- Due to the achievable size reduction and high capacities, the installation of RPBs in PCC processes is considered to be promising
- Short residence time in RPBs require fast reaction kinetics
- The casing in RPBs can have a major share in the overall mass transfer
- The enhanced absorption of CO_2 in RPBs by use of fast reacting solvents is linked with an increased release of heat, which must be removed. RPBs with integrated cooling coils might be a solution and a subject to current research activities
- Having a small footprint and no dependency on changes in orientation, RPB technology offers many advantages for offshore applications such as oil platforms.
- 40000 t a^{-1} of captured CO_2 is the largest reported RPB CO_2 absorption application
- An improved selectivity towards H_2S could be achieved for the cleaning of sulfuric flue gas by reducing the CO_2 absorption from 80% (column) to 9% in an RPB. At the same time, a volume reduction by a factor of 10 and a high gas capacity factor was reported. Hence, H_2S absorption is a promising field of application for RPBs.

– The largest industrial scale RPB has been reported for a SO_2 absorption process. There, a 3.5 m diameter rotor is used to treat gas flow rates up to 200000 $m^3\,h^{-1}$.
– Depending on the field of application, volume reductions in the range of 3–10 are achievable by the use of RPBs.
– Up to now, there are no mature modeling approaches for RPBs available as correlations being valid for different RPB designs are still subject of investigations.

4.6 Quiz

Question 1. Why are there different zones of gas-liquid mass transfer in RPBs. How do they affect the absorption of flue gases along the radial position in the packing?

Question 2. Does the mass transfer in RPBs always take place exclusively in the packing? If not, why are there any further contributors to the overall mass transfer?

Question 3. Why is the use of amine blends, e.g., PZ/MEA, considered to be beneficial in RPBs?

Question 4. Why are 30 wt.% MEA solutions often chosen for CO_2 absorption packed columns? What are the advantages and disadvantages of using concentrations larger than 30 wt.%? Why do RPBs facilitate the use of higher-concentrated MEA solutions?

Question 5. How does an increasing/decreasing liquid-to-gas ratio in CO_2 absorption affect the position of the temperature bulge in RPBs?

Question 6. Draw the factor of volume reduction (RPB compared with packed column). Why does this factor change along the radial positon/axial height of the column?

Question 7. What are the major advantages of RPBs to exploit remote natural gas and oil fields? Why is this application considered to gain more attention within the next years?

Question 8. Which premises determine the validity of a correlation derived for a mass transfer parameter? How is this linked with current rate-based modeling approaches in RPBs?

Question 9. How do gas flow rate and rotational speed affect the effective interfacial area and the gas-side mass transfer in RPBs? What can be done to further enhance the mass transfer? For answering these questions, also take a look at Chapter 7.

Question 10. Why are low lean solvent loadings beneficial for RPB applications?

4.7 Solutions

All answers can be found within this chapter.

References

[1]   Sandilya P, Rao DP, Sharma A, Biswas G. Gas-Phase Mass Transfer in a Centrifugal Contactor. Ind. Eng. Chem. Res. 2001;40:384–92.

[2]   Rao, D.P., Chandra, A., Goswami, P.S. Characteristics of Flow in a Rotating Packed Bed (HIGEE) with Split Packing. Industrial & Engineering Chemistry Research. 2005; 44, 11:4051–60.

[3]   Burns JR, Ramshaw C. Process Intensification: Visual Study of Liquid Maldistribution in Rotating Packed Beds. Chem. Eng. Sci. 1996;51:1347–52.

[4]   Guo K, Guo F, Feng Y, Chen J, Zheng C, Gardner NC. Synchronous visual and RTD study on liquid flow in rotating packed-bed contactor. Chem. Eng. Sci. 2000;55:1699–706.

[5]   Luo Y, Chu GW, Zou HK, Zhao ZQ, Dudukovic MP, Chen JF. Gas–Liquid Effective Interfacial Area in a Rotating Packed Bed. Ind. Eng. Chem. Res. 2012;51:16320–25.

[6]   Yang K, Chu G, Zou H, Sun B, Shao L, Chen JF. Determination of the Effective Interfacial Area in Rotating Packed Bed. Chem. Eng. J. 2011;168:1377–82.

[7]   Burns JR, Jamil JN, Ramshaw C. Process Intensification: Operating Characteristics of Rotating Packed Beds – Determination of Liquid Hold-up for a high-voidage Structured Packing. Chem. Eng. Sci. 2000;55:2401–15.

[8]   Yang Y, Xiang Y, Chu G, Zou H, Luo Y, Arowo M, Chen JF. A noninvasive X-ray technique for determination of liquid holdup in a rotating packed bed. Chem. Eng. Sci. 2015;138:244–55.

[9]   Chen YH, Chang CY, Su WL, Chen CC, Chiu CY, Yu YH, Chiang PC, Chiang SIM. Modeling Ozone Contacting Process in a Rotating Packed Bed. Ind. Eng. Chem. Res. 2004;43:228–36.

[10]   Sang L, Luo Y, Chu GW, Zhang JP, Xiang Y, Chen JF. Liquid flow Pattern Transition, Droplet Diameter and Size Distribution in the Cavity Zone of a Rotating Packed Bed: A visual Study. Chem. Eng. Sci. 2017a;158:429–38.

[11]   Sang L, Luo Y, Chu G-W, Liu Y, Liu XZ, Chen JF. Modeling and Experimental Studies of Mass Transfer in the Cavity Zone of a Rotating Packed Bed. Chem. Eng. Sci. 2017b;170:355–64.

[12]   IPCC: Climate Change 2014 Synthesis Report. Intergovernmental Panel on Climate change (IPCC) (2014)

[13]   Wang M, Lawal A, Stephenson P, Sidders J, Ramshaw C. Post-combustion CO_2 capture with chemical absorption. Chem. Eng. Res. Des. 2011;89:1609–24.

[14]   Harbold G, Park J: Using the GasTran® Deaeration System to achieve low dissolved oxygen levels for superior line speed and product quality: A case study in carbonated soft drink bottling (2008); last checked 08. 06.2021

[15]   Wang M, Joel AS, Ramshaw C, Eimer D, Musa NM. Process Intensification for Post-Combustion CO_2 Capture with Chemical Absorption: A Critical Review. Applied Energy. 2015;158:275–91.

[16]   Joel AS, Wang M, Ramshaw C, Oko E. Process analysis of intensified absorber for post-combustion CO_2 capture through modelling and simulation. International Journal of Greenhouse Gas Control. 2014;21:91–100.

[17]   Pan SY, Wang P, Chen Q, Jiang W, Chu YH, Chiang PC. Development of high-gravity technology for removing particulate and gaseous pollutant emissions. Journal of Cleaner Production. 2017;149:540–56.

[18] Wang Y-M, Chen Y-S. Capture of CO_2 by highly concentrated alkanolamine solutions in a rotating packed bed. Environ. Prog. Sustainable Energy. 2019;38:1–7.

[19] Zhang W, Peng X, Li Y, Teng L, Zhu J. Hydrodynamic characteristics and mass transfer performance of rotating packed bed for CO_2 removal by chemical absorption: A review. J. Nat. Gas Sci. Eng. 2020a;79:2–13.

[20] Zhao B, Tao W, Zhong M, Su Y, Cui G. Process, Performance and Modeling of CO_2 Capture by Chemical Absorption using High Gravity: A Review. Renewable and Sustainable Energy Reviews. 2016;65:44–56.

[21] Lin CC, Chen YW. Performance of a cross-flow rotating packed bed in removing carbon dioxide from gaseous streams by chemical absorption. International Journal of Greenhouse Gas Control. 2011;5(4):668–75

[22] Cheng HH, Tan CS. Removal of CO_2 from indoor air by alkanolamine in a rotating packed bed. Sep. Purif. Technol. 2011;82:155–56.

[23] Jassim MS, Rochelle G, Eimer D, Ramshaw C. Carbon Dioxide Absorption and Desorption in Aqueous Monoethanolamine Solutions in a Rotating Packed Bed. Ind. Eng. Chem. Res. 2007;46:2823–33.

[24] Lin CC, Liu WT, Tan CS. Removal of Carbon Dioxide by Absorption in a Rotating Packed Bed. Ind. Eng. Chem. Res. 2003;42:2381–86.

[25] Lee J, Toluwanimi K, Attidekou P. Carbon capture from a simulated flue gas using a rotating packed bed adsorber and mono ethanol amine (MEA). Energy Procedia. 2017;114:1834–40.

[26] Tan CS, Chen JE. Absorption of carbon dioxide with piperazine and its mixtures in a rotating packed bed. Sep. Purif. Technol. 2006;49:174–80.

[27] Wu T-W, Hung Y-T, Chen M-T, Tan C-S. CO_2 capture from natural gas power plants by aqueous PZ/DETA in rotating packed bed. Sep. Purif. Techn. 2017;186:309–17.

[28] Leimbrink M, Tlatlik S, Salmonc S, Kunze A-K, Limberg T, Spitzer R, Gottschalk R, Górak A, Skiborowski M. Pilot scale testing and modeling of enzymatic reactive absorption in packed columns for CO_2 capture. Int. J. Greenh. Gas Control. 2017a;62:100–12.

[29] Schäffer A: Amine und Aminmischungen zur CO_2-Absorption aus Kraftwerksrauchgasen und ihr Energiebedarf zur Regeneration. Ph.D. Dissertation, Stuttgart University, Stuttgart (2013)

[30] Zhang LL, Wang JX, Liu Z-P, Lu Y, Chu GW, Wang W-C, Chen J. Efficient Capture of Carbon Dioxide with Novel Mass-Transfer Intensification Device Using Ionic Liquids. AIChE J. 2013;59(8):2957–65

[31] Rajan S, Kumar M, Ansari MJ, Rao DP, Kaistha N. Limiting Gas Liquid Flows and Mass Transfer in a Novel Rotating Packed Bed (HiGee). Ind. Eng. Chem. Res. 2011;50:986–97.

[32] Cheng HH, Lai CC, Tan CS. Thermal regeneration of alkanolamine solutions in a rotating packed bed. International Journal of Greenhouse Gas Control. 2013;16:206–16.

[33] Yu CH, Tan CS. CO_2 Capture by Aqueous Solution Containing Mixed Alkanolamines and Diethylene Glycol in a Rotating Packed Bed. Energy Procedia. 2014;63:758–64.

[34] Cheng HH, Tan CS. Carbon dioxide capture by blended alkanolamines in rotating packed bed. Energy Procedia. 2009;1:925–32.

[35] Yu CH, Cheng HH, Tan CS. CO_2 capture by alkanolamine solutions containing diethylenetriamine and piperazine in a rotating packed bed. International Journal of Greenhouse Gas Control. 2012;9:136–47.

[36] Sheng M, Xie C, Zeng X, Sun B, Zhang L, Chu G, Lue Y, Chen J-F, Zou H. Intensification of CO_2 capture using aqueous diethylenetriamine (DETA) solution from simulated flue gas in a rotating packed bed. Fuel. 2018;234:1518–27.

[37] Chen H, Tsai TC, Tan CS. CO_2 capture using amino acid sodium slat mixed with alkanolamines. International Jpurnal of Greenhouse Gas Control. 2018;79:127–33.

[38] Lu X, Xie P, Ingham DB, Ma L, Pourkashanian M. Modelling of CO_2 absorption in a rotating packed bed using an Eulerian porous media approach. Chem. Eng. Sci. 2019;199(18)302–18.

[39] Jafari B, Rahimi MR, Ghaedi M, Dashtian K, Mosleh S. CO_2 capture by amine-based aqueous solution containing atorvastatin functionalized mesocellular silica foam in a counter-current rotating packed bed: Central composite design modeling. Chem. Eng. Res. D. 2018;120:64–74.

[40] Kvamsdal HM, Rochelle GT. Effects of the Temperature Bulge in CO_2 Absorption from Flue Gas by Aqueous Monoethanolamine. Ind. Eng. Chem. Res. 2008;47:867–75.

[41] Gaspar J, Solms N, Thomsen K, Fosbøl PL. Multivariable Optimization of the Piperazine CO_2 Post-Combustion Process. Energy Procedia. 2016;86:229–38.

[42] Lee U, Burre J, Caspari A, Kleinekorte J, Schweidtmann AM, Mitsos A. Techno-economic Optimization of a Green-Field Post-Combustion CO_2 Capture Process Using Superstructure and Rate-Based Models. Ind. Eng. Chem. Res. 2016;55:12014–26.

[43] Plaza JM, Rochelle GT. Modeling pilot plant results for CO_2 capture by aqueous piperazine. Energy Procedia. 2011;4:1593–600.

[44] Meeuwse M, van der Schaaf J, Schouten JC. Multistage rotor-stator spinning disc reactor. AIChE J. 2012;58:247–55.

[45] Chen YS, Lin CC, Liu HS. Mass Transfer in a Rotating Packed Bed with Viscous Newtonian and Non-Newtonian Fluids. Ind. Eng. Chem. Res. 2005a;44:1043–51.

[46] Yu CH, Wu TW, Tan CS. CO_2 capture by piperazine mixed with non-aqueous solvent diethylene glycol in a rotating packed bed. International Journal of Greenhouse Gas Control. 2013;19:503–09.

[47] Oko E, Ramshaw C, Wang M. Study of absorber intercooling in solvent-based CO_2 capture based on rotating packed bed technology. Energy Procedia. 2017b;142:3511–16.

[48] Oko E, Ramshaw C, Wang M. Study of intercooling for rotating packed bed absorbers in intensified solvent-based CO_2 capture process. Appl. Energy. 2018;223:302–16.

[49] Ramshaw C, Mallinson RH: Mass transfer processUS 4283255 (1981)

[50] Fowler R. Higee – A status report. Chem. Eng. London 1989;456:35–37.

[51] Ramshaw C. Opportunites for exploiting centrifugal fields. Chem. Eng. London 1987;437:17–21.

[52] Short H. New Mass-Transfer Find is a Matter of Gravity. Chem. Eng. London 1983;90:23–29.

[53] Reay D, Harvey A, Ramshaw C. Process Intensification. Amsterdam, London: Elsevier, 2008.

[54] Sudhoff D, Leimbrink M, Schleinitz M, Górak A, Lutze P. Modelling, Design and Flexibility Analysis of Rotating Packed Beds for Distillation. Chem. Eng. Res. Des. 2015;94:72–89.

[55] Kelleher T, Fair JR. Distillation Studies in a High-Gravity Contactor. Ind. Eng. Chem. Res. 1996;35:4646–55.

[56] Zhang J, Liu B, Li Q, Zhang X: CO_2 Capture from Coal Fired Power Plant and EOR in Shengli oilfield of Sinopec China, Shengli Petroleum Administrative Bureau (2011a)

[57] Wood DA, Mokhatab S, Economides MJ: Technology options for securing markets for remote gas. Proceedings of 87th Annual Convention of the Gas Processors Association (2008)

[58] Baldea M, Edgar TF, Stanley BL, Kiss AA. Modular manufacturing processes. AIChE J. 2017;75:4262–72.

[59] Xiang L, Wua L, Gao L, Chena J-F, Liu Y, Zhao H. Pilot scale applied research on CO_2 removal of natural gas using a rotating packed bed with propylene carbonate. Chem. Eng. Res. D. 2019;150:33–39.

[60] Chang EE, Chen TL, Pan SY, Chen YH, Chiang PC. Kinetic modeling on CO_2 capture using basic oxygen furnace slag coupled with cold-rolling wastewater in a rotating packed bed. Journal of Hazardous Materials. 2013;260:937–46.

[61] Guo J, Jiao W, Guisheng Q, Yuan Z. Applications of high-gravity technologies in gas purifications: A review. Chinese Journal of Chemical Engineering. 2019;27:1361–73.

[62] Yildirim Ö, Kiss AA, Hüser N, Leßmann K, Kenig EY. Reactive absorption in chemical process industry: A review on current activities. Chem. Eng. J. 2012;213:371–91.

[63] Liu YZ, Li P, Li Y, Kang RC, Diao JX. Pilot test on treatment of high concentration nitrogen oxides by high gravity technology. Chem. Ind. Eng. Prog. 2007;26:1058–61. in Chinese.

[64] Li Y, Liu Y, Zhang L, Su Q, Jin G. Absorption of NOx into nitric acid solution in rotating packed bed. Chin. J. Chem. Eng. 2010;18:244–48.

[65] Zhang LL, Wang JX, Sun Q, Zeng XF, Chen JF. Removal of nitric oxide in rotating packed bed by ferrous chelate solution. Chem. Eng. J. 2012;181–182:624–29.

[66] Sun BC, Sheng MP, Gao WL, Zhang LL, Arowo M, Liang Y, Shao L, Chu GW, Zou HK, Chen JF. Absorption of nitrogen oxides into sodium hydroxide solution in a rotating packed bed with preoxidation by ozone. Energy Fuel. 2017;31:11019–25.

[67] Chen T-L, Chen Y-H, Chiang P-C. Enhanced performance on simultaneous removal of NOx-SO2-CO2 using a high-gravity rotating packed. Chemical Engineering Journal. 2020;393. https://doi.org/10.1016/j.cej.2020.124678

[68] Neumann K, Gladyszewski K, Groß K, Qammar H, Wenzel D, Górak A, Skiborowski M. A guide on the industrial application of rotating packed beds. Chemical Engineering Research and Design. 2018;134:443–62.

[69] Savage DW, Funk EW, Yu WC, Astarita G. Selective absorption of hydrogen sulfide and carbon dioxide into aqueous solutions of methyldiethanolamine. Ind. Eng. Chem.Fund. 1986;25 (3):326–30.

[70] MacKenzie DH, Prambil FC, Daniels CA, Bullin JA. Design & operation of a selective sweetening plant using MDEA. Energy Prog. 1987;7:31–36.

[71] Chen YS, Lin FY, Lin CC, Tai CYD, Liu HS. Packing characteristics for mass transfer in a rotating packed bed. Ind.Eng. Chem. Res. 2006;45(20):6846–53.

[72] Trent DL, Tirtowidjojo D, Quarderer GJ: Reactive stripping in a rotating packed bed for the production of hypochlorous acid. In: Process Intensification for the Chemical Industry – Smaller, Cheaper and Safer Production (1999). 4th International Conference on Process Intensification for the Chemical Industry (BHR Group Conference), Antwerp, Belgium.

[73] Qian Z, Xu LB, Li ZH, Li H, Guo K. Selective Absorption of H2S from a Gas Mixture with CO_2 by Aqueous N-Methyldiethanolamine in a Rotating Packed Bed. Ind. Eng. Chem. Res. 2010;49:6196–203.

[74] Qian Z, Li ZH, Guo K. Industrial Applied and Modeling Research on Selective H2S Removal Using a Rotating Packed Bed. Ind. Eng. Chem. Res. 2012;51:8108–16.

[75] Zou H, Chu GW, Zhao H, Xiang Y, Chen JF. Process Intensification of High-Gravity Reactor for Enviromental Engineering: From Fundamental to Industrialization. Sci. Sin. Chim. 2014;44:1413.

[76] Chu G-W, Luo Y, Shan C-Y, Zou H-K, Xiang Y, Shao L, Chen J-F. Absorption of SO2with ammonia-based solution in a co-current rotating packed bed. Ind. Eng. Chem. Res. 2014;53 (40):15731–37.

[77] Jiang X, Liu Y, Gu M. Absorption of sulphur dioxide with sodium citrate buffer solution in a rotating packed bed. Chin. J. Chem. Eng. 2011;19(4):687–92.

[78] Song YH, Chen JM, Fu J, Chen JF. Research on particle removal efficiency of the rotating packed bed. Chem. Ind. Eng.Prog. 2003;22:499–502.

[79] Zhang Y, Liu LS, Liu Y. Experimental study in flue gas dedusting by hyper gravity rotary bed. Environ. Eng. 2003;21(58):42–43.

[80] U.S. Environmental Protection Agency: Air Pollution Control Technology – Fact Sheet: Flue Gas Desulfurization – Wet, Spray Dry and Dry Scrubbers (2003).

[81] Kunze AK, Lutze P, Kopatschek M, Maćkowiak JF, Maćkowiak J, Grünewald M, Górak A. Mass Transfer Measurements in Absorption and Desorption. Chem. Eng. Res. Des. 2015a;104:440–52.

[82] Yi F, Zou HK, Chu GW, Shao L, Chen JF. Modeling and experimental studies on absorption of CO_2 by Benfield solution in rotating packed bed. Chem. Eng. J. 2009;145:377–84.

[83] Qian Z, Xu L, Cao H, Guo K. Modeling Study on Absorption of CO_2 by Aqueous Solutions of N -Methyldiethanolamine in Rotating Packed Bed. Ind. Eng. Chem. Res. 2009;48:9261–67.

[84] Kang JL, Sun K, Wong DSH, Jang SS, Tan CS. Modeling studies on absorption of CO_2 by monoethanolamine in rotating packed bed. International Journal of Greenhouse Gas Control. 2014a;25:141–50.

[85] Kang JL, Luo ZJ, Liu JL, Sun K, Wong DSH, Jang SS, Tan CS, Shen JF. Experiment and Modeling Studies on Absorption of CO_2 by Dilute Ammonia in Rotating Packed Bed. Energy Procedia. 2014b;63:1308–13.

[86] Joel AS, Wang M, Ramshaw C. Modelling and simulation of intensified absorber for post-combustion CO_2 capture using different mass transfer correlations. Applied Thermal Engineering. 2015;74:47–53.

[87] Chamchan N, Chang JY, Hsu HC, Kang JL, Wong DSH, Jang SS, Shen JF. Comparison of rotating packed bed and packed bed absorber in pilot plant and model simulation for CO_2 capture. Journal of the Taiwan Institute of Chemical Engineers. 2016;73:20–26.

[88] Kang JL, Wong DSH, Jang SS, Tan CS. A comparison between packed beds and rotating packed beds for CO_2 capture using monoethanolamine and dilute aqueous ammonia solutions. International Journal of Greenhouse Gas Control. 2016;46:228–39.

[89] Thiels M, Wong DSH, Yu CH, Kang JL, Jang SS, Tan CS. Modelling and Design of Carbon Dioxide Absorption in Rotating Packed Bed and Packed Column. IFAC-PapersOnLine. 2016;49:895–900.

[90] Oko E, Wang M, Ramshaw C. Study of mass transfer correlations for rotating packed bed columns in the context of solvent-based carbon capture. Int. J. Of Greenhouse Gas Control. 2019;91:1–9.

[91] Oko E, Wang M, Ramshaw C. Study of Mass Transfer Correlations for Intensified Absorbers in Post-combustion CO_2 Capture Based on Chemical Absorption. Energy Procedia. 2017a;114:1630–36.

[92] Sang L, Ch G-W, Sund B-C, Zhang -L-L, Chen J-F. A three-zone mass transfer model for a rotating packed bed. AIChE J. 2019;65:1–11.

[93] Peng X, Lu X, Yang X, Ingham D, Ma L, Pourkashanian M. Characteristics of liquid flow in a rotating packed bed for CO_2 capture: A CFD analysis. Chem. Eng. Sci. 2017;172:216–29.

[94] Zarei F, Rahimi M, Razavi R, Baghband R. Insight into the experimental and modeling study of process intensification for post-combustion CO_2 capture by rotating packed bed. Journal of Cleaner Production. 2019;211:953–61.

[95] Zhao B, Su Y, Tao W. Mass Transfer Performance of CO_2 Capture in Rotating Packed Bed: Dimensionless Modeling and Intelligent Prediction. Applied Energy. 2014;136:132–42.

[96] Yang Y-C, Ouyang Y, Zhang N, Yua Q-J, Arowoc M. A review on computational fluid dynamic simulation for rotating packed beds. J. Chem Technol. Biotechnol. 2019;94:1017–31.

[97] Zhang W, Peng X, Li Y, Teng L, Zhu J. CFD analysis of the hydrodynamic characteristics in a rotating packed bed with multi-nozzles. Chem. Eng. Pro. 2020b;158:1–16.

[98] Danckwerts PV. Gas-Liquid Reactions. New York: McGraw-Hill, 1970.

[99] Lewis WK, Whitman WG. Principles of Gas Absorption. Ind. Eng. Chem. 1924;16:1215–20.

[100] Hoffmann A, Maćkowiak JF, Górak A, Haas M, Löning J-M, Runowski T, Hallenberger K. Standardization of Mass Transfer Measurements: A Basis for the Description of Absorption Processes. Chem. Eng. Res. Des. 2007;85:40–49.

[101] Onda K, Takeuchi H, Okumoto Y. Mass Transfer Coefficients Between Gas and Liquid Phases in Packed Columns. J. Chem. Eng. Japan. 1968;1:56–62.

[102] Saha D. Prediction of mass transfer coefficient in rotating bed contactor (Higee) using artificial neural network. Heat Mass Transfer. 2009;45:451–57.

[103] Chen YS. Correlations of Mass Transfer Coefficients in a Rotating Packed Bed. Ind. Eng. Chem. Res. 2011;50:1778–85.

[104] Lin CC, Liu WT. Mass transfer characteristics of a high-voidage rotating packed bed. J. Ind. Eng. Chem. 2007;13:71–78.

[105] Sun BC, Wang XM, Chen JM, Chu GW, Chen J, Shao L. Simultaneous Absorption of CO_2 and NH3 into Water in a Rotating Packed Bed. Industrial and Engineering Chemistry Research. 2009;48:11175–80.

[106] Sun BC, Zou HK, Chu GW, Shao L, Zeng Z-Q, Chen J. Determination of mass-transfer coefficient of CO_2 in NH3 and CO_2 absorption by materials balance in a rotating packed bed. Ind. Eng. Chem. Res. 2012;51:10949–54.

[107] Liu HS, Lin CC, Wu SC, Hsu HW. Characteristics of a Rotating Packed Bed. Ind. Eng. Chem. Res. 1996;35:3590–96.

[108] Lin CC, Wei TY, Liu WT, Shen KP. Removal of VOCs from Gaseous Streams in a High-Voidage Rotating Packed Bed. J. Chem. Eng. Japan. 2004;37:1471–77.

[109] Shivhare MK, Rao DP, Kaistha N. Mass Transfer Studies on Split-Packing and Single-Block Packing Rotating Packed Beds. Process Intensification by Alternative Energy Forms and Transfer Mechanisms. 2013;71:115–24.

[110] Brauer H, Gaebelein W. Fluiddynamik und flüssigkeitsseitiger Stoffübergang in einer für Anwendungen in Weltraumfahrzeugen konzipierten Stoffaustauschmaschine. Forschung Im Ingenieurwesen. 1984;50:67.

[111] Chen YH, Chiu CY, Chang CY, Huang YH, Yu YH, Chiang PC, Shie JL, Chiou CS. Modeling Ozonation Process with Pollutant in a Rotating Packed Bed. Ind. Eng. Chem. Res. 2005b;44:21–29.

[112] Munjal S, Duduković MP, Ramachandran P. Mass-Transfer in Rotating Packed Beds – II. Experimental Results and Comparison with Theory and Gravity Flow. Chem. Eng. Sci. 1989;44:2257–68.

[113] Luo Y, Luo JZ, Chu GW, Zhao ZQ, Arowo M, Chen JF. Investigation of effective interfacial area in a rotating packed bed with structured stainless steel wire mesh packing. Chem. Eng. Sci. 2016;170:347–54.

[114] Guo K, Zhang Z, Luo H, Dang J, Qian Z. An Innovative Approach of the Effective Mass Transfer Area in the Rotating Packed Bed. Ind. Eng. Chem. Res. 2014;53:4052–58.

[115] Leimbrink M, Neumann K, Kupitz K, Górak A, Skiborowski M. Enzyme Accelerated Carbon Capture in different Contacting Equipment – A Comparative Study. Energy Procedia. 2017b;114:795–812.

[116] Wojtasik J, Gładyszewski K, Skiborowski M, Górak A, Piątkowski M. Enzyme-enhanced CO_2 absorption process in rotating packed bed. Chemical Papers. 2019;73:861–69.

[117] Kunze AK, Dojchinov G, Haritos VS, Lutze P. Reactive absorption of CO_2 into enzyme accelerated solvents: From laboratory to pilot scale. Applied Energy. 2015b;156:676–85.

[118] Zhang Y, Que H, Chen CC. Thermodynamic modeling for CO_2 absorption in aqueous MEA solution with electrolyte NRTL model. Fluid Phase Equilibria. 2011b;311:67–75.

[119] Austgen DM, Rochelle GT, Peng X, Chen CC. Model of vapor-liquid equilibria for aqueous acid gas-alkanolamine systems using the electrolyte-NRTL equation. Ind. Eng. Chem. Res. 1989;28:1060–73.

[120] Kucka L: Modellierung und Simulation der reaktiven Absorption von Sauergasen mit Alkanolaminlösungen. Ph.D. Dissertation, TU Dortmund University, Dortmund, Germany (2003)

[121] Pinsent BRW, Pearson L, Roughton FJW. The kinetics of combination of carbon dioxide with hydroxide ions. Transactions of the Faraday Society. 1956;52:1512–20.

[122] Tobiesen FA, Svendsen HF, Juliussen O. Experimental validation of a rigorous absorber model for CO_2 postcombustion capture. AIChE J. 2007;53:846–65.

[123] Jassim MS: Process Intensification: Absorption and Desorption of Carbon Dioxide from Monoethanolamine Solutions Using Higee Technology. Ph.D. thesis, University of Newcastle Upon Tyne, Newcastle, England (2002)

[124] Aboudheir A, Tontiwachwuthikul P, Chakma A, Idem R. Kinetics of the reactive absorption of carbon dioxide in high CO_2-loaded, concentrated aqueous monoethanolamine solutions. Chem. Eng. Sci. 2003;58:5195–210.

[125] Nagel O, Hegner B, Kürten H. Kriterien für die Auswahl und die Auslegung von Gas/Flüssigkeits-Reaktoren. Chem. Ing. Tech. 1978;50:934–44.

[126] Brauer H. Stoffaustausch-Maschinen. Chancen für den Gerätebau und die stoffwandelnde Industrie. Chem. Ing. Tech. 1986;58:97–107.

[127] Sulzer Chemtech Ltd: Structured Packings Energy-efficient, innovative and profitable, http://www.sulzer.com; last checked 08.06.2021

Tobias Pyka, Jörg Koop

5 Rotating packed beds in distillation: rotor configurations

5.1 Fundamentals

Distillation is one of the most investigated and most established separation processes in chemical industry. The conventional distillation column consists of a vertical vessel with internals like trays or packings providing the interfacial area between gas and liquid. The cross-sectional area of a conventional column is, in most cases, constant. One of the major differences to rotating packed beds (RPBs), at least to RPBs with standard isotropic internals, is that the cross-sectional area increases with radial length. Because of this, the comparability of a conventional distillation column and an RPB regarding mass transfer efficiency is not trivial. Different fundamental equations describe the operation parameters and mass transfer efficiency in distillation. These equations enable comparison between conventional columns and RPBs since the conceptual models are valid for both. Nevertheless, such comparison is only possible under certain assumptions which will be elucidated and discussed in detail.

The first quantity for the operation is the F-factor F_G, which is a measure of the gas load of the equipment. It is defined as the product of the superficial gas velocity u_s and the square root of the gas density ρ_V [eq. (5.1)]: [1]

$$F_G = u_s \cdot \sqrt{\rho_V} \tag{5.1}$$

In the case of an isotropic packing in an RPB, the gas velocity decreases with an increasing radius, which results in a decreasing F_G. The question arises which cross-sectional area is used to determine the F_G within an RPB. F_G is calculated at the eye of the RPB because there the highest velocities and according F_G occur, potentially initiating the flooding. Neumann et al. [2] proposed a modified equation for calculation of F_G, where the gas velocity is integrated to account for its change along the radius. When the nonconstant F_G long the radial depth of RPBs is discussed, it shall be mentioned that there is the possibility to achieve a constant F_G in RPB. That is achieved by tailor-made packings with a constant cross-sectional area along the radius. A detailed discussion of such packing designs is given in Chapter 7.

To quantify the mass transfer efficiency of a conventional distillation column the HETP (height equivalent to a theoretical plate) is a frequently used key performance indicator. It is calculated by dividing the axial height of the packing in the

Tobias Pyka, Jörg Koop, TU Dortmund University

https://doi.org/10.1515/9783110724998-005

column by the achieved number of theoretical stages N_{th}. It can be directly derived from mass transfer performance tests [1, 3]. This concept is adapted to RPBs by replacing the axial height with the radial depth of the packing, with r_o being the outer and r_i the inner radius (eq. (5.2)) [4]:

$$\text{HETP}_{\text{RPB}} = \frac{(r_o - r_i)}{N_{th}} \tag{5.2}$$

To use this quantity for comparison of different RPBs or even RPBs with conventional distillation columns, the assumption is made that the mass transfer occurs only within the packing and is evenly distributed along the radius. These assumptions are highly questionable for the following reasons: First, **the mass transfer does also take place within the casing of the RPB and not only within the packing** [5, 6]. For this reason, the HETP$_{\text{RPB}}$ overestimates the mass transfer efficiency within the packing, especially for small rotors [7]. Second, **the mass transfer efficiency within an isotropic packing along the radius is not linear.** Usually, the highest mass transfer efficiency is situated at the inner radius of the packing, the so-called end zone [8]. Therefore, the HETP$_{\text{RPB}}$ of RPBs with a smaller rotor tends to be smaller because the contribution of the end zone is more pronounced. Thus, a linear scale-up by increasing the packing length like in conventional distillation columns is, due to the aforementioned reasons, not possible. A detailed overview of the mass transfer zones of an RPB is further provided in Chapters 3 and 6, and the scalability of the packing is further discussed in Chapter 7.

Another approach to quantify the mass transfer efficiency of columns is the *HTU-NTU* concept [9]. This concept describes the required height of a transfer unit (HTU) to obtain one mass transfer unit [9]. It is based on a material balance of a differential volume element [9]. The assumptions which are made in this approach are: mass transfer can be described by film theory and constant vapor and liquid flows [9]. Since the geometry of an RPBs differs from a conventional distillation column, Singh [10] referred to the area of a transfer unit (ATU), which is known as the *ATU-NTU* concept. The resulting equation is presented in eq. (5.3):

$$\pi\left(r_o^2 - r_i^2\right) = \frac{\dot{m}_G}{\rho_G h K_G a_e} \int_{y_{in}}^{y_{out}} \frac{dy}{(y^* - y)} = ATU_g \cdot NTU \tag{5.3}$$

For the evaluation, the parameters, gas mass flow $\dot{m}_G$, gas density ρ_G, axial height h, and overall volumetric gas side mass transfer coefficient $K_G a_e$ are assumed as constant along the radius of the packing. For operation under total reflux ($x = y$), the integral can be simplified to [eq. (5.4)]:

$$NTU = \int\limits_{y_{in}}^{y_{out}} \frac{dx}{(y^*(x) - x)} \qquad (5.4)$$

Since the dedicated aim of HiGee (high-gravity) technology, and especially RPBs, is to increase the mass transfer, either by increasing the mass transfer coefficient K_G or the effective interfacial area a_e or both, the assumption of constant values for those parameters is questionable. That becomes obvious by considering the following: The centrifugal acceleration is the main contributor to the increase of each parameter and the centrifugal acceleration linearly increases with the radius. Furthermore, the cross-sectional area increases linearly with the radius, leading to decreasing gas and liquid loadings at the outer packing regions. Thus, it seems reasonable to discretize the rotor into smaller elements which allow using individual values for each discrete element. That is a known procedure for conventional columns where the assumption of constant mass transfer coefficients along with the height also does not always hold [11]. For RPBs, the discretization was done by Sudhoff et al. [12]. They applied equiareal discretization ending up with rings of the same ground area and, thus, same volume and same absolute interfacial area [12]. Based on specific loading and integrated centrifugal acceleration, the $K_G a_e$ values for each element were calculated and used for subsequent calculations. However, oftentimes, the varying conditions along the radius of the rotor are neglected by applying the ATU-NTU concept for the integral performance of the overall equipment. It has to be noted that additional effects happen inside in an RPB affecting the mass transfer. The most prominent of those are the mass transfer in the casing or the end effect in the inner packing region. Furthermore, Groß et al. [13] showed that for large rotational speeds, severe liquid maldistribution in isotropic metal foam packings occurs for certain gas and liquid loads (see Chapter 3).

Consequently, a concept for evaluating the mass transfer efficiency, which takes all aforementioned effects within an RPB into account, is still missing.

5.2 Rotor configurations

In the last two decades, the rotor configuration of the conventional RPB with an isotropic packing and a single rotor like in Fig. 5.1 was modified in various ways [14–20]. The conventional RPB consisting of one rotor equipped with an isotropic packing, for instance, metal foam or wire meshes, was modified in different ways to improve the mass transfer and enlarge the operating window. The modifications mainly concern the internals of the rotor and the number of rotors connected. The aim was to improve the design to intensify the process further and to overcome constructional challenges. The subsequent chapter will first introduce the design features of single- and two-staged rotor configurations. Subsequently, the idea about the concepts will be explained and discussed.

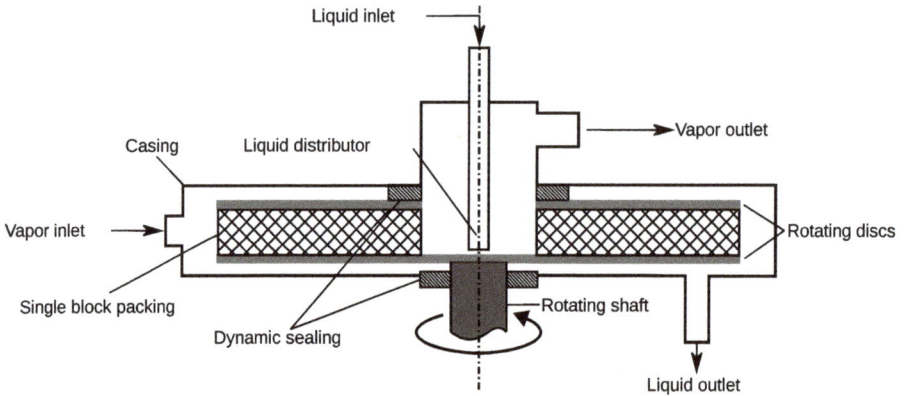

Fig. 5.1: Schematic drawing of RPB.

Before the different rotor modifications are presented, the design of a conventional RPB with an isotropic packing is explained especially regarding its constructional disadvantages. A conventional RPB is equipped with a rotor consisting of two interconnected discs with a packing between them (see Fig. 5.1). To avoid a vapor bypass to the eye of the rotor, a dynamic seal is used at the upper disc. The packing commonly consists of metal wire mesh, glass spheres, or metal foam [21], offering a high specific surface area. By increasing the height of the packing, the gas and liquid capacity can be easily increased. In the eye of the rotor, the liquid is sprayed into the rotating packing via liquid distributors to apply the liquid evenly. Then the liquid flows radially outwards and the vapor flows radially inwards. Based on the flow pattern, the liquid residence time is quite short, which is dependent on the further application, advantageous or disadvantageous [14]. Furthermore, at least two RPB are needed to conduct a continuous distillation with a rectifying and stripping section (Chapter 1, Fig. 1.3) because an intermediate feed is challenging to implement in a single RPB. Contrary to expectations, Sandilya et al. [22] discovered that in metal wire mesh or metal foam packing, the gas-side mass transfer is in a similar range compared to fixed bed columns, and it is not enhanced by rotational speed [22]. These drawbacks have motivated ongoing investigations of new rotor configurations for RPBs.

5.2.1 Rotating zig-zag bed

The rotating zig-zag bed (RZB) is one of the first modified rotors and was patented in 2001 by Ji et al. [23]. The simplified sketch of the rotor of the RZB is illustrated in Fig. 5.2. Compared to the conventional RPB, only the lower disc is rotating, while the upper disc is stationary and attached to the housing of the device. The discs are equipped with concentric circular sheets, while the upper part of the rotating baffles is perforated. The radial distance between the adjacent rotational and stationary baffles is kept constant. Due to this arrangement of baffles, the gas and liquid phase flow in a zig-zag shape, which results in two different contact mechanisms: a crosscurrent and countercurrent contact. Despite these modifications, the inlet and outlet for gas and the liquid outlet are the same as a conventional RPB [14, 24].

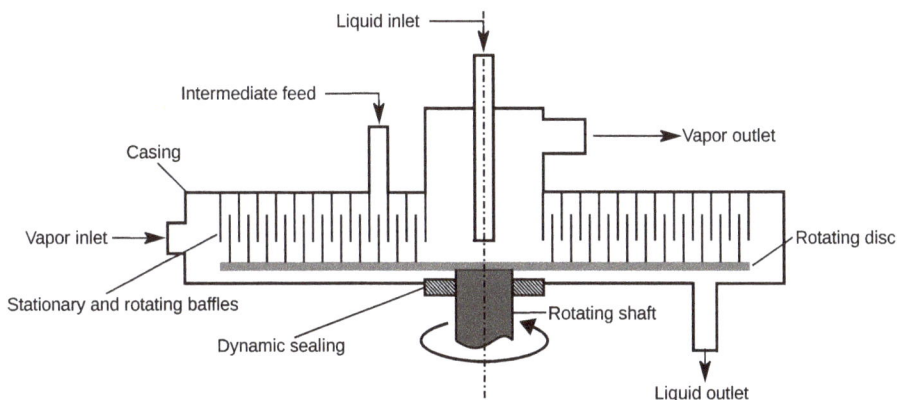

Fig. 5.2: Schematic drawing of RZB [24].

The following *advantages* over a conventional RPB of an RZB are stated:
- The manufacturing is simplified because only a static seal is needed for the stationary upper disc instead of a dynamic seal [14, 24]. (A dynamic sealing for the rotating shaft is still needed.)
- A major advantage is the possibility to install an intermediate feed to perform continuous distillation with stripping and rectifying section in a single machine. To realize this with a conventional RPB, two apparatuses or a costly intermediate feed concept are needed, which increases the apparatus expenditure [14, 24].
- Initial liquid distributors are not required as the liquid flows by centrifugal force from the rotational baffles and forms fine droplets by impinging the stationary baffles [24].
- Since the rotational baffles are only perforated at the upper half, the liquid hold-up is increased which leads to a longer residence time [14, 24].
- As a consequence of the aforementioned advantage of the saving of a dynamic seal for the upper disc and an initial liquid distributor, the manufacturing complexity of multiple rotors within an RZB is reduced [14, 24].
- The RZB can be operated at higher F_G, which results in an enlarged operating window [14, 24].

Due to the modified rotor design also *disadvantages* are coming up:
- The pressure drop and the power consumption are higher compared to conventional RPBs because of the repeated acceleration-deceleration of the fluids along the zig-zag path [25].
- The axial height of the rotor and the diameter of a conventional column cannot be compared directly. Thus, a capacity expansion is not trivial.
- Liquid can be entrained in the clearance between rotational baffles and the stationary disc, which results in backmixing [16].

Due to the different rotor designs, similar issues arise when comparing the mass transfer efficiency, as addressed in the previous section. In the following, only the number of theoretical stages will be listed to classify the mass transfer efficiency into context. Wang et al. [14] performed distillation studies in an RZB with inner and outer diameter of the rotor of 100 and 520 mm and with a methanol/water mixture. The F_G varied from about 1.6 to 7 Pa$^{0.5}$, and they achieved about three to five theoretical stages. Lin et al. [26] could achieve about one to three theoretical stages within an RPB with methanol/ethanol as a chemical system. The RPB had an inner and outer diameter of 122 and 294 mm and was operated up to F_G of 2.2 Pa$^{0.5}$. Hilpert et al. [27] used an ethanol/water mixture in a distillation study and an RPB with inner and outer diameter of the rotor of 120 and 450 mm. They varied the F_G from 2.3 to 4.9 Pa$^{0.5}$ and obtained two to six theoretical stages [27]. It is important to note that the studies mentioned above are indicative and that the machines have different dimensions. **Nevertheless, the mass transfer efficiency of an RZB and RPB is within a similar range** [14, 24, 26].

The RZB presents the only rotor design which is reportedly applied on an industrial scale for distillation. In 2011, Wang et al. [28] reported about 200 commercialized distillation processes where separation of binary and ternary mixtures, such as ethanol/water, acetone/water, DMSO/water, DMF/water, ethyl acetate/water, methanol/*tert*-butanol, dichloromethane/silyl-ethers, methanol/formaldehyde/water, methanol/toluene/water, ethyl acetate/toluene/water, methanol/methylal/water, and methanol/DMF/water, was performed in an RZB. Besides the application as a conventional distillation, Wang et al. [28] also reported the application as an extractive distillation to separate anhydrous ethanol by adding ethylene glycol (see Fig. 5.3) [28].

This process consists of two RZBs with two stages each. One RZB is used to separate anhydrous ethanol and the other to recover the ethylene glycol. The fluid flow of a two-stage RZB differs from a single-stage RZB as follows: After passing the upper rotor, the liquid is collected at the bottom of the upper housing, and there is also an external feed introduced. These two liquid streams are mixed and flow to the lower rotor by gravity. Therefore, in a two-stage RZB, an intermediate feed can be made either between the rotors, at the upper rotor, at the lower rotor, or combinations thereof. The vapor flows countercurrently to the liquid. It enters the lower rotor going to the bottom of the upper housing and finally passing the upper rotor [28].

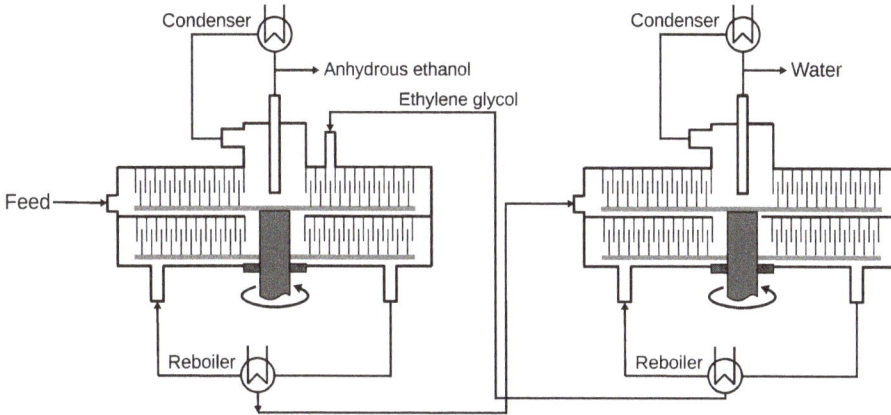

5.2.2 Crossflow concentric-baffle rotating bed

To further develop the rotor design of the RZB, the crossflow concentric-baffle rotating bed (CRB) was designed (see Fig. 5.4), which is supposed to overcome the disadvantages of the RZB regarding power consumption and liquid phase backmixing. The main constructional difference is in the baffle design. All concentric baffles are attached to the lower rotating disc reaching the upper disc. Originally, the upper disc was static, like in the RZB. To avoid a gas bypass between the baffles and the static disc, a labyrinth seal was used, which failed to function properly. Afterwards, a gland seal was used instead which connected both rotor discs and successfully hindered a gas bypass. Furthermore, in the vertical direction, the baffles can be divided into three zones: gas-hole zone, liquid-hole zone, and non-hole zone. To control the flow direction, the diameter of the holes in the gas-hole zone is about two times greater. The baffles are arranged in a way that the vertical orientation of the zones is alternating. This leads to a crossflow: the vapor flows radially inwards through the gas-hole zone in a zig-zag shape, while crossing the straight line liquid flow through the liquid-hole zone. In early designs, a liquid distributor was installed at the lower rotor disc. However, it proved to be disadvantageous, and the baffle design itself provided a better liquid distribution [16, 29].

The positive impact on this rotor design especially compared to an RZB is as follows:

- Lower energy consumption, as the kinetic energy loss is reduced with all baffles rotating [16, 29].
- The pressure drop is also lower because of less change of flow direction and sudden expansion/ contraction along the radius [16].
- As a consequence of the gas-hole zone, liquid entrainment can be prevented [16].

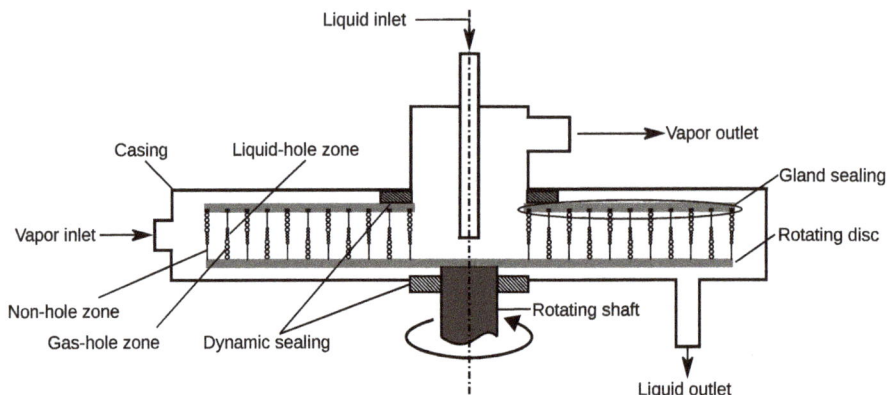

Fig. 5.4: Schematic drawing of CRB [16].

Since, in the optimized design, both discs are rotating, an intermediate feed is challenging to implement. Improvements in the form of lower required shaft power, lower pressure drop, and lower backmixing were achieved, but mass transfer performance remained in a similar range to RZB. **Thus, the implemented rotor modifications did not lead to a significant improvement compared to the conventional RZB regarding mass transfer performance** [16, 29].

5.2.3 Counterflow concentric ring rotating bed

Another development of the RZB represents the counterflow concentric ring rotating bed (CFCR-RB) [15]. The CFCR-RB has similar objectives as the CRB, aiming to diminish the drawbacks of the RZB, such as low liquid entrainment, high gas-liquid throughput, and low pressure drop. The constructional implementation also resembles the CRB. The rotor consists of a lower rotating disc equipped with concentric rings reaching up to the upper stationary disc (see Fig. 5.5). The upper stationary disc compromises concentric grooves which function combined which the concentric ring as a labyrinth seal to avoid a vapor bypass. This design enables the implementation of an intermediate feed with relatively low effort, as is exemplarily shown in Fig. 5.5. Besides that, the major difference to a CRB is the perforation of the concentric rings being uniform over the entire height of the concentric ring. Thus, both fluids flow countercurrently through these. Due to the perforation of the concentric rings, a repeated droplet formation and film flow of the liquid are supposed to happen [15].

These constructional changes have a positive impact regarding:
- The pressure drop could significantly be reduced especially at high F_G compared to an RZB rotor because of more free cross-sectional area and lesser direction changes of the vapor [15].

Liquid inlet

Intermediate feed

Vapor outlet

Casing

Vapor inlet

Rotating disc

Perforated concentric rings

Rotating shaft

Dynamic sealing

Liquid outlet

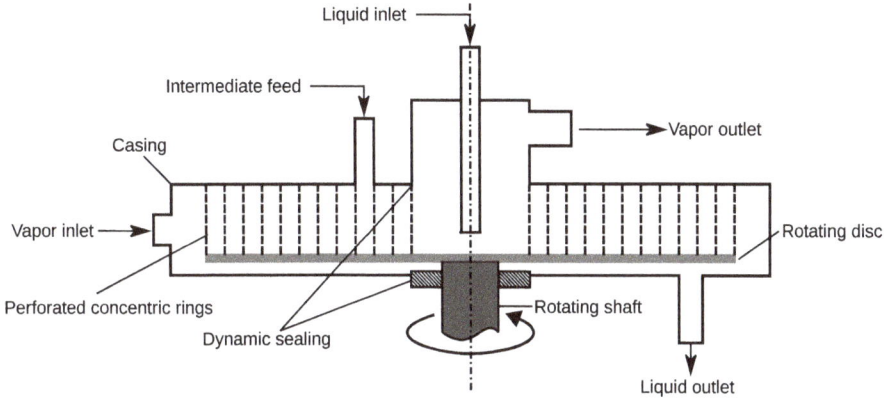

Fig. 5.5: Schematic drawing of a CFCR-RB [15].

- The liquid entrainment could be prevented at high F_G as compared with an RZB rotor. That is an indicator that the turndown ratio and the gas-liquid through-put are increased within a CFCR-RB. [15]
- Since the upper disc is stationary, an intermediate feed and a multirotor design can be easily implemented [15].

The negative impacts are as follows:
- The effective interfacial area (a_e) is lower than in a single-block packing [15]. One reason for this phenomenon could be that the perforated rings offer less specific surface area than single-block packings.
- Furthermore, the mass transfer performance is lower compared to an RZB rotor at low F_G [15].

5.2.4 Rotating packed bed with split-packing

Sandilya et al. [22] discovered that the limiting vapor side mass transfer coefficient in conventional RPBs does not differ significantly compared to conventional columns. Therefore, Chandra et al. [30] suggested a different rotor design which is called a RPB with split-packing (SP-RPB). Their approach was to enhance the vapor side mass transfer coefficient by increasing the tangential slip velocity within the packing along the radius [30, 31]. Another expectation of this rotor modification was to enhance the throughputs [30, 31].

The constructional implementation takes place by using two discs rotating in opposite directions, which is illustrated in Fig. 5.6 [30, 31]. Each disc is equipped with packing in form of four concentric rings consisting of wire mesh [32]. In contrast to a conventional RPB, the vapor does not leave the packing through the eye, but through an opening over the third packing ring [17]. This different outlet is inspired by the crosscurrent flow RPB and intends to enhance the throughput [30, 31]. The liquid is applied to the innermost ring with a liquid distributor [30, 31].

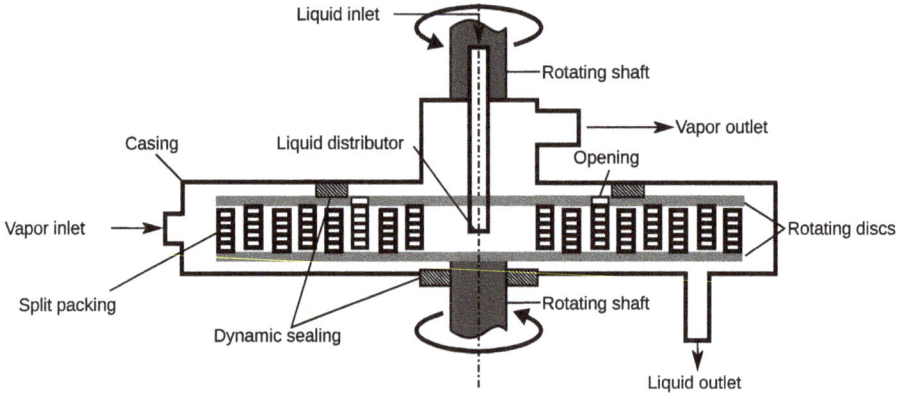

Fig. 5.6: Schematic drawing of SP-RPB [17].

This design modification leads to the following advantages:
- a_e is increased due to better liquid distribution [31, 33, 34].
- Intensive liquid redistribution at every packing ring.

The drawbacks are as follows:
- Based on the two independent rotors, two motors are needed [30, 31]. As a by-effect of this, a coaxial rotor design is quite challenging to build.
- Intermediate feed is, like in a conventional RPB, difficult to implement [30, 31].
- Split-packings are more prone to deformation than single-block packings [35]. Additional supports could compensate for it, but these obstructions would lead to liquid maldistribution and should be avoided [35, 36].

The intended enhancement of the vapor side mass transfer coefficient by increasing the tangential slip and of the throughput could not be proved by experiments [17, 37]. Mondal et al. [17] conducted distillation experiments with F_G up to 0.6 Pa$^{0.5}$. They assumed entrainment of liquid at $F_G = 0.6$ Pa$^{0.5}$. Therefore, the throughputs could not be increased. Unfortunately, the effect of co- and counter-rotation was not investigated to confirm the enhancement of the tangential slip [17]. Shivhare et al. [37] and Reddy et al. [31] determined the overall volumetric gas-side mass transfer coefficient $K_G a_e$ which characterizes gas-limited mass transfer processes like distillation at co- and counter-rotation. Both experimental studies revealed that the direction of rotation had a minor effect on $K_G a_e$ [31, 37]. **Consequently, it can be summarized that the tangential slip cannot significantly enhance the $K_G a_e$** [31, 37]. Shivhare et al.

[37] additionally equipped the rotor with a single-block packing and observed significantly lower $K_G a_e$ for single-block packing compared to split-packing in general. **They concluded that the split-packing itself due to a repeated liquid redistribution leads to better wetting of the packing, which results in an enhanced mass transfer** [37].

5.2.5 Rotating packed bed with packing and blades

Another approach to intensify the mass transfer by changing the rotor was invented by Luo et al. [18] which is a RPB with packing and blades (BP-RPB). The idea behind it is to generate multiple end zones (see Chapter 6) and thus enhance the mass transfer along the whole radius. The implementation of this idea is realized by arranging alternating rings of packing and blades. As shown in Fig. 5.7, the rotor is equipped with three packing and two blade rings. The blades should create droplets that are sheared by the next packing ring [18].

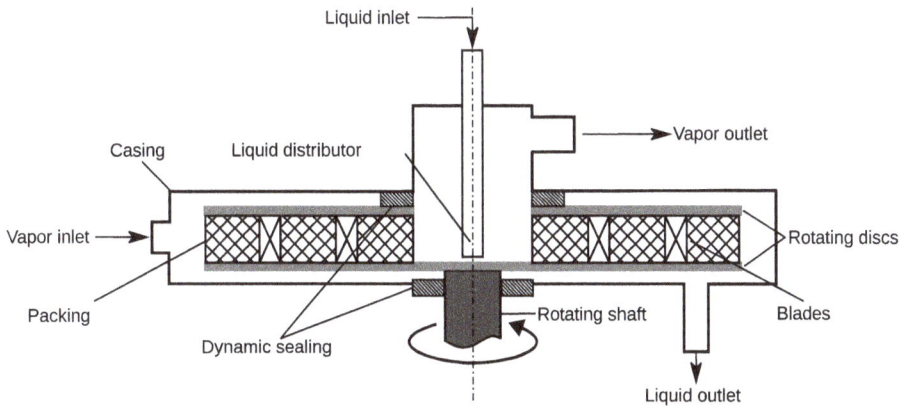

Fig. 5.7: Schematic drawing of BP-RPB [18].

Compared to a single-block packing, the rotor with packing and blades results in the following improvements:
- The a_e is enhanced due to the repeated collisions of the liquid to the packing [18].
- Also, the $K_L a_e$ increases as a consequence of the creation of small droplets [18].

On the contrary, this rotor modification results in the following challenges:
- Since the blades have to be attached to the disc, the construction effort of this rotor is higher than for a conventional RPB [18].
- Just as in a conventional RPB, an intermediate feed is challenging to implement.

Since so far, no distillation studies were conducted in this rotor configuration, a conclusion about the mass transfer performance compared to a conventional RPB cannot be made yet. Nevertheless, this rotor design was adapted to a two-staged RPB design in which distillation experiments were carried out successfully (see Section 5.2.7) [20].

5.2.6 Two-stage counter-current rotating packed bed

Luo et al. [19] combined the rotor designs of a conventional RPB and an RZB to exploit a synergy in a two-stage counter-current RPB (TSCC-RPB), which is illustrated in Fig. 5.8. The upper disc of each rotor is stationary to facilitate a coaxial multirotor design and includes concentric sheets with pores, which differs from the upper rotor design of an RZB. The lower rotating disc of each rotor is not equipped with concentric circular perforated baffles rather with concentric circular casings into which conventional packing of wire mesh is located. The liquid is introduced in the eye of the upper rotor and is collected at the bottom of the upper housing after passing the packing. Afterwards, it flows to the eye of the lower rotor, passes the packing, and leaves the TSCC-RPB via the liquid outlet. By contrast, the vapor enters the RPB at the bottom of the lower rotor and leaves it in the eye of the upper rotor. Thus, the fluids are contacted countercurrently [19].

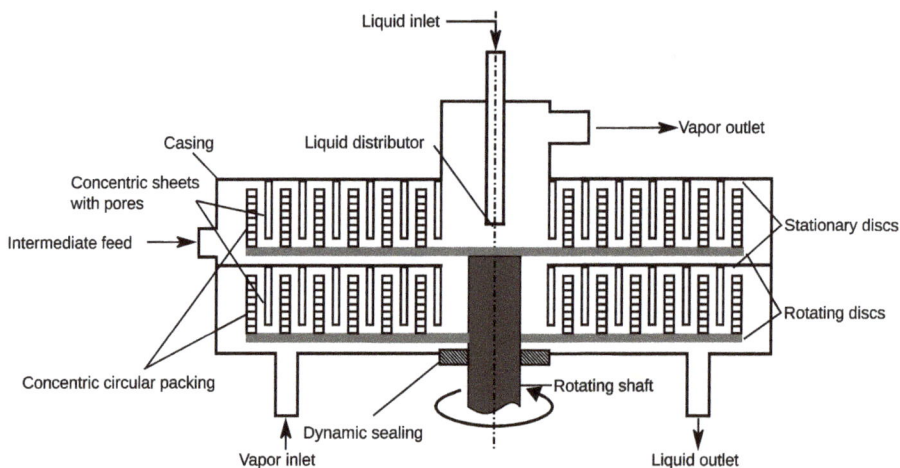

Fig. 5.8: Schematic drawing of a TSCC-RPB [19].

The synergies arising from the RZB and RPB concept are as follows:
–   No dynamic sealing for the upper disc of each rotor is necessary for the two-stage design because of the RZB rotor design [14, 24, 28].

- Compared to a conventional RPB, an intermediate feed can be implemented between the two rotors to achieve a continuous operation. [19]
- The rotor design is more compact than two single-stage RPBs which possibly results in lower heat loss since the flow path of the fluid is shorter [19].
- Due to modified packing and rotor design, the pressure drop and energy consumption decrease, and the mass transfer performance increases compared to an RZB [19, 38].

The main drawback represents the manufacturing of the packing and the rotor design needs high manufacturing precision, and the costs of a TSCC-RPB are higher than a conventional RPB [20].

It is worth mentioning that the mass transfer performance of the TSCC-RPB could be lower at some operation conditions than of two single-stage RPBs [19, 39]. Furthermore, the mass transfer in the so-called cavity zone (see Chapter 6) could be more pronounced in the two-staged design than in a single-stage design since the surface area of the casing is higher.

5.2.7 Two-stage blade-packing rotating packed bed

The most recent rotor modification represents the two-stage blade-packing RPB (TSBP-RPB) [20]. The TSBP-RPB combines the design of the TSCC-RPB and the BP-RPB (see Fig. 5.9). This modification seeks for simplification of the structure and maintaining the high mass transfer performance of a TSCC-RPB by equipping the rotors with a blade-packing instead of a static and rotational disc. The blade-packing is adapted from the BP-RPB. Similar to a TSCC-RPB, the liquid flows through the upper rotor and, subsequently, through the lower rotor while the vapor flows countercurrently [20].

Most of the advantages and disadvantages of the TSCC-RPB are also valid for the TSBP-RPB if not explicitly addressed in the following comparison:
- The TSBP-RPB can maintain the mass transfer performance of the TSCB-RPB at a lower rotational speed which results in lower energy consumption [20].
- The rotor of the TSBP-RPB is easier to construct than the rotor of the TSCC-RPB [20]. Nevertheless, the construction effort of a blade-packing is still higher than a single-block packing [18]. Furthermore, additional dynamic sealing is needed for each rotor equipped with a blade packing which increases the construction effort to avoid gas bypasses [14, 24].

To quantify the statement of similar mass transfer performance of the TSCC-RPB and the TSBP-RPB, a comparison is given in Tab. 5.1. Exceptionally, the $HETP_{RPB}$ of the devices are compared since the radial packing depth of both devices is quite similar (radial packing depth of TSCC-RPB: 210 mm, radial packing depth of TSBP-RPB: 206 mm) [20, 38].

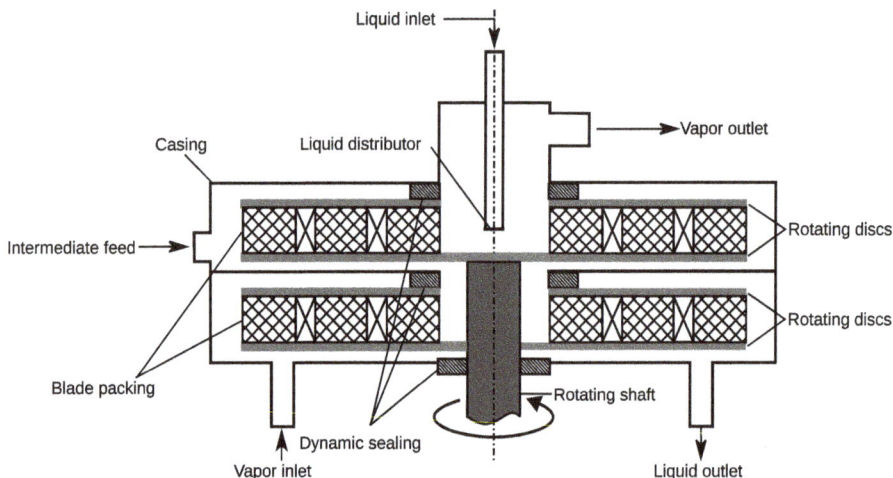

Fig. 5.9: Schematic drawing of a TSBP-RPB [20].

Tab. 5.1: Comparison of TSCC-RPB and TSBP-RPB.

Device	TSCC-RPB [38]	TSBP-RPB [20]
Chemical system	Ethanol/water	Ethanol/water
d_o [mm]	0.355	0.348
d_i [mm]	0.145	0.128/0.156
h_r [mm]	46	40
Optimal rotational speed [min^{-1}]	800	700
HETP$_{RPB}$ [mm]	3.1–4.7	1.9–10

As shown in Tab. 5.1, the HETP$_{RPB}$ are in a similar range; therefore, it is possible to achieve the mass transfer performance of a TSCC-RPB in a TSBP-RPB [20]. The simplification of the rotor design is questionable since additional manufacturing effort of the blade packing and the dynamic sealing is needed. However, the concept of the TSBP-RPB could be easily modified by using a single-block isotropic packing for each rotor as in the basic RPB concept (Fig. 5.1). That would omit the need for the blades but, of course, the dynamic seal in between the rotors is still needed.

For further information, an overview of the limited experimental and theoretical studies on distillation with RPBs is given in Tab. 5.2. This overview includes the rotor configuration, the packing type, the rotor dimensions, the chemical system, and a short description of the respective study.

Tab. 5.2: List of international high-gravity distillation studies which are available in English.

Year	Author	Rotor configuration	Packing type	Rotor dimensions (d_i, d_o, h_r) [mm]	Chemical system	Description	Reference
1983	Ramshaw, Short	RPB	Wire mesh	200/380, 800, 25/300	Ethanol/propanol	Experimental investigation of continuous distillation	[4, 39]
1996	Kelleher et al.	RPB	Metal foam	175, 600, 300	Cyclohexane/ n-heptane	Experimental investigation at total reflux	[40]
2002	Lin et al.	RPB	Wire mesh	61, 147, 95/25	Methanol/ethanol	Experimental investigation at total reflux	[26]
2004	Rao et al.	–	–	–	–	Review and appraisal of RPBs	[21]
2008	Li et al.	Two RPB	Corrugated disk packing, cross meshwork packing, wave thread packing	60, 110 63	Methanol/water	Experimental investigation of continuous distillation	[41]
2008	Wang et al.	RZB	Concentric rings with and without perforation	100, 520, 78	Methanol/water	Experimental investigation at total reflux	[14]
2008	Wang et al.	RZB	Concentric rings with and without perforation	R_1: 200, 630, 80 R_2: 200, 630, 80	Ethanol/water	Experimental investigation at total reflux	[24]
2009	Nascimento et al.	RPB	Random rashing ring, wire mesh	22, 80, 40	n-Hexane/n-heptane	Experimental investigation at total reflux	[42]
2010	Li et al.	Two RPB	Fin baffle packing	60, 110, 30	Ethanol/water	Experimental investigation of continuous distillation	[43]

(continued)

Tab. 5.2 (continued)

Year	Author	Rotor configuration	Packing type	Rotor dimensions (d_i, d_o, h_r) [mm]	Chemical system	Description	Reference
2010	Rahimi and Karimi	–	–	–	–	Theoretical investigation on neural network model for prediction of mass transfer coefficient in RPBs	[44]
2010	Agarwal et al.	SP-RPB	–	–	n-Butane/isobutane, benzene/cumene	Theoretical investigation and design procedure for continuous distillation	[35]
2012	Mondal et al.	SP-RPB	Wire mesh, split packing	60, 310, 27	Methanol/ethanol	Experimental investigation at total reflux	[17]
2012	Luo et al.	TSCC-RPB	Concentric rings with pores or wire mesh	R_1: 145, 355, 58 R_2: 145, 355, 58	Acetone/water	Experimental investigation of continuous distillation	[19]
2012	Prada et al.	RPB	–	200, 800, 80	Ethanol/water	Computational study of RPB in distillation	[45]
2012	Krishna et al.	SP-RPB	–	–	Synthesis of methyl acetate	Theoretical study of reactive distillation in RPBs	[46]
2013	Chu et al.	TSCC-RPB	Concentric rings with pores or wire mesh	R_1: 145, 365, 58 R_2: 145, 365, 58	Methanol/water	Experimental investigation of continuous distillation	[38]
2014	Li et al.	CFCR-RPB	Perforated concentric rings	140, 272, 15/45	Ethanol/water	Experimental investigation at total reflux	[15]
2014	Wang et al.	CRB	Perforated concentric baffles	100, 242, 80	Ethanol/water	Experimental investigation at total reflux	[16, 29]

Year	Author	Type	Packing		Mixture	Description	Ref.
2015	Sudhoff et al.	RPB	–	–	–	Mathematical tool to simulate distillation performance	[12, 47]
2016	Luo et al.	TSBP-RPB	Blade packing	R_1: 128, 348, 40 R_2: 158, 348, 40	Methanol/water	Experimental investigation of continuous distillation	[20]
2017	Garcia et al.	–	–	–	–	Review on HiGee distillation	[7]
2017	Li et al.	RPB	Wave thread stainless steel wire mesh	180, 285, 310	Ethanol/water	Modeling of vacuum distillation in RPB	[48]
2018	Qammar et al.	RPB	Knit mesh	146, 400, 10	Ethanol/water	Experimental investigation at total reflux	[49]
2019	Wang et al.	–	–	–	–	Review of RPB	[50]
2019	Qammar et al.	RPB	3D-printed zig-zag packing	146, 400, 10	Ethanol/water	Experimental investigation at total reflux	[51]
2019	Wang et al.	RZB	Different configuration of concentric rings with and without perforation	118, 285, 50	Ethanol/water	Experimental investigation at total reflux	[52]
2022	Hilpert et al.	RPB	Metal foam	120, 450, 10	Ethanol/water, ethanol/n-propanol, ethanol/n-propanol/water	Experimental investigation at total reflux, first-time ternary distillation, rate-based modeling	[27]

5.3 Take-home messages

- The comparison regarding mass transfer performance of a conventional column and different rotor configurations themselves are challenging due to the different geometries.
- At the state-of-the-art, only RZB and CFCR are suitable for continuous distillation with a single rotor since an intermediate feed can be applied to them.
- A high tangential slip velocity does not necessarily enhance the mass transfer.
- The different rotor configurations have been modified in the number of rotors, the internals of the rotor, or both.
- No rotor configuration is superior in every aspect like operating window, mass transfer performance, energy demand, and construction complexity. Usually, an advantage achieved in one place goes along with at least one disadvantage elsewhere.
- Only RZB is applied for industrial distillation processes.

5.4 Quiz

Question 1. How is the F-factor changing along the radius?

Question 2. Why is using HETP for comparing a conventional column and an RPB not suitable?

Question 3. Which assumptions have been made in the *ATU-NTU* concept?

Question 4. Which geometrical parameter needs to be changed to increase the capacity of a conventional RPB?

Question 5. How can a continuous distillation be realized in different rotor configurations?

Question 6. What are the differences in construction between an RZP and a conventional RPB?

Question 7. Name one advantage and one disadvantage of each rotor configuration? Which one is the best of all?

Question 8. What is the idea of the concept of packings and blades?

Question 9. Is a coaxial multi-rotor design of a conventional RPB feasible?

Question 10. Is a distillation gas- or liquid-side limited?

Question 11. Why is an intermediate feed challenging to implement in a conventional RPB?

Question 12. List commonly used chemical systems for distillation studies.

Question 13. What is an application of a distillation process with two RZBs? What is the function of each RZB? Sketch of the process.

5.5 Exercises

5.5.1 Continuous distillation with RPBs

A mixture consisting of a lower boiling component A and a higher boiling component B has to be separated in a process consisting of two conventional RPBs. Each RPB is equipped with metal foam with an inner diameter of 150 mm, an outer diameter of 450 mm, and an axial height of 10 mm. The process operates under adiabatic conditions and at a pressure of 1 bar. The feed flow is at liquid boiling conditions. The vapor-liquid equilibrium data and the operating lines for the process are given in the diagram below (Fig. 5.10). $\rho_{V,out}$ and $\dot{m}_{V,out}$ are the vapor density and mass flow of the leaving distillate stream.

Details:
- $\rho_{V,out} = 1\,kg/m^3$
- $\dot{m}_{V,out} = 50\,kg/h$

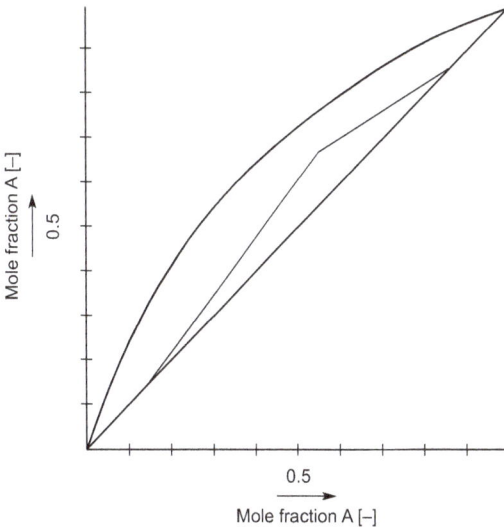

Fig. 5.10: McCabe/Thiele diagram of A/B.

Exercise 1. Draw a flow sheet of the process. Label the streams and apparatuses.

Exercise 2. Calculate the F-factor of the RPB which is used as the rectifying section.

Exercise 3. What are the mole fractions of component A in the feed, distillate, and bottom product?

Exercise 4. Calculate the HETP$_{\text{RPB}}$ for each RPB.

5.6 Solutions

5.6.1 Quiz

Solution (Question 1). The F_G is decreasing from the inner to the outer packing since the cross-sectional is increasing linearly: $\rho_{V,\text{out}} = u_s \cdot \sqrt{\rho_V} = \frac{\dot{V}_G}{A_p(r)} \cdot \sqrt{\rho_V}$

Solution (Question 2). The HETP for RPBs is solely related to the packing length, but the separation efficiency is different over the packing length since different mass transfer zones are present. Furthermore, the mass transfer in the outer cavity of the RPB is not included. Therefore, the HETP for RPBs is in general a poor measure for separation efficiency.

Solution (Question 3). For the *ATU-NTU* concept, the following parameters are assumed as constant along the radius of the packing: gas mass flow $\dot{m}_G$, gas density ρ_G, axial height h, and overall volumetric gas side mass transfer coefficient $K_G a_e$. Furthermore, it must be valid that the mass transfer can be described by film theory.

Solution (Question 4). The capacity of conventional RPB increases by increasing the axial height of the packing.

Solution (Question 5). Two conventional RPBs are needed to conduct a continuous distillation with a rectifying and stripping section (see Fig. 5.10) because an intermediate feed is challenging to implement in a single RPB. A continuous distillation can be realized with solely one apparatus if an intermediate feed is present.

Solution (Question 6). Compared to a conventional RPB, at the RZB, only the lower disc is rotating, while the upper disc is stationary and attached to the housing of the device. The discs are equipped with concentric circular sheets, while the upper part of the rotating baffles is perforated and while the rotor of RPB is equipped with a packing. Additionally, the RZB has an intermediate feed.

Solution (Question 7). Tab. 5.3 lists one advantage and one disadvantage of each rotor configuration.

Tab. 5.3: Advantages and disadvantages of rotor configurations.

Rotor configuration	Advantage		Disadvantage	
RPB	–	Easy increase of fluid capacity	–	No intermediate feed
RZB	–	Simplified manufacturing	–	High pressure drop
CRB	–	Low pressure drop	–	Low mass transfer performance
CFCR-RB	–	High F_G feasible	–	Low a_e
SP-RPB	–	High a_e	–	Tow motors are needed
BP-RPB	–	High a_e	–	No intermediate feed
TSCC-RPB	–	High mass transfer performance	–	High manufacturing precision
TSBP-RPB	–	Low energy consumption	–	High construction effort

The superiority of the rotor configurations among each other depends strongly on the application; thus, no rotor configuration is superior in every respect.

Solution (Question 8). The idea behind it is to generate multiple end zones and thus enhance the mass transfer along the whole radius.

Solution (Question 9). It is feasible, but a dynamic seal for each rotor is needed.

Solution (Question 10). It is gas-side limited.

Solution (Question 11). To realize an intermediate feed, the fluid needs to be introduced to the rotating packing which requires an elaborate seal system.

Solution (Question 12). The most common chemical systems used for distillation in RPBs are ethanol/water and methanol/water.

Solution (Question 13). An application including two RZBs is the extractive distillation of ethanol/water by adding ethylene glycol (see Fig. 5.3). One RZB is used to separate anhydrous ethanol and the other to recover the ethylene glycol.

5.6.2 Continuous distillation with RPBs

Solution (Exercise 1) Fig. 5.11 shows a flow diagram of a distillation process with two RPBs.

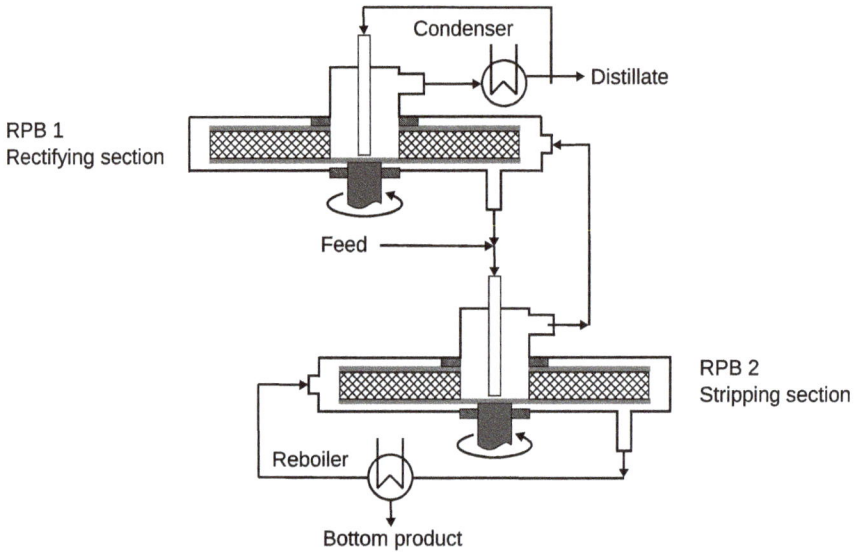

Fig. 5.11: Solution of exercise 1, process sketch.

Solution (Exercise 2)

Initial equation: $F_G = u_s \cdot \sqrt{\rho_V}$

Modification to: $F_G = \dfrac{\dot{m}_{V,out}}{\rho_{V,out} \cdot A_p} \cdot \sqrt{\rho_{V,out}}$

$$A_{P,eye} = \pi \cdot d_i \cdot h$$

➜ $F_G \approx 2.95\,Pa^{0.5}$

Solution (Exercise 3) Reading of Fig. 5.10.

➜ $x_{feed} \approx 0.55, \; x_{bottom\,product} \approx 0.15, \; x_{distillate} \approx 0.87$

Solution (Exercise 4) Use Fig. 5.10 to do the stage construction (see solution in Fig. 5.12). Reboiler equals one theoretical stage. Therefore, in each RPB (rectifying and stripping section), two theoretical stages are achieved. In the last step use:

$$HETP_{RPB} = \frac{(r_o - r_i)}{N_{th}}$$

➜ $HETP_{RPB} = 0.075\,m$

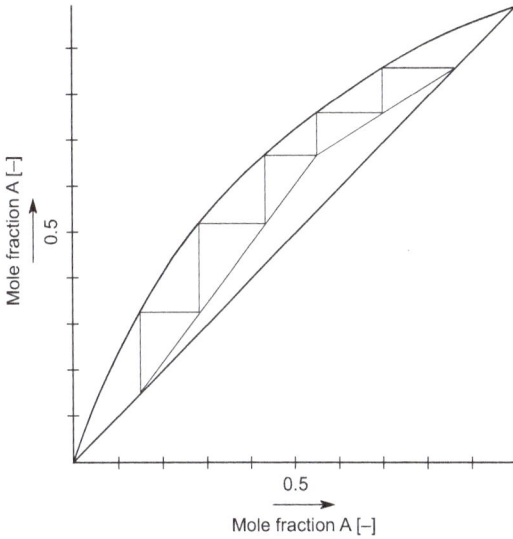

Fig. 5.12: Stage construction in McCabe/Thiele-diagram of A/B.

List of symbols

a_e	effective interfacial area	$m^2\ m^{-3}$
A_p	cross-sectional area of the packingr	
ATU	area of a transfer unit	m^2
d_i	inner diameter of bed packing/internal	mm
d_o	outer diameter of bed packing/internal	mm
F_G	F-factor	$Pa^{0.5}$
h	axial height	m
HETP	height equivalent to a theoretical plate	m
h_r	height of the rotor	mm
HTU	height of a transfer unit	m
$K_G a_e$	overall volumetric gas- or vapor-side mass transfer coefficient	s^{-1}
$K_L a_e$	overall volumetric liquid-side mass transfer coefficient	s^{-1}
$\dot{m}_G$	gas mass flow	$kg\ s^{-1}$
N_{th}	number of theoretical stages	–
NTU	number of a transfer unit	–
r	radius	m
u_s	superficial gas velocity	$m\ s^{-1}$
$\dot{V}$	volume flow	$m^3\ s^{-1}$
x	mole fraction of liquid	–
y	mole fraction of gas	–

Greek letters

ρ	density	kg m^{-3}
ν	reflux ratio	–

Subscripts

eye	eye of tde RPB
G	gas
i	inner
in	inflowing
L	liquid
o	outer
out	outflowing
P	packing
R	rotor
RPB	rotating packed bed
s	superficial
V	vapor

Superscript

*	**at equilibrium conditions**

List of abbreviations

CFCR-RB	counterflow concentric ring rotating bed
CRB	crossflow concentric-baffle rotating bed
HiGee	high-gravity
BP-RPB	packed bed with packing and blades
R_1	upper rotor
R_2	lower rotor
RPB	rotating packed bed
RZB	rotating zig-zag bed
SP-RPB	rotating packed bed with split-packing
TSCC-RPB	two-stage counter-current rotating packed bed
TSBP-RPB	two-stage blade-packing rotating packed bed

References

[1]   Düssel RSJ, Dssel R, Stichlmair J, Groebel M, eds. Rektifikation. In: Fluidverfahrenstechnik. Hoboken: Wiley-VCH; 2008.

[2]   Neumann K, Hunold S, Skiborowski M, Górak A. Dry pressure drop in rotating packed beds – systematic experimental studies. Ind. Eng. Chem. Res. 2017;56(43):12395–405.

[3]   Cai TJ, ed. Column Performance Testing Procedures. In: Distillation: Operation and Applications. Salt Lake City: Academic Press, 2014.

[4]   Ramshaw C. 'HIGEE' distillation – An example of process intensification. Chem. Eng. (London). 1983;389:13–14.

[5]   Guo K, Wen J, Zhao Y, Wang Y, Zhang Z, Li Z, et al. Optimal packing of a rotating packed bed for H(2)S removal. Environ. Sci. Technol. 2014;48(12):6844–49.

[6]   Yang K, Chu G, Zou H, Sun B, Shao L, Chen J-F. Determination of the effective interfacial area in rotating packed bed. Chem. Eng. J. 2011;168(3):1377–82.

[7]   Cortes Garcia GE, van der Schaaf J, Kiss AA. A review on process intensification in HiGee distillation. J. Chem. Technol. Biotechnol. 2017;92(6):1136–56.

[8]   Guo K A Study on Liquid Flowing Inside the Higee Rotor. Ph.D. Dissertation. Beijing, China; 1996.

[9]   Chilton TH, Colburn AP. Distillation and absorption in packed columns: a convenient design and correlation method. Ind. Eng. Chem. 1935;27:255–60.

[10] Singh SP. Air Stripping of Volatile Organic Compounds from Groundwater: An Evaluation of a Centrifugal Vapor-Liquid Contactor. Ph.D. Dissertation. United States; 1989.

[11] Sattler K, Feindt HJ. Thermal Separation Processes: Principles and Design. Weinheim: VCH; 1995.

[12] Sudhoff D, Neumann K, Lutze P. An integrated design method for rotating packed beds for distillation. Comput. Aided Chem. Eng. 2014;33:1303–08.

[13] Groß K, Bieberle A, Gladyszewski K, Schubert M, Hampel U, Skiborowski M, et al. Analysis of flow patterns in high-gravity equipment using gamma-ray computed tomography. Chem. Ing. Technik. 2019;91(7):1032–40.

[14] Wang GQ, Xu OG, Xu ZC, Ji JB. New HIGEE-rotating Zigzag bed and its mass transfer performance. Ind. Eng. Chem. Res. 2008;47(22):8840–46.

[15] Li Y, Li X, Wang Y, Chen Y, Ji J, Yu Y, et al. Distillation in a counterflow concentric-ring rotating bed. Ind. Eng. Chem. Res. 2014;53(12):4821–37.

[16] Wang GQ, Guo CF, Xu ZC, Li YM, Ji JB, New Crossflow A. Rotating Bed, Part 1: Distillation Performance. Ind. Eng. Chem. Res. 2014;53(10):4030–37.

[17] Mondal A, Pramanik A, Bhowal A, Datta S. Distillation studies in rotating packed bed with split packing. Chem. Eng. Res. Des. 2012;90(4):453–57.

[18] Luo Y, Chu G-W, Zou H-K, Wang F, Xiang Y, Shao L, et al. Mass transfer studies in a rotating packed bed with novel rotors: Chemisorption of CO 2. Ind. Eng. Chem. Res. 2012;51 (26):9164–72.

[19] Luo Y, Chu G-W, Zou H-K, Xiang Y, Shao L, Chen J-F. Characteristics of a two-stage counter-current rotating packed bed for continuous distillation. Chem. Eng. Process. Process Intensif. 2012;52:55–62.

[20] Luo Y, Chu G, Sang L, Zou H, Xiang Y, Chen J. A two-stage blade-packing rotating packed bed for intensification of continuous distillation. Chin. J. Chem. Eng. 2016;24(1):109–15.

[21] Rao DP, Bhowal A, Goswami PS. Process intensification in rotating packed beds (HIGEE): An appraisal. Ind. Eng. Chem. Res. 2004;43(4):1150–62.

[22] Sandilya P, Rao DP, Sharma A, Biswas G. Gas-phase mass transfer in a centrifugal contactor. Ind. Eng. Chem. Res. 2001;40(1):384–92.

[23] Ji JB, Wang LH, Xu ZC, Bao TH. Equipment of zigzag high-gravity rotating beds: China patent 01134321.4, Zhejiang University of Technology (China), 2001.

[24] Wang GQ, Xu ZC, Yu YL, Ji JB. Performance of a rotating zigzag bed – A new HIGEE. Chem. Eng. Process. Process Intensif. 2008;47(12):2131–39.

[25] Li Y, YuLi Y, XuLi Z, LiLi X, Liu X, Ji J. Rotating zigzag bed as trayed HIGEE and its power consumption. Asia-Pac. J. Chem. Eng. 2013;8(4):494–506.

[26] Lin CC, Ho TJ, Liu WT. Distillation in a rotating packed bed. J Chem Eng Jpn. 2002;35:1298–304.

[27] Hilpert M, Calvillo Aranda GU, Repke J-U. Experimental analysis and rate-based stage modeling of multicomponent distillation in a rotating packed bed. Chem. Eng. Process. Process Intensif. 2021;171:108651.

[28] Wang GQ, Xu ZC, Ji JB. Progress on Higee distillation – Introduction to a new device and its industrial applications. Chem. Eng. Res. Des. 2011;89(8):1434–42.

[29] Wang GQ, Guo CF, Xu ZC, Yu YL, Ji JB. A new crossflow rotating bed, part 2: Structure optimization. Ind. Eng. Chem. Res. 2014;53(10):4038–45.

[30] Chandra A, Goswami PS, Rao DP. Characteristics of flow in a rotating packed bed (HIGEE) with split packing. Ind. Eng. Chem. Res. 2005;44(11):4051–60.

[31] Reddy KJ, Gupta A, Rao DP, Rama OP. Process intensification in a HIGEE with split packing. Ind. Eng. Chem. Res. 2006;45(12):4270–77.

[32] Bhattacharya S, Mondal A, Bhowal A, Datta S. Evaporative cooling of water in a rotating packed bed (Split Packing). Ind. Eng. Chem. Res. 2010;49(2):847–51.

[33] Liu Y, Gu D, Xu C, Qi G, Jiao W. Mass transfer characteristics in a rotating packed bed with split packing. Chin. J. Chem. Eng. 2015;23(5):868–72.

[34] Rajan S, Kumar M, Ansari MJ, Rao DP, Kaistha N. Limiting gas liquid flows and mass transfer in a novel rotating packed bed (HiGee). Ind. Eng. Chem. Res. 2011;50(2):986–97.

[35] Agarwal L, Pavani V, Rao DP, Kaistha N. Process intensification in higee absorption and distillation: Design procedure and applications. Ind. Eng. Chem. Res. 2010;49(20):10046–58.

[36] Hacking JA, Delsing N, de Beer MM, van der Schaaf J. Improving liquid distribution in a rotating packed bed. Chem. Eng. Process. Process Intensif. 2020;149:107861.

[37] Shivhare MK, Rao DP, Kaistha N. Mass transfer studies on split-packing and single-block packing rotating packed beds. Chem. Eng. Process. Process Intensif. 2013;71:115–24.

[38] Chu G-W, Gao X, Luo Y, Zou H-K, Shao L, Chen J-F. Distillation studies in a two-stage counter-current rotating packed bed. Sep. Purif. Technol. 2013;102:62–66.

[39] Short H. New mass-transfer find is a matter of gravity. Chem. Eng. 90.4. 1983;90(4):23–29.

[40] Kelleher T, Fair JR. Distillation studies in a high-gravity contactor. Ind. Eng. Chem. Res. 1996;(35):4646–55.

[41] Li X, Liu Y, LI Z, WANG X. Continuous distillation experiment with rotating packed bed. Chin. J. Chem. Eng. 2008;16(4):656–62.

[42] Nascimento JVS, Ravagnani TMK, Pereira JAFR. Experimental study of a rotating packed bed distillation column. Braz. J. Chem. Eng. 2009;26:219–26.

[43] Xiuping LI, Youzhi LIU. Characteristics of fin baffle packing used in rotating packed bed. Chin. J. Chem. Eng. 2010;18:55–60.

[44] Rahimi MR, Karimi H. A mega-trend diffusional neural network model for prediction of mass transfer coefficient of rotating packed bed distillation column. In: CHISA: Conference Proceedings, 2010.

[45] Prada RJ, Martínez EL, Wolf Maciel MR. Computational study of a rotating packed bed distillation column. Comput. Aided Chem. Eng. 2012;30:1113–17.

[46] Krishna G, Min TH, Rangaiah GP. Modeling and analysis of novel reactive higee distillation. Comput. Aided Chem. Eng. 2012;31:1201–05.

[47] Sudhoff D, Leimbrink M, Schleinitz M, Górak A, Lutze P. Modelling, design and flexibility analysis of rotating packed beds for distillation. Chem. Eng. Res. Des. 2015;94:72–89.

[48] Li W, Song B, Li X, Liu Y. Modelling of vacuum distillation in a rotating packed bed by Aspen. Appl. Therm. Eng. 2017;117:322–29.

[49] Qammar H, Hecht F, Skiborowski M, Górak A. Experimental investigation and design of rotating packed beds for distillation. Chem. Eng. Trans. 2018;69:655–60.

[50] Wang Z, Yang T, Liu Z, Wang S, Gao Y, Wu M. Mass transfer in a rotating packed bed: A critical review. Chem. Eng. Process. Process Intensif. 2019;139:78–94.

[51] Qammar H, Gładyszewski K, Górak A, Skiborowski M. Towards the development of advanced packing design for distillation in rotating packed beds. Chem. Ing. Technik. 2019;91 (11):1663–73.

[52] Wang GQ, Zhou ZJ, Li YM, Ji JB. Qualitative relationships between structure and performance of rotating zigzag bed in distillation. Chem. Eng. Process. Process Intensif. 2019;135:141–47.

Hina Qammar

6 Rotating packed beds in distillation: binary distillation

While high-gravity technology is mostly treated as new technology, the concept of centrifugal field separation exists for several decades. Podbielniak [1] was one of the first to file a patent using a spirally wound tube on rotating support [1]. Over the years, different design improvements for rotor and packing were investigated and patented, while most of the current designs of rotating packed beds (RPBs) are based on the patent of Pilo and Dahlbeck [2]. Over time, several investigations have been published showing different mass transfer and performance models of a PRB [3–8]. Recently, few studies have been initiated focusing on the systematic design of RPB [5, 6, 9–12] to help pave the way toward the industrial implementation of this technology. For the systematic design of RPBs, it is very important to understand the hydrodynamics (Chapter 3) as well as the mass transfer of different sections of the RPB, that is, how much each section is contributing to the total mass transfer. The understanding of these fundamentals will not only help to develop reliable scale-up and design rules but will also help to improve the separation efficiency of these sections.

The basics of distillation in RPBs with a comprehensive literature overview of different rotor configurations and advances in the design of a rotor and an RPB are described in Chapter 5. The current chapter focuses on binary distillation in an RPB, with a detailed discussion of an example application. First, some basic concepts are discussed in Section 6.1 followed by reported applications of this technology in Section 6.2. Section 6.3 explains the detailed example of ethanol dehydration in an RPB where experimental setup, procedure, and results are discussed. The comparison of RPBs with the conventional column and other variants of HiGee technology improves the understanding of this technology and is further pursued in Section 6.4. Section 6.5 further presents some advanced concepts about the basics of mass transfer distribution using telemetry techniques in an RPB, differentiating the rotating and non-rotating sections and their contributions. Some fundamentals about the available modeling approaches for gas-side mass transfer limited processes in RPBs are finally given in Section 6.6.

Hina Qammar, TU Dortmund University

https://doi.org/10.1515/9783110724998-006

6.1 Fundamentals

In the chemical process industry, distillation is one of the most important and commonly used separation techniques. In traditional distillation columns, vapor flows upward while counter-currently the liquid flow is directed downward due to the gravitational force. For several decades, a number of techniques have been investigated to intensify the contact between the vapor and the liquid phase from which the application of centrifugal forces in a rotating packed bed (RPB) is one of the most promising ones [9, 13]. The concept allows controlling the separation efficiency through a change of the applied rotational speed. In RPBs high shear forces are supplied by rotation of a rotor, containing a packed bed, in which gas and liquid phases are contacted in a counter-current operation. The rotor consists of an annular-shaped packing enclosed by the casing as discussed in further detail in Chapter 5, Fig. 5.1. In an RPB, the liquid is supplied at the inner periphery of the annular packing and is accelerated radially outward by an applied centrifugal field into the stationary casing. The vapor phase is introduced into the casing and flows radially inward counter-currently to the liquid due to an enforced pressure gradient. Thin films and fine droplets are generated by the application of centrifugal forces, 100–1,000 times the gravitational force, consequently leading to intensified contact between the vapor and liquid phase [14]. The potential benefits include intense mixing, reduced footprint, shorter residence time, use of high specific surface area packings, and an enlarged operating window [15] compared to the traditional gravity-based separations. Some of the basic terminologies and equations to characterize the distillation operation and mass transfer in an RPB have been described in the previous chapters. The following description presents the basis for the interpretation of the experimental results presented in Section 6.3.4.

To characterize the vapor flow rate in RPBs the F-factor is used. It is determined by the superficial gas/vapor velocity and the density of the gas/vapor. As the packing area decreases from the outer to the inner radius, the velocity increases as the vapor transit the packing. With that, the F-factor increases until it reaches its maximum value at the eye of the rotor [9]. Therefore, when the F-factor is mentioned in this chapter it refers to the maximum value at the inner radius, calculated with the following equation:

$$F_G = \frac{\dot{m}_{v,out}}{\rho_{v,out} \cdot A_{c,i}} \cdot \sqrt{\rho_{v,out}} \tag{6.1}$$

$A_{c,i}$ in the above equation is the cylindrical surface area at the inner radius of the packing and is defined as:

$$A_{c,i} = 2 \cdot \pi \cdot r_i \cdot h \tag{6.2}$$

To evaluate the mass transfer efficiency of conventional columns as well as RPBs, the concept of the number of theoretical stages (N_{th}) is applied commonly. To determine the achieved number of theoretical stages, knowledge about the equilibrium line is necessary. An analytical approach for the system, ethanol-water, presented as an example in the experimental results section is provided by Li et al. [16]:

$$y^* = \frac{\alpha \cdot x}{1 + (\alpha - 1) \cdot x} \tag{6.3}$$

where x_E is the mole fraction of ethanol in the bulk liquid phase, y^* is the mole fraction of ethanol in the vapor phase being in equilibrium with the bulk liquid and α is the relative volatility of ethanol in water. The relative volatility is a function of the ethanol composition, which can be approximated by the following correlation at atmospheric pressure [16]

$$\alpha = \begin{cases} 1.1213 \cdot (x + 0.2)^{-1.5235} & x \leq 0.292 \\ 0.8938 \cdot x^{-1.062} & x > 0.292 \end{cases} \tag{6.4}$$

When operated at total reflux the mole fraction in the reflux liquid, x, and the mole fraction in the distillate vapor, y, are equal [17]. Therefore, using the composition data and the equilibrium data together with eqs. (6.3) and (6.4), the number of theoretical stages can be calculated.

6.2 Application in the field of distillation

Over recent years, a number of applications of RPBs have been reported [7, 18]. Some applications benefit from the short residence time and intense mixing of the contacting phases in the RPB while others take advantage of the reduced equipment size. Neumann et al. [7] summarized some of the industrial applications of RPBs containing gas-liquid [19–21], vapor-liquid [22–24], and liquid-liquid mixtures [25, 26] in a recent review [7]. In a single RPB, supplying the feed along the radial axis of the rotor is challenging as the rotor rotates along with the shaft. Therefore, either two single-stage RPBs or an RPB with a multirotor is needed to perform the continuous distillation process. However, the recent geometrical modifications of the rotor have brought other options, like Rotating Zigzag Bed (RZB), to perform continuous distillation [27]. The details of the design and structure of an RZB are discussed in Chapter 5, but it must be mentioned that most of the industrial applications of RZBs for distillation are found in China [23]. They are also called rotating distillation machines, supergravity separators, or Higee distillation system [24]. There are almost 30,000 chemical companies in China where RZB is potentially applied in product separation and solvent recovery processes [23, 24]. In recent years, the most prominent applications for RZBs in the field of vapor-liquid contacting processes have been the recovery of solvents from products or waste streams. Xu et al. [22] studied the separation of THF-methanol-water system in an RZB and in a conventional column and showed that a significant amount of space can be reduced using RZB. Moreover, they showed that using water as an extract in addition to the feed shifted the mass balance and reduced the methanol concentration in the top product and 94.9 wt.% THF can be obtained as a product with a small methanol contamination of only 0.0024 wt.% [22].

Methanol, for instance, is one of the largest commodity chemicals in the world that is mostly used as a base material in the production of other chemicals like formaldehyde, acetic acid, or products like synthetic fabrics, paints, and adhesives, plastics, or as a chemical agent in pharmaceuticals. Owing to the health and environmental hazards associated with methanol, it is often necessary to recover it from the waste liquid or exhaust gas streams [28]. Solvent recovery is also necessary for the economic feasibility of an industrial process. The recovery lowers not only the fresh solvent requirement but also a possibly expensive waste treatment.

An interesting example of solvent recovery is the separation of methanol from an aqueous stream containing viscous heat-sensitive materials, which are prone to decomposition. Wang et al. [23] used this application and compared the performance of a distillation column with that of an RZB. However, it is not a plain comparison of equipment exchange only as two different operations, i.e., batch distillation in the packed column and continuous steam stripping in RZB are compared. As this makes the comparison of the different equipment difficult, it needs to be considered that an application of the steam stripping operation in the conventional column would be ineffective due to higher residence time, since the longer contact time with the high-temperature steam would result in product decomposition. Therefore, batch distillation must be performed at low temperatures under vacuum conditions which reduce the mass transfer rates due to the increased viscosity of the liquid mixture at these process conditions.

A distinct advantage permitted by the use of RZB is the significantly reduced residence time that allows the continuous steam stripping mode in an RZB [23]. Therefore, it becomes reasonable to compare both the equipment while considering the mode of operation and the particular operating conditions as they are only applicable through the use of RZB. The RZB application offers reduced utility as well as equipment costs due to the higher operating temperatures and reduced equipment size. Moreover, the applied centrifugal forces in an RZB provide intense turbulence and higher liquid velocities for the viscous liquid, thus leading to thinner liquid films and an increased mass transfer rate. Although the power consumption of operating an RZB is generally additional to that of a reboiler and condenser, in this comparison the total power consumption of operating a packed column is 35 kW including refrigeration and vacuum while that of the RZB is 18 kW since cooling water is used in the condenser [23]. By the use of RZB the reduction in the area occupied is four times less than that of the packed column [23].

A similar example where the use of an RZB not only reduced the occupied space for the equipment but also improved the product purity is the binary separation of methanol-water mixture [7, 23]. An RZB of a diameter of 0.83 m and a total height of 0.8 m, excluding the motor, is reported for the separation of a binary methanol-water mixture with a total capacity of 500 kg/h [24]. Furthermore, with these dimensions, 99.8 wt.% pure methanol is produced at the top and 0.2 wt.% in the bottom product. Neumann et al. [7] performed a theoretical study to enable a potential comparison between the packed column and the RZB. The separation of the methanol-water mixture is taken as a reference for a simulation of the process in Aspen Plus® using a built-in RADFRAC model and Sulzer BX500 structured packing. All the input data like feed, top, and bottom compositions, and reflux ratio make the comparison on the same basis and are given in Tab. 6.1.

Tab. 6.1: Simulation data for methanol-water distillation.

Item	Feed	Top	Bottoms
Flowrate [kg/h]	500	350.4	149.6
Methanol [wt./wt.]	0.7	0.998	0.002
Water [wt./wt.]	0.3	0.002	0.998
Column Operating Conditions			
Pressure [bar]	1.01	Reflux ratio	1.5
Feed condition	Saturated liquid	Distillate to feed ratio (mass)	0.7
Packing	Sulzer BX500		

The simulation results of column sizing showed that a packed column with a height of 13 m would be required to achieve a top product purity of 99.8 wt.% like that reported for the RZB. However, the reported packed column for this case has a height of 11 m, a diameter of 0.6 m, and 3 wt.% methanol in the bottom product [23, 24]. **Consequently, the application of RZB leads to an equipment size reduction of an order of magnitude** as compared to the conventional distillation column. Furthermore, a sensitivity study conducted as part of the theoretical study mentioned above [7] showed the effect of variations in feed and product compositions on the column height. The authors [7] showed that when the composition of the feed is varied from 70 wt.% methanol to 50 wt.% methanol keeping the column height (13 m), distillate to feed ratio (0.7), and reflux ratio (1.5) constant then the top product purity dropped from 99.8 wt.% methanol to 71.4 wt.%. Moreover, the changes during operation could be adjusted in an RZB by changing the rotational speed and without changing the design [7, 23], which offers an additional degree of freedom [3–6]. **This implies that the rotational speed of an RZB can be altered to react to changes in feed or product compositions, providing a degree of flexibility that is absent for the static column equipment used as a benchmark.**

6.3 Detailed example: ethanol dehydration

One way to develop reliable models is to rigorously perform well-designed experiments and gather experimental data. Information on how a particular variable is affected by changing any design or operational parameter is very important in the development of the design rules. In this regard a set of well-designed experiments were performed to evaluate the effect of rotational speed, packing type, and packing length on the separation efficiency (measured in terms of number of theoretical stages) of an RPB.

6.3.1 Experimental setup

Figure 6.1 shows the pilot-scale RPB, having a diameter of 1 m and a height of 1.15 m, used for the experiments in this study. This RPB has a maximum rotational speed of 1,500 rpm and the motor connected to it has a power consumption of up to 2.2 kW [6]. It has a vapor inlet, vapor outlet, and liquid inlet at the top and has four liquid outlets at the bottom. The simplified flow diagram of the experimental setup for distillation at total reflux under atmospheric pressure is further shown in Fig. 6.2. All experiments are performed at counter-current operation and the liquid is distributed via three flat fan nozzles with a 120° spray angle at the eye. A labyrinth seal at the top prevents the vapors from bypassing the rotor and a liquid seal at the bottom of the rotating shaft prohibits any leakages. Two types of conventional packing, namely, metal foam and knit mesh, that are used for the experiments are also shown. The metal foam packing, NCX1116, consists of five packing rings that fit smoothly into one another. This packing is investigated with all five rings as well as with only two rings with a rotor of an outer diameter of 400 mm.

The dehydration of ethanol is the most studied chemical system in RPBs and also one of the standard test mixtures for distillation recommended by Onken and Arlt [29]. The feed mixture of 10 wt.% ethanol is charged into the reboiler before each experiment. The reboiler is electrically heated and has a total heating power of 20 kW and a capacity of 65 l. The reboiler heating duty can be varied between 1–20 kW. Deionized water (998 kg m^{-3} at 20 °C) and absolute ethanol (789 kg m^{-3} at 20 °C) are used to prepare the feed mixture. Density measurements are used to evaluate the composition of this mixture. For this purpose, a Densito 30PX by Mettler-Toledo GmbH is used, which has a density accuracy of ± 1 kg m^{-3}. The reflux flow rate is measured with a Coriolis flowmeter from SIEMENS having a SITRANS FC300 sensor with MASS 6000 transmitter and a flow accuracy of 0.1% of the rate. Furthermore, there are several temperature sensors installed at the plant that will help in the characterization of mass transfer as described in Section 6.5.

The design specifications of the investigated RPB, rotor, and packing types with dimensions are listed in Tabs. 6.2 and 6.3 respectively. For the experimental investigation, two types of conventional packings and an empty rotor were used. The major properties of the conventional packings like porosity and specific surface area are listed in Tab. 6.2. In Tab. 6.3 dimensions of five metal foam rings are mentioned which are used to determine the effect of radial packing length on the separation efficiency of the RPB. The results of these experiments are presented in Section 6.3.3.

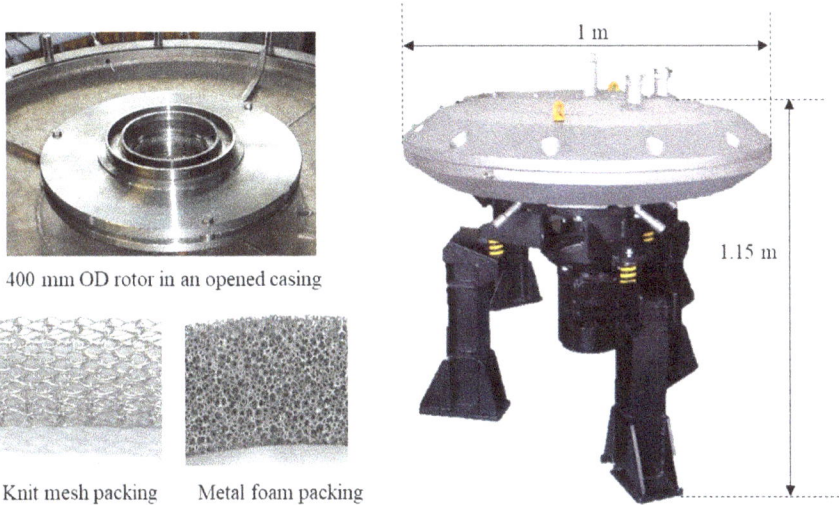

400 mm OD rotor in an opened casing

Knit mesh packing   Metal foam packing

1 m

1.15 m

Fig. 6.1: Pilot-scale RPB on the right with the rotor shown in an opened casing on the top left and conventional RPB packing types on the bottom left [6].

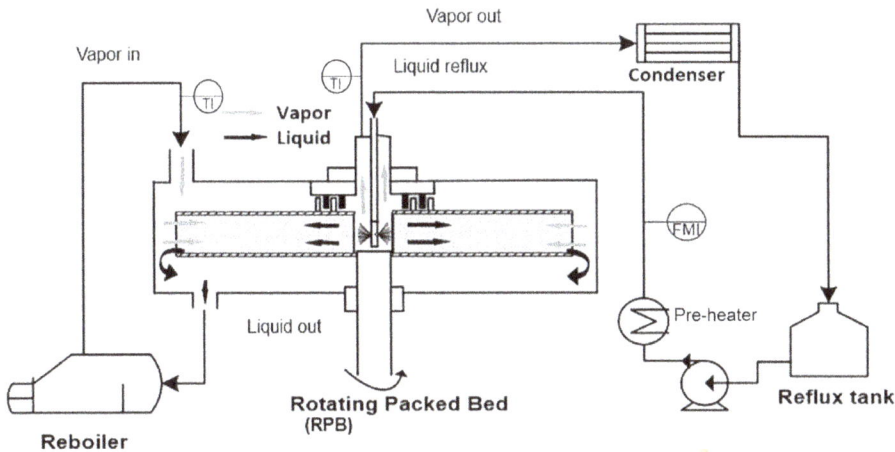

Fig. 6.2: Flow scheme of RPB pilot plant under total reflux at atmospheric pressure [6].

Tab. 6.2: Design specifications of the investigated RPB and internals [6].

RPB and internals		Investigated conventional packing types			
Specification	Dimension		ε [%]	ap [m2 m-3]	Ø [mm]
RPB casing inner diameter	860 mm	Metal knit mesh (KM)	87	2,496	–
Rotor outer diameter	400 mm	Metal foam NCX1116	92	1,000	1.4
Packing inner and outer diameter	di = 146 mm, do = 380 mm				
Axial height of packing	h = 10 mm				

Tab. 6.3: Metal foam rings inner and outer diameters.

	Ring 1	Ring 2	Ring 3	Ring 4	Ring 5
Inner diameter [m]	0.146	0.165	0.200	0.250	0.300
Outer diameter [m]	0.165	0.199	0.250	0.300	0.380

6.3.2 Experimental procedure

To investigate the influence of different design and operational parameters the same experimental procedure is used. First, the rotational speed of the RPB was set to a value between 1 s^{-1}–20 s^{-1}, and a 10 wt.% ethanol-water mixture of approximately 55 kg was pumped into the reboiler. Before starting the reboiler, nitrogen that is used as inert gas was supplied into the reboiler to ensure safe operation. The electrically heated reboiler was set to a power resulting in an F-factor of 2.1 Pa$^{0.5}$ at the eye of the rotor. As the feed mixture boils in the reboiler, a constant stream of vapor was introduced into the RPB through the vapor inlet located at the outer edge of the casing. It flows radially inward through the packing and leaves the RPB through the eye of the rotor. The vapor stream leaving the RPB was condensed in the overhead condenser and the condensate was collected in the reflux tank. A constant mass of the liquid in the reflux tank was used to control the reflux pump and to ensure operation under total reflux. Online measurement of the flow rate and density of the liquid reflux was obtained using a Coriolis mass flow meter. Once a steady state was reached, liquid samples were drawn from the liquid reflux, the bottom stream of the RPB, and from the reboiler. A steady state was assumed once the temperature change was less than 0.3 °C, and the density change of the liquid reflux was less than 0.5 kg m^{-3} over 10 min. The liquid samples were further analyzed with the Mettler Toledo-Densito 30 PX. The first steady state is achieved in approximately two hours and further, it took approximately 20–30 min to achieve a new steady-state operation after modifying the rotational speed. Though the longer initial period for obtaining the steady state can be assigned to the heat uptake by the solid steel casing of the RPB, the short periods for reaching the new steady state indicate the capability of the RPB to quickly adjust to changes during operation. Each experiment was performed twice. Initially, the rotational speed was increased stepwise from $1\ s^{-1}$ to $20\ s^{-1}$, while afterward the rotational speed was reduced stepwise in the same intervals to complete the cycle and examine potential hysteresis effects.

6.3.3 Discussion of experimental results

Influence of rotational speed and packing

As often said about RPBs that rotational speed is an additional degree of freedom that can be used to manipulate the operating window or the separation efficiency [6, 11, 30]. A validation of this can be seen in Fig. 6.3 where the number of theoretical stages for knit mesh packing and empty rotor as a function of rotational speed is plotted.

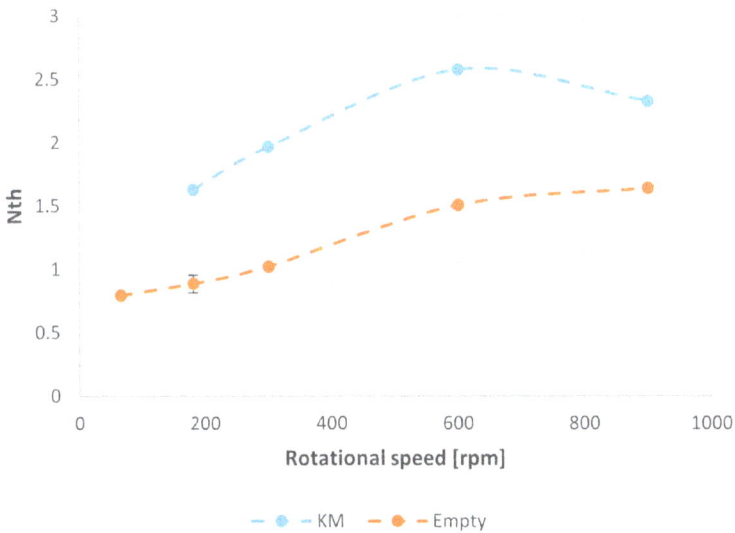

Fig. 6.3: Number of theoretical stages for the knit mesh (KM) packing and the empty rotor (400 mm OD) at an F-factor 2.1 $Pa^{0.5}$.

In many publications, mostly for absorption/desorption, the separation efficiency is shown to be always increasing with the rotational speed [4]. However, this is not generally the case with distillation as shown in Fig. 6.3 where the separation efficiency initially increases and then decreases with rotational speed (the case of the packed rotor). This maximum in separation efficiency was also reported by other authors [5, 6, 30, 31]. Qammar et al. [5] interpreted this maximum in separation efficiency as a balance between the benefits of high turbulence and low residence time. The increase of centrifugal acceleration with the rotational speed changes the hydrodynamics in the packing; the liquid flow in the packing changes from pore flow to droplet flow and finally to film flow [30]. This more uniform liquid distribution leads to larger interfacial areas between the vapor and the liquid phase at higher rotational speeds which increases the mass transfer efficiency. However, after a certain rotational speed, the separation efficiency starts to decrease with increasing rotational speed. This limitation in separation efficiency can be explained due to lower residence times in the apparatus for rising velocities which lead to a decrease in mass transfer. **The maximum separation efficiency for distillation in an RPB is reached when the benefits due to the higher interfacial area balance out the loss of separation efficiency caused by the lower time available for mass transfer in the packing.** Additionally, it must be considered that the higher the rotational speed, the higher the electricity cost, so an optimum point must be investigated for the given system and operating conditions.

Figure 6.3 also shows how beneficial packing can be in the separation of a particular test mixture. It is considered that the overall separation efficiency in an RPB is because of the packing mounted on the rotor. However, in general, all the areas in an RPB having gas-liquid contact can contribute to the overall fluid separation in an RPB. The question arises, how much is the contribution of packing and other areas? To determine the contribution of all the areas, specialized techniques are required, e.g., sample withdrawal from different points of the rotor or measurement of temperature profile on the rotor (Section 6.5). However, in this section, only the contribution of the packed rotor is evaluated by conducting two experiments: one without packing (empty rotor) and one with packing. These experiments will help in understanding the benefits of packing as well as its contribution to the total separation in an RPB. It also will eventually help in developing reliable models (Chapter 8) and design rules for RPBs.

Figure 6.3 shows that the separation efficiency of the whole device i.e., the packing and the rotor is higher than the empty rotor. An interesting fact is that the empty rotor also provides certain separation efficiency almost equivalent to 1.6 theoretical stages at 900 rpm. It seems to be resulting from the surface area provided by the rotating plates and the casing. As the liquid is sprayed on the empty rotor plates, it might form films, rivulets, or droplets depending on the flow regime and hence mass transfer takes place even without packing [5, 6, 30]. Furthermore, these droplets/film/rivulets pass through the space between the static casing and the rotating plates after leaving the rotor and eventually splash against the static casing walls. Since the casing is filled with vapor, mass transfer takes place in the casing as well.

When the packing is added on the empty rotor plates then the surface area provided for mass transfer increases as well as packing introduces more turbulence to the flow of liquid/vapor that most likely results in smaller droplet size and hence the overall separation efficiency of RPB increases. Based on the given conditions (knit mesh packing and flat fan nozzles) a packed rotor provides only one additional theoretical stage as compared to an empty rotor. This is an important point to be taken into account while designing the packings for RPBs since there seems to be a lot of potential to enhance the separation efficiency of the packing. To this point, it appears that the potential of the high surface area of packings is still not fully exploited.

Influence of radial packing depth

In the previous section, it is shown that the empty rotor also contributes to the overall mass transfer. An interesting question that may arise is does each radial segment of packing length contribute equally to the mass transfer and if not then how is the contribution of various packing segments from the inner to outer side of the rotor? Moreover, the general question is how to scale up radial packing length? One way could be to perform multiple experiments with the varying radial length of the packing or by withdrawing samples at different radial positions of the rotor. However, due to rotation, withdrawing samples from the rotor could be a very challenging task.

Chen et al. [17] studied the mass transfer efficiency of an RPB with various inner and outer radii. They investigated mass transfer coefficients as a function of liquid flow rate, rotational speed, the inner and outer radius of the packed bed, and radial position of the bed during various experiments with deaeration of water. Experimental results showed that the liquid-side mass transfer coefficient, k_La, increased with increasing rotational speed and liquid flow rate. Furthermore, they correlated the increased k_La at decreased packing volume with the more prominent end effects (see Chapter 3) at reduced packing volume and proposed a correlation predicting k_La taking end effects into account.

Similarly, Groß et al. [11] evaluated the scale-up of packing in an RPB by investigating the deaeration of water by nitrogen using two different radial packing lengths. They also showed that increasing the packing volume improves the overall mass transfer coefficient in a way that a specific degassing value can be achieved with a significantly lower rotational speed. However, the k_La value reduces significantly for larger radial packing lengths showing the flow nonuniformities along the radial length of the packing. Such a study is rarely found for distillation. Therefore, to evaluate the effect of increasing packing length on the separation efficiency of distillation, several experiments were conducted with the following packing lengths: 0 m (empty rotor), 0.1 m, 0.125 m, and 0.19 m. Chromium-nickel metal foam NCX1116 by

Recemat is used as the packing. The inner radius of the packing is kept constant at 0.073 m while the outer radius of the packing was varied as mentioned above. The whole metal foam packing is cut into rings to facilitate the investigation with different packing lengths as shown in Fig. 6.4.

Fig. 6.4: Chromium-Nickel metal foam NCX1116 cut into five rings.

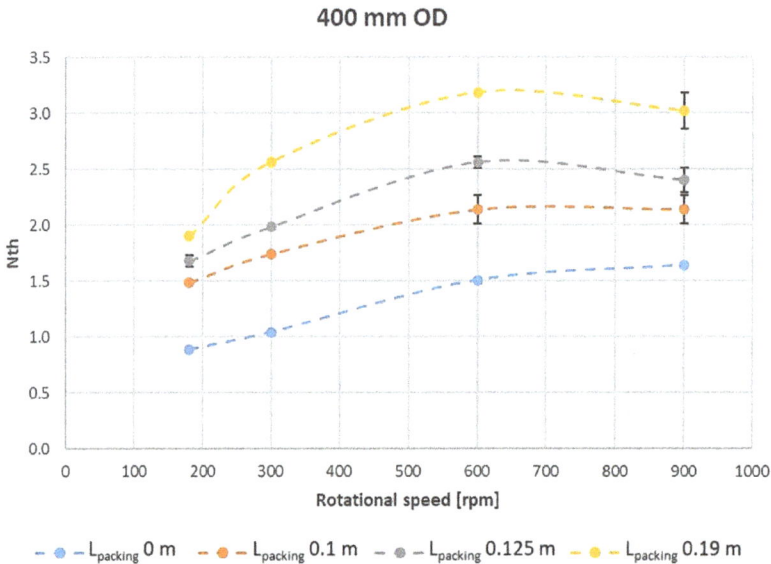

Fig. 6.5: Effect of different radial packing lengths on the separation efficiency of metal foam NCX1116 at an F-factor of 2.1 $Pa^{0.5}$ and 400 mm OD rotor.

Figure 6.5 shows the number of theoretical stages gained for different packing lengths. For all the investigated packing lengths, the separation efficiency increases with increasing rotational speed and reaches a maximum value. Furthermore, larger radial packing lengths lead to higher separation efficiencies, the maximum value of about 3.2 theoretical stages is achieved for the full metal foam ($L_{packing}$ 0.19 m) at 600 rpm. Larger radial packing lengths mean larger packing volume inside the rotor which provides a larger surface area for the mass transfer and thereby enhances the fluid separation efficiency. **Even though the mass transfer improves for larger radial packing depths, it seems possible to reach the same separation efficiency in some cases with less or even without packing by increasing the rotational speed.** For example, in Fig. 6.5, the separation efficiency of 1.5 theoretical stages can be achieved with each rotor configuration at a different rotational speed. For a radial packing length of 0.19 m, this efficiency can be achieved at a rotational speed of about 100 rpm; for the smaller radial packing length of 0.1 m, the required rotational speed is around 200 rpm whereas for the empty rotor a speed of 600 rpm is needed. These results also demonstrate one of the promising advantages of RPB which is the flexibility of operation. The results indicate that **in some of the rotor configurations the same separation efficiency can be reached with a small amount of packing by spinning it faster.**

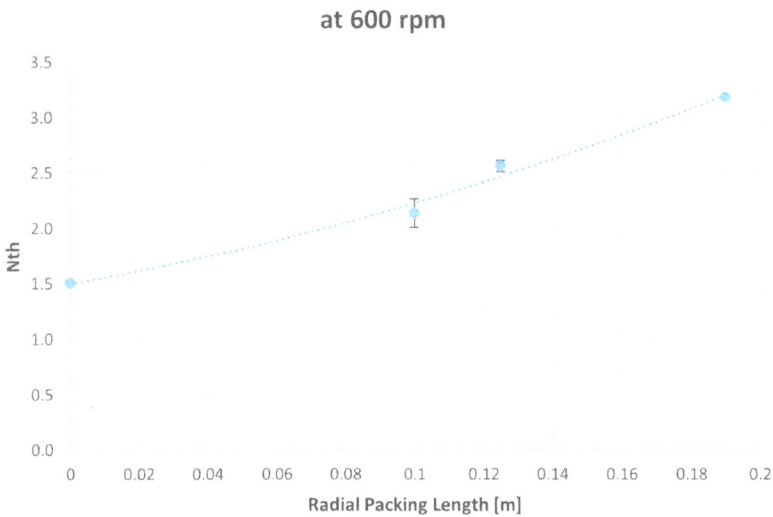

at 600 rpm

Fig. 6.6: Dependence of number of theoretical stages on radial packing length at 600 rpm for a 400 mm OD rotor at an F-factor of 2.1 $Pa^{0.5}$.

Figure 6.6 illustrates how the radial packing length effects the separation efficiency measured in terms of the number of theoretical stages at 600 rpm. With a radial packing length of 0.1 m, 2.1 theoretical stages are achieved, and by almost doubling the radial packing length, i.e., 0.19 m, an improvement of approx. 50% in the number of theoretical stages is observed. These results indicate that the increase in the radial packing length contributes to the overall mass transfer not very effectively. The increasing cross-sectional area, decreasing liquid load, and maldistribution of liquid seem to be the factors making the outer part of the packing not very effective [32] (see Chapter 3). There is room for improvement of packing design to balance the loss of mass transfer due to variations in the outer parts of the packing as also addressed in Chapter 7.

6.4 Comparison with the conventional column and RZB

One of the major advantages of rotating packed beds compared to conventional columns is the possible size reduction. To be able to evaluate how large this potential is, the separation efficiency of an RPB can be compared to the separation efficiency of a conventional column. For conventional distillation columns, HETP values are a common parameter used to compare the performance of different types of packings along with pressure drop. However, **the same HETP concept may not be reasonable to use for the packings where the fluid loading is changing continuously** like isotropic RPB packings. In this regard, a new type of packing called Zickzack (ZZ) packing was introduced [6] that uses the concept similar to the tray of a distillation column. The concept of Zickzack packing is also similar to the zigzag design in a rotating zig-zag bed (RZB) with the difference that in an RPB the whole packing rotates with the rotor like a solid block. Additive manufacturing was used [33] to print this packing with a constant cross-section area achieved by reducing the radial distance between two consecutive baffles from the inner to the outer side of the rotor, and hence the fluid loading also remains constant along the radial packing length. Details of the additive manufacturing procedure as well as the Zickzack packing design are further discussed in Chapter 7.

The initial distillation trials with the Zickzack packing showed an improved mass transfer performance as compared to the random RPB packings, namely, wire mesh and metal foam due to higher liquid holdup and more uniform hydrodynamic conditions [6]. Moreover, they showed comparable mass transfer performance to the industrially applied variant in China, i.e., RZB, at a reduced pressure drop. With these types of packing having constant cross-sectional and consequently constant fluid loading, the concept of HETP values can be used. As also described in Chapter 3, **packing is not the only part of an RPB providing surface area for mass transfer,** but the casing also contributes to the overall mass transfer. Qammar et al. [6] estimated that the casing contribution to the overall mass transfer is equivalent to one theoretical stage and redefined the HETP value [6] for RPBs as

$$\text{HETP} = \frac{(r_o - r_i)_{\text{packing}}}{N_{\text{th}} - 1} \tag{6.5}$$

where r_o and r_i are the outer and inner radius of the packing respectively. Figure 6.7 shows the comparison of different column packings (Raschig Super-Ring made of metal, HOLPACK) [34] with RPB packing (ZZ) [6] and two rotor variations of RZB [35]. The experimental data for the column resulted from the binary distillation of ethanol-water mixture under total reflux conditions in a column with an inner diameter of 0.213 m and a packing height of 2.8 m [34]. Figure 6.7 shows that **compared to a conventional column a reduction of the HETP value of 4–10-fold can be achieved with an RPB or RZB. The reduction of HETP values relates to reduced packing volumes required in an RPB as compared to the column** which opens the possibility of using dedicated packing for certain processes regardless of the high cost of this specific packing.

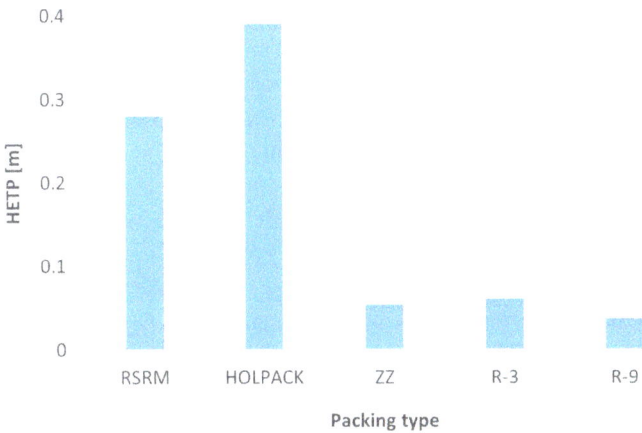

Fig. 6.7: Comparison of HETP values of static column with different packing and rotor types in an RPB; F-factor = 0.6 Pa$^{0.5}$ for RSRM and HOLPACK; F-factor = 1.6 Pa$^{0.5}$ for ZZ, R-3 and R-9 [6].

Using eq. (6.7), HETP values are calculated for ZZ, R-3, and R-9 and as can be seen from Fig. 6.7, an HETP value of 5.3 cm is calculated for the ZZ packing at the maximum separation efficiency, which is comparable to the two variants of RZB. However, in an RZB an opposing set of rotating and static baffles are used that causes increased friction in an RZB and results in a higher pressure drop and power consumption as compared to the single block RPB. For the two rotor variations of the RZB shown in the above figure a pressure drop of approx. 138 – 158 Pa per theoretical stage is reported at an F-factor of 1.6 Pa$^{0.5}$ [35], as compared to a pressure drop of only approx. 14 Pa per theoretical stage, for the ZZ packing [6]. Additionally, the electrical power consumption of an RPB at the rotational speed giving the highest separation efficiency, i.e., 181 rpm for ZZ is 83 W, while the estimated power consumption from the published data of the RZB (R-3 and R-9) is approx. 250 W [6].

Recently, Li et al. [36] introduced **a modified version of RZB called compound RZB (CRZB)** which consists of several perforated concentric circular ripple disks mounted on a perforated rotating ring at the inner part of the rotor. Moreover, a stationary four-armed liquid distributor is used in CRZB. The construction of the outer part of the rotor is similar to that of the RZB. It is reported that **the end effects of CRZB are stronger than the RZB thus leading to the higher interfacial area and enhanced mass transfer [36].** The modification at the inner part of the rotor leads to lower gas flow resistance and eventually lower pressure drop as compared to RZB. The authors propose it as the **second-generation RZB** that can be used commercially [36].

6.5 Distribution of mass transfer in RPBs: end effects

In an RPB, the liquid droplets leave the rotor at high speed into the casing that is filled with the vapor phase. Therefore, the two phases are also in contact with the casing. Similarly, the reflux liquid sprayed on the packing at the inner periphery contacts the vapor phase leaving the rotor at the inner periphery, i.e., the eye of the rotor. Therefore, as stated in Chapters 3 and 5, these end effects have to be accounted for, and assigning the total mass transfer achieved in an RPB solely to the packing results in an overestimation of the packing performance. Hence the mass transfer correlations and scale-up rules should be developed accordingly. The mass transfer in an RPB can be divided into three zones (Chapter 3): the central zone, the packing zone, and the casing zone. An experimental study was conducted by Qammar et al. [6] on the distribution of mass transfer in RPBs by using the temperature measurements at the vapor inlet, vapor outlet as well as in casing [5]. However, due to the lack of temperature measurements on the rotor, the mass transfer could only be measured in two zones, i.e., the packing and the casing. Since for binary distillation at total reflux and atmospheric pressure, composition data can be estimated using temperature measurements. The authors illustrated some very interesting trends for a low F-factor of 0.35 $Pa^{0.5}$ and 0.65 $Pa^{0.5}$, noticing that at higher rotational speeds almost 80% of the mass transfer in the specific experiments took place in the casing. Furthermore, they showed that the mass transfer decreases in the packing and increases in the casing with increasing rotational speed. Basically, the mass transfer is shifted from the packing to the casing at a high rotational speed [5]. The reasons for the poor packing performance at high rotational speeds are not fully understood; however, the decrease of liquid holdup or residence time in the packing with increasing rotational speed seems to be one of the obvious reasons [5].

For an improved insight, temperature measurements using a tailored telemetry setup for distillation in an RPB were developed for the first time [37]. The online measurement of temperature alongside the radial packing depth is not only helpful to understand the contribution of the packing to the total mass transfer but will also help determine the changes in the radial concentration profile.

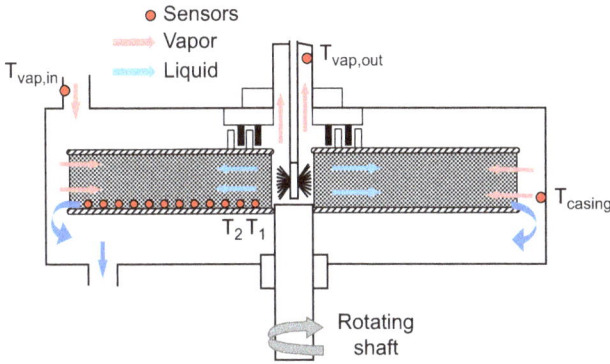

Fig. 6.8: Schematic of RPB showing temperature measurement points. T1–T12 are the wireless temperature sensors and $T_{vap,in}$, $T_{vap,out}$, and T_{casing} are the thermocouples used at the vapor inlet, vapor outlet, and in the casing [37].

Figure 6.8 shows the schematic of the telemetry setup. Twelve temperature sensors were mounted on the rotor for wireless transmission of signals that allow monitoring of the temperature profile in the rotor for different operational (e.g., rotational speed, F-factor) and design variables (e.g., packing types, nozzle types). The PT1000 sensors used in this setup have an accuracy class A ($\pm 0.15°C + 0.002*|T|$). The outer diameter of the rotor used for telemetry setup was 600 mm. The remaining details of the experimental setup for total-reflux distillation at atmospheric pressure in a single-stage RPB are the same as described in Section 6.3.2 [6]. Figure 6.9 shows the total number of theoretical stages achieved in an RPB with an empty rotor having a 600 mm outer diameter rotor at an F-factor of 2.1 $Pa^{0.5}$. As it can be seen clearly that initially the separation efficiency increases and then after 600 rpm it decreases with rotational speed. Figure 6.10 shows the radial temperature profile of the same experiment at 600 rpm and 900 rpm at atmospheric pressure. The temperature reading (T_{12}) at a radial length of 253 mm shows the temperature of the ethanol-water vapor mixture entering the rotor of the RPB from the casing. The radial temperature profile of 900 rpm has an overall slightly higher temperature than the temperature profile at 600 rpm which indicates a slightly poor packing performance which is also reflected in the total separation efficiency results in Fig. 6.9. A higher vapor temperature at 900 rpm indicates a lower ethanol composition in the vapor as compared to the liquid, therefore showing relatively poor separation as compared to 600 rpm. The temperature of vapor entering the casing of the RPB is measured to be approx. 95 °C using the $T_{vap,in}$ temperature sensor. As the vapor enters the casing of the RPB, the temperature changes because of the mass transfer in the casing, and it is also evident from Fig. 6.10.

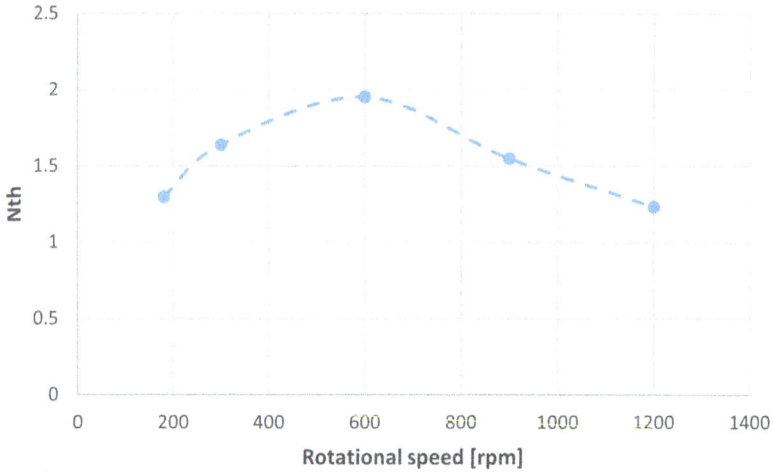

Fig. 6.9: Total separation efficiency (rotor + casing) of an RPB with an empty rotor having twelve temperature sensors at an F-Factor of 2.1 $Pa^{0.5}$, atmospheric pressure, and 600 mm OD rotor.

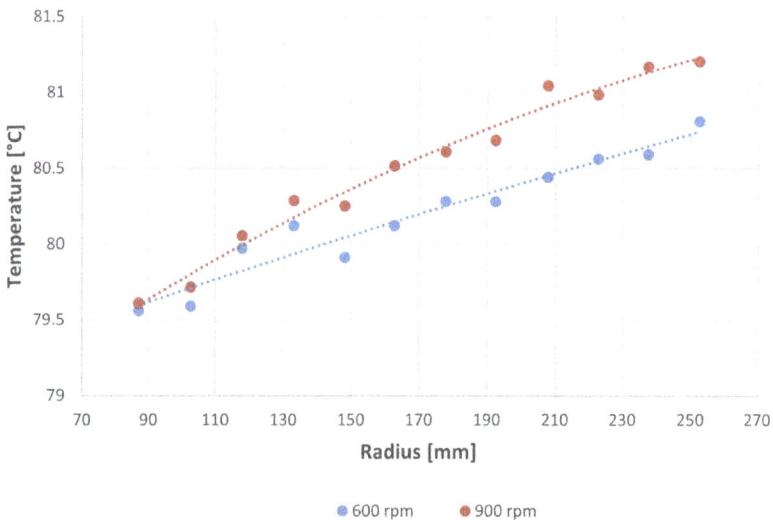

Fig. 6.10: Measured temperature profile at 600 rpm and 900 rpm in the empty rotor at 2.1 $Pa^{0.5}$ F-factor, 600 mm OD rotor.

Depending on the rotational speed the vapor entering the rotor has a temperature between 80.7–81.2 °C. The temperature of vapors entering the rotor is measured with a T_{12} sensor mounted on the outermost part of the rotor at a radial distance of 253 mm. The temperature-composition data of the ethanol-water system at atmospheric pressure gives an estimation of the change in the composition of ethanol in different sections of RPB. For example, the difference in composition of vapors determined at $T_{vap,in}$ and T_{12} gives an idea of how much mass transfer takes place in the casing. Similarly, the change in vapor composition from T_{12} to T_1 can be related to mass transfer in the packing. These calculations reveal that for an empty rotor at an F-factor of 2.1 Pa$^{0.5}$ majority of the mass transfer takes place in the casing at 600 rpm and 900 rpm. Hence, this supports the theory that the rotor is not the only part contributing to the mass transfer in an RPB. Furthermore, the **temperature profile in the rotor can also help in understanding the so-called end-zone effects.** On the inner radius of the rotor, close to the eye, the temperature profile for 900 rpm is steeper as compared to the outer radius of the rotor. This could potentially be an indication that **the inner few centimeters of the rotor (or packing) are contributing to the mass transfer more effectively than the rest of the packing.** Moreover, these kinds of measurements also reflect how the rotational speed affects the mass transfer in the rotor as well as the in the casing. Similarly, radial temperature profiles for different kinds of packing will be helpful in the development of packing design as well as in developing reliable design and scale-up rules for RPBs.

6.6 Mass transfer modeling

Modeling of mass transfer in an RPB is a challenging task as the mass transfer depends on many variables, for instance, the radial packing length, the fluid loads, the rotational speed, and the characteristics of the packing. Parameters such as the cross-sectional area and the centrifugal force vary along the radius of the packing while they remain constant in conventional columns. Increasing the rotational speed leads to stronger centrifugal fields, thus the liquid flows through the packing mainly as thin liquid film or tiny droplets and provides a larger interfacial area and decreased mass transport resistance. On the contrary, high centrifugal forces cause lower residence times because the liquid velocity increases. These shorter contact times between the vapor and the liquid phase may lead to a decreased overall mass transfer in the RPB [3, 6]. These kinds of irregularities make the modeling of mass transfer in an RPB a very challenging task. Most of the correlations on the mass transfer modeling in an RPB for distillation are derived from absorption or stripping processes. There are very limited studies dealing with distillation in RPBs. The few available studies mostly focus on gas-side mass transfer coefficient as this is limiting for most distillation processes [38–41].

The HETP concept in a conventional column assumes that the mass transfer is equally distributed over the radius of the column; however, in an RPB the cross-sectional area increases with the radius, thereby reducing the mass transport efficiency in a fixed radial length. To overcome this problem Singh et al. [41] proposed a concept similar to the height of a transfer unit (HTU) and a number of transfer units (NTU) concept for columns.

$$\pi \left(r_o^2 - r_i^2\right) = ATU \cdot NTU \qquad (6.6)$$

The height is replaced by the area of the packing that is equal to a transfer unit. The area in the ATU value refers to the area of a circular ring the gas passes if the packing is considered from the top view and not the cross-sectional area.

Later Kelleher and Fair [3] adapted the concept and defined ATU_g and NTU_g for the vapor-side according to eqs. (6.7) and (6.8), respectively, deriving from a material balance over a differential volume element of the rotor in the RPB:

$$ATU_g = \frac{\dot{m}_g}{\rho_g \cdot h \cdot K_g a} \qquad (6.7)$$

$$NTU_g = \int_{y_{in}}^{y_{out}} \frac{dy}{(y^* - y)} \qquad (6.8)$$

At total reflux the composition of the top vapor coming out of the RPB, and the reflux liquid are the same, i.e., x equals y, thus the NTU_g value can be calculated with eq. (6.8) using the sample-based distillate and bottom concentrations. The concept can also be adapted for the liquid-side mass transfer resulting in ATU_l and NTU_l and given by the following equations:

$$ATU_l = \frac{\dot{m}_l}{\rho_l \cdot h \cdot K_l a} \qquad (6.9)$$

$$NTU_l = \int_{x_{in}}^{x_{out}} \frac{dx}{(x - x^*)} \qquad (6.10)$$

Kelleher and Fair [3] also presented a correlation for the vapor-side mass transfer coefficient, $k_g a$, based on the model for packed columns by Onda et al. [38]:

$$k_g a = c_1 \cdot \left(\frac{a_p \cdot D_g}{d_p}\right) \cdot Re_g^{c_2} \cdot Gr_g^{c_3} \cdot Sc_g^{c_4} \qquad (6.11)$$

where a_p is the specific surface area of the packing, D_g represents the diffusion coefficient of the vapor phase and d_p is the effective diameter of the packing [4]. The latter can be determined as follows [4]

$$d_p = \frac{6 \cdot \left(1 - \varepsilon_p\right)}{a_p} \qquad (6.12)$$

Therein ε_p is the packing porosity. Mondal et al. [4] used the ATU-NTU concept to calculate the overall volumetric mass transfer coefficient $K_g a$ and proposed a mass transfer correlation that relates the vapor load and the centrifugal acceleration to the overall volumetric mass transfer coefficient $K_g a$

$$K_g a = c_1 \cdot (F - factor)^{c_2} \cdot (\omega^2 \cdot r_{av})^{c_3} \tag{6.13}$$

where r_{av} is the average radius of the packing and $\omega^2 r_{av}$ denotes the centrifugal acceleration. They compared the experimental $K_g a$, determined by eq. (6.7) using a split packing and distillation of methanol-ethanol mixture in an RPB, with the theoretical $K_g a$ estimated by eq. (6.13) and found the results to lie within a range of ± 30 % deviation.

Sudhoff [9] presented a generalized correlation for the volumetric mass transfer coefficient as used in many of the studies. For the liquid-side mass transfer coefficient, the index "j" becomes "l" and for the vapor-/gas-side the index "j" becomes "g" [9]

$$k_j a = C \, Re_j^{z_1} Gr_j^{z_2} Sc_j^{z_3} We_j^{z_4} \left(\frac{a_p \, D_i}{d_p} \right)^{z_5} \tag{6.14}$$

The Reynolds number Re_j, the Grashof number Gr_j, the Schmidt number Sc_j, and the Weber number We_j are dimensionless.

Due to limited information on the distribution of mass transfer in an RPB almost all these correlations were applied attributing the total mass transfer in an RPB to the packing only. However, as discussed in Section 6.5 that the total mass transfer in an RPB can be divided into different sections. Furthermore, **the separation efficiency of distillation in an RPB is not always increasing with rotational speed rather a maximum in separation exists that needs to be addressed in the mass transfer correlations.** Another rate-based modeling approach based on non-equilibrium stages is presented in Chapter 8 addressing the above-mentioned issues. In this regard, a few studies [6, 11, 30, 42] have also recently started developing the mass transfer correlations accordingly. With this development, the correlations can be used to predict the mass transfer performance more reliably and hence can pave the way to the industrial applications of this technology.

6.7 Take-home messages

- The mass transfer efficiency of an RPB can be altered by manipulating the rotational speed.
- The mass transfer efficiency does not always increase with rotational speed for distillation; a maximum exists.
- Fluid separation also takes place without packing in an RPB, i.e., in an empty rotor

- The static casing also contributes to the total mass transfer in an RPB, and the contribution of the static casing increases with increasing rotational speed.
- The inner few centimeters of the packing contribute more effectively to the mass transfer as compared to the rest of the packing; "end-effects."
- It is not reasonable to define the mass transfer efficiency of a random packing in RPB using parameters like HETP.
- The concept of constant cross-sectional area along with the radial packing depth as used in the Zickzack packing is helpful for the scalability purpose as well as for comparison with conventional columns and RZB.
- Telemetry is a useful technique to understand the fundamentals of mass transfer in an RPB as well as the mass transfer zones.
- The temperature profile in the rotor varies with the rotational speed.
- The radial temperature profile in the rotor helps in explaining the variations of the total mass transfer efficiency of an RPB with the rotational speed.
- The increasing rotational speed causes a steeper temperature profile, especially at the inner side of the rotor.
- The available mass transfer correlations need to be modified considering the maximum in separation efficiency as well as the end effects and casing contribution.

6.8 Quiz

1. Can you name a few random and structured packing used in an RPB?
2. Why in random RPB packings the fluid loading change from the inner to the outer side of the rotor?
3. What is the main concept behind the development of Zickzack packing?
4. How does the separation efficiency of distillation in an RPB changes with rotational speed?
5. It is found that in some cases after a certain rotational speed the mass transfer in the "packing" decreases with further increase of rotational speed. What could be the reason for this?
6. What could be the reason for the decrease in overall separation efficiency after the maximum in an RPB?
7. Can you name at least one technique that can be used to analyze the mass transfer in the "rotor" of an RPB?
8. Which parameter(s) can be used for comparing the efficiency of conventional columns and RPBs and under what conditions?
9. The formula used for one of the parameters for comparing the separation efficiency of conventional columns and RPBs is updated recently. Write down the modified form of the formula.

10. In most of the available correlations for estimating the mass transfer in an RPB, which two important considerations are missing?

6.9 Exercise

Exercise 1. Distillation is to be carried out in an RPB with 10 wt.% ethanol-water solutions to determine the temperature distribution in the RPB for a specific rotor configuration. A few temperature sensors are installed at various positions in the RPB. The temperature of the vapors at the vapor inlet is measured to be 95 °C. As the vapors enter the casing of the RPB, the temperature drops to 81.5 °C because of intense mixing and mass transfer. Three temperature sensors are installed in the rotor. The temperature sensor installed at the eye of the rotor shows a reading of 79 °C, whereas sensors at the middle and outer radial length show a temperature of 80 °C, and 80.5 °C respectively. Qualitatively draw the temperature profile in the RPB indicating all the temperatures at different positions.

Exercise 2. A binary mixture of 5 wt.% ethanol and water containing a small amount of high boiling corrosive additive need to be separated using distillation. To obtain the required ethanol purity in the product, seven theoretical stages are required. After some preliminary investigations, two types of equipment are shortlisted, i.e., conventional column and RPB. It appears that the cost of packing will have a significant impact on the choice of the equipment as the packing needs to be coated with a very special type of corrosion-resistant material due to the highly corrosive nature of the additive in the process fluids. Assume that RSRM packing is chosen for the column and ZZ is selected for the RPB. Irrespective of the type of packing, the price of applying the protective coating on the packing is 5,000 €/m^3. Assume that the actual price of packing is negligible as compared to the price of the coating material. Initial estimations show that the diameter of the conventional column needed to fulfill this task is 0.2 m and the inner diameter of the rotor of RPB needed for this process is 0.146 m. The axial height of the rotor is 10 mm. Using Fig. 6.11, estimate the cost of packing for both the equipment.

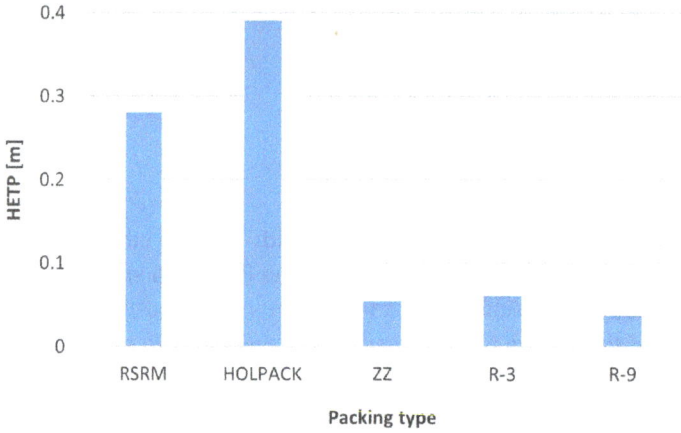

Fig. 6.11: HETP values of different packings used in the conventional column and RPB (cf. Figure 6.7).

6.10 Solution

Solution to Questions 1–10: The answer to questions can be found from the main body of this chapter.

Solution to Exercise 1: The temperature profile can be represented as shown in Fig. 6.12.

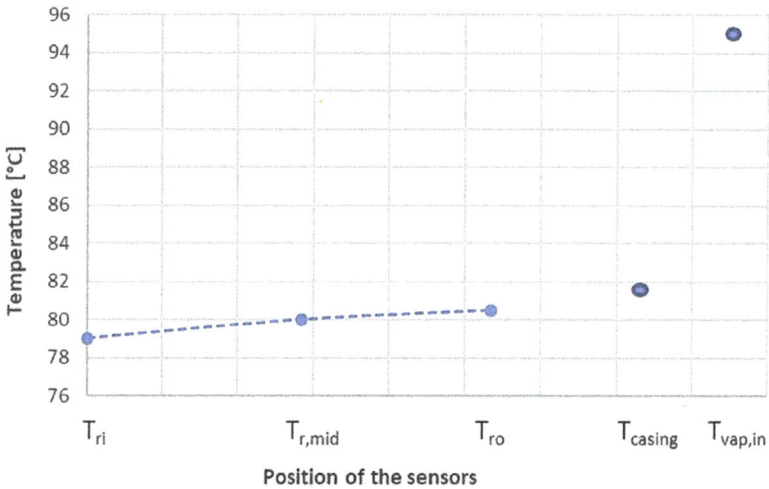

Fig. 6.12: Temperature profile in an RPB.

Solution to Exercise 2: To estimate the cost of the packing, the volume of the packing needs to be calculated using the HETP values given in Fig. 6.11. The volume of the packing for the column can be estimated as follows:

$$V_{col,packing} = \pi \frac{(d)^2}{4} h_{col,packing} \tag{6.15}$$

$$h_{col,packing} = HETP_{RSRM} \times Nth = 0.28m \times 7 = 1.96m \tag{6.16}$$

$$V_{col,packing} = \pi \frac{0.2^2 m^2}{4} 1.96m = 0.062m^3 \tag{6.17}$$

As known, that for RPB the height of packing needed for separation is the radial packing length from the inner to the outer radius of the rotor. Therefore, using eq. (6.5) and Fig. 6.11:

$$\frac{(r_o - r_i)}{Nth - 1} = 0.053m \rightarrow (r_o - r_i) = 0.32m \tag{6.18}$$

Since r_i is 0.073 m, therefore

$$r_o = 0.393m \rightarrow d_o = 0.786m$$

$$V_{RPB,packing} = \pi \frac{(d_o^2 - d_i^2)}{4} h_{RPB,packing} \tag{6.19}$$

$$V_{RPB,packing} = \pi \frac{(0.786^2 - 0.146^2)m^2}{4} 0.01m = 0.005m^3 \tag{6.20}$$

$$\text{Cost of column packing} = V_{col,packing} \times \frac{€}{m^3} \cong 308 € \tag{6.21}$$

$$\text{Cost of RPB packing} = V_{RPB,packing} \times \frac{€}{m^3} \cong 24 € \tag{6.22}$$

The calculations show that the cost of RPB packing will be lower.

List of symbols

Latin letters

A	surface area	m^2
a_p	specific geometric packing surface area per unit	$m^2\,m^{-3}$
d	diameter	m
D	diffusion coefficient	$m^2\,s^{-1}$
d_p	equivalent spherical diameter, particle diameter	m
F-factor	gas capacity factor	$Pa^{0.5}$
F_G	gas capacity factor at the inner radius of the rotor	$Pa^{0.5}$
g	gravitational acceleration	$m\,s^{-2}$

h	axial length	m
K_ga	overall volumetric gas- or vapor-side mass transfer coefficient	s^{-1}
k_ga	volumetric gas side mass transfer coefficient	s^{-1}
$L_{packing}$	radial packing length	m
$\dot{m}$	mass flow rate	$kg\ s^{-1}$
r	radius	m
T	temperature	°C
u	Velocity	$m\ s^{-1}$
x	liquid molar fraction	–
y^*	equilibrium vapor composition	–

Greek letters

α	relative volatility	–
ε	porosity	–
ρ	density	$kg\ m^{-3}$
Ø	pore diameter of foam packing	mm
ω	angular velocity	$rad\ s^{-1}$

Subscripts

av	average
c	cylindrical
col	column
empty	without packing installed
g	gas
i	inner
l	liquid
o	outer
v	vapor

Dimensionless numbers

Gr	Grashof number	–
Re	Reynolds number	–
Sc	Schmidt number	–
We	Weber number	–

List of abbreviations

ATU	area of a transfer unit
CRZB	compound rotating zig-zag bed
HETP	height equivalent to a theoretical plate
HTU	height of a transfer unit
NTU	number of transfer unit

OD	outer diameter
RPB	rotating packed bed
RSRM	Raschig super ring metal
RZB	rotating zig-zag bed
ZZ	Zickzack packing

References

[1] Podbielniak WJ. Centrifugal counter current contact apparatus(US2004011); 1935; Available from: https://patents.google.com/patent/US2004011A/en.

[2] Wilhelm PC, Wilhelm DS. Apparatus for intimate contacting of two fluid media having different specific weight(US2941872A); 1960; Available from: https://patents.google.com/patent/US2941872A/en.

[3] Kelleher T, Fair JR. Distillation studies in a high-gravity contactor. Ind. Eng. Chem. Res. 1996;35(12):4646–55.

[4] Mondal A, Pramanik A, Bhowal A, Datta S. Distillation studies in rotating packed bed with split packing. Chem. Eng. Res. Des. 2012;90(4):453–57.

[5] Qammar H, Hecht F, Skiborowski M, Górak A. Experimental investigation and design of rotating packed beds for distillation. Chem. Eng. Trans. 2018;69:655–60.

[6] Qammar H, Gładyszewski K, Górak A, Skiborowski M. Towards the development of advanced packing design for distillation in rotating packed beds. Chem. Ing. Technik. 2019;91(11):1663–73.

[7] Neumann K, Gladyszewski K, Groß K, Qammar H, Wenzel D, Górak A, et al. A guide on the industrial application of rotating packed beds. Chem. Eng. Res. Des. 2018;134:443–62.

[8] Wenzel D, Górak A. Review and analysis of micromixing in rotating packed beds. Chem. Eng. J. 2018;345:492–506.

[9] Sudhoff D, Leimbrink M, Schleinitz M, Górak A, Lutze P. Modelling, design and flexibility analysis of rotating packed beds for distillation. Chem. Eng. Res. Des. 2015;94:72–89.

[10] Sudhoff D, Neumann K, Lutze P. An integrated design method for rotating packed beds for distillation. Comput. Aided Chem. Eng. 2014;33:1303–08.

[11] Groß K, de Beer M, Dohrn S, Skiborowski M. Scale-up of the radial packing length in rotating packed beds for deaeration processes. Ind. Eng. Chem. Res. 2020;59(23):11042–53.

[12] Agarwal L, Pavani V, Rao DP, Kaistha N. Process intensification in HiGee absorption and distillation: Design procedure and applications. Ind. Eng. Chem. Res. 2010;49(20):10046–58.

[13] Sudhoff D, Lutze P, Górak A Multi-stage-counter-current rotating-packed-bed for distillation. In: AIChE Spring Meeting 2013; 2013.

[14] Rao DP. The Story of "HIGEE". Indian Chem. Eng. 2015;57(3-4):282–99.

[15] Neumann K, Hunold S, Groß K, Górak A. Experimental investigations on the upper operating limit in rotating packed beds. Chem. Eng. Process. Process Intensif. 2017;121:240–47.

[16] Li Y, Li X, Wang Y, Chen Y, Ji J, Yu Y, et al. Distillation in a counterflow concentric-ring rotating bed. Ind. Eng. Chem. Res. 2014;53(12):4821–37.

[17] Chen Y-S, Lin -C-C, Liu H-S. Mass transfer in a rotating packed bed with various radii of the bed. Ind. Eng. Chem. Res. 2005;44(20):7868–75.

[18] Cortes Garcia GE, van der Schaaf J, Kiss AA. A review on process intensification in HiGee distillation. J. Chem. Technol. Biotechnol. 2017;92(6):1136–56.

[19] Zheng C, Guo K, Song Y, Zhou X, Al D, Xin Z, et al. Industrial practice of HIGRAVITEC in water deaeration. 2nd International Conference on Process Intensification in Practice / applications and opportunities 1997(28):273–88.

[20] Peel J, Howarth CR, Ramshaw C. Process intensification: Higee seawater deaeration. Chem. Eng. Res. Des. 1998;76(5):585–93.

[21] Harbold G, Park J. Using the GasTran® Deaeration System to achieve low dissolved oxygen levels for superior line speed and product quality: A case study in carbonated soft drink bottling. In: 7th International Conference on Process Intensification 2008.

[22] Xu ZC, Ji JB, Wang G, Li XH, Li Y. Rotating zigzag bed application in extractive distillation process of THF-methanol-water system. Mod. Chem. Ind. (China). 2012;32(6):94–96.

[23] Wang GQ, Xu ZC, Ji JB. Progress on Higee distillation – Introduction to a new device and its industrial applications. Chem. Eng. Res. Des. 2011;89(8):1434–42.

[24] Hangzhou Hecci Technology Co., Ltd. Distillation expert in the fine chemical and pharmaceutical industries: Manufacturer. [July 13, 2021]; Available from: https://www.global sources.com/si/AS/Hangzhou-Hecci/6008850191753/Homepage.htm.

[25] Zhao H, Shao L, Chen J-F. High-gravity process intensification technology and application. Chem. Eng. J. 2010;156(3):588–93.

[26] Wenzel D, Nolte K, Górak A. Reactive mixing in rotating packed beds: On the packing's role and mixing modeling. Chem. Eng. Process. Process Intensif. 2019;143:107596.

[27] Wang GQ, Xu OG, Xu ZC, Ji JB. New HIGEE-rotating zigzag bed and its mass transfer performance. Ind. Eng. Chem. Res. 2008;47(22):8840–46.

[28] Chen J, Liu YZ, Li XP. Distillation process of methanol in rotating packed bed. Chem. Ind. Eng. Prog. 2009;28(8):1333.

[29] Onken U, Arlt W. Recommended Test Mixtures for Distillation Columns. 2nd editon, Warwickshire: The institution of Chemical Engineers; 1990.

[30] Hilpert M, Repke J-U. Experimental investigation and correlation of liquid-side mass transfer in pilot-scale rotating packed beds. Ind. Eng. Chem. Res. 2021;60(14):5251–63.

[31] Chu G-W, Gao X, Luo Y, Zou H-K, Shao L, Chen J-F. Distillation studies in a two-stage counter-current rotating packed bed. Sep. Purif. Technol. 2013;102:62–66.

[32] Groß K, Bieberle A, Gladyszewski K, Schubert M, Hampel U, Skiborowski M, et al. Analysis of flow patterns in high-gravity equipment using gamma-ray computed tomography. Chem. Ing. Tech. 2019;134(6):443.

[33] Gładyszewski K, Skiborowski M. Additive manufacturing of packings for rotating packed beds. Chem. Eng. Process. 2018;127:1–9.

[34] Darakchiev S, Semkov K. A study on modern high effective random packings for ethanol-water rectification. Chem. Eng. Technol. 2008;31(7):1039–45.

[35] Wang GQ, Zhou ZJ, Li YM, Ji JB. Qualitative relationships between structure and performance of rotating zigzag bed in distillation. Chem. Eng. Process. Process Intensif. 2019;135:141–47.

[36] Li Y, Liu P, Wang G, Ji J. Enhanced mass transfer and reduced pressure drop in a compound rotating zigzag bed. Sep. Purif. Technol. 2020;250:117188.

[37] Hampel U, Schubert M, Döß A, Sohr J, Vishwakarma V, Repke J-U, et al. Recent advances in experimental techniques for flow and mass transfer analyses in thermal separation systems. Chem. Ing. Technik. 2020;92(7):926–48.

[38] Onda K, Takeuchi H, Okumoto Y. Mass transfer coefficients between gas and liquid phases in packed columns. J. Chem. Eng.Jpn. 1968;1:56–62.

[39] Keyvani M, Gardner NC. Operating Characteristics of Rotating Beds: Office of Scientific and Technical Information (OSTI). United States: U.S. Department of Energy. Office of Scientific and Technical Information. 1988.

[40] Vivian JE, Brian PLT, Krukonis VJ. The influence of gravitational force on gas absorption in a packed column. AIChE J. 1965;11(6):1088–91.

[41] Singh SP, Wilson JH, Counce RM, Lucero AJ, Reed GD, Ashworth RA, et al. Removal of volatile organic compounds from groundwater using a rotary air stripper. Ind. Eng. Chem. Res. 1992;31(2):574–80.

[42] Hilpert M, Calvillo Aranda GU, Repke J-U. Experimental analysis and rate-based stage modeling of multicomponent distillation in a rotating packed bed. Chem. Eng. Process. Process Intensif. 2021;108651.

Rouven Loll, Jörg Koop

7 3D printed packings for rotating packed beds

7.1 Fundamentals

In recent years, 3D printing of packings has become an integral part of packing development for rotating packed beds (RPBs) [1–5]. Since there are many possibilities arising from the 3D printing technologies, this chapter first explains the aims of packing design for RPBs. Afterward, a selection of the most relevant technologies for this purpose is presented. Furthermore, an overview of recent 3D printed packings in literature is given. As examples, the purposes of the specific structures of the Zickzack packing and the spiral-packing are identified. The corresponding design process of these packings and the dependence of their design on the 3D printer type used are explained in detail. Finally, two exemplary packings that have been successfully tested in pilot-scale RPBs are presented.

7.1.1 Motivation

The technical maturity of available packings for RPBs is still rather low compared to the packed columns. This gap can be attributed to the novelty of the RPB technology. While numerous research studies have been carried out on packed columns since the 1930s [6], the idea of exploiting high-gravity fields in rotating packed beds has only gained special attention over the last three decades [7]. Over time, a deep understanding of hydrodynamics and mass transfer in packed columns has been achieved and is available in almost each form and detail level. On the other hand, knowledge about packed RPBs is still quite fresh and far less distributed. Thus, mass transfer in RPBs is in most cases not as deeply understood as it is for conventional packed columns.

Unfortunately, the knowledge and techniques for engineering and construction available for packed columns cannot simply be transferred to RPBs. Partially some ideas and concepts may be re-used, but there are limitations arising from the RPB's geometry. As the cross-sectional area of RPBs increases linearly with increasing radii for isotropic porous packings like the most prominent knitted-mesh or metal foam packings (see Section 1.3.2) liquid and gas loadings are decreasing from the inner radius of a packing to the outer radius. Assuming that the gas inlet is at the outer radius of a packing and the liquid phase is distributed onto the packing at the

Rouven Loll, Jörg Koop, TU Dortmund University

https://doi.org/10.1515/9783110724998-007

inner radius, the gas load and its velocity increase while the liquid load decreases on their path through the packing. In contrast to that, the mean velocity of the liquid phase rises while the liquid phase moves through the packing due to the centrifugal field generated by the rotation of the packed bed.

Investigations about the liquid holdup (see Section 3.1.1) demonstrate that the holdup distributes unevenly over the radius of a packing. While there is a relatively high liquid holdup in the inner 2–3 centimeters of an isotropic packing, it is vastly reduced in the remaining packing. It is often reported that the mass transfer in an isotropic porous packing can be divided into an inner ring called the end zone with a high mass transfer and the rest of the packing called the bulk zone with significantly lower mass transfer [7]. Thus, even gas and liquid loading and holdup along the radius cannot be realized with an isotropic packing in an RPB [8]. That makes it difficult to tailor the dimensions of a packing to a specific separation task, especially when many theoretical stages need to be provided by the packing. **New kinds of packing structures should therefore be developed to provide a scalable packing design.**

Additive manufacturing (AM) offers the possibility to quickly manufacture new types of packings for RPBs. Using current state of the art technology, it is possible to model and print an entirely new kind of packing within a few days. That way, different packing designs can be tested and evaluated efficiently leading to an accelerated rate, in which RPB-technology is developed in research institutions and in industry. Furthermore, it provides a tool to quickly test packings that are tailored for a specific application in lab scale or pilot plant experiments. Depending on the morphology and the technology used, it may also be more cost-effective to manufacture a custom packing for a specific application using additive manufacturing rather than with traditional manufacturing processes.

7.1.2 Overview of suitable 3D printing technologies for packings production

In recent years, several publications have discussed the characteristics of different additive manufacturing (AM) technologies in the context of fluid separations. **The different AM technologies can be categorized into four types: liquid-based, powder-based, wire-based, and foil-based technologies** [9]. In addition to the manufacturing of packings [5, 10], they have also been used for manufacturing various equipment like microreactors, pumps, or separation membranes [9]. In this subchapter a summary of AM technologies suitable for the manufacturing of packings for RPBs is given.

The **liquid-based** technology provides the highest resolutions for printing [11] and was the first commercially available AM technology, starting with stereolithography (SLA) which was introduced in 1988 [9]. In SLA, a UV laser light is tracing the shape of the generated piece layer after layer onto a built platform in a bulk of photosensitive liquid resin. This induces a selective photopolymerization process in a discrete volume element of the liquid resin, leaving the hardened polymer in the desired shape. The maximum resolution is restricted by the laser spot diameter of the 3D printer [12]. Another liquid-based technology based on SLA is digital light processing (DLP). This technology differs from SLA in the light source and the way it is projected on the work piece. During DLP the digital light projector projects the entire image of one single layer simultaneously. Thus, the printing time only depends on the height of the printed model. The AM technology with the highest resolution is two-photon polymerization (TPP), where a photochemical or physical transformation within a transparent resin is induced by simultaneous absorption of two photons. With this technology, a reaction can be induced in a volume as small as a few atto-liters. In Inkjet 3D printing, light-curable resins are deposited on a table in a similar way as in a customary paper printer ink is deposited on paper [11].

In **powder-based** AM technologies, powders of polymers, metals or ceramics are used instead of a liquid raw material for printing. The particles are selectively bonded together with a liquid binder (binder jetting), sintered together in selective laser sintering (SLS), or melted together in selective laser melting (SLM) or electron beam melting (EBM). After finishing one layer, a powder delivery system adds new loose particles for the next layer on top the previously processed layer. The residual loose powder in the bed acts as support material [11].

Wire-based technologies, such as fused deposition modeling (FDM), use coiled printing material which is continuously fed to an extrusion head. The extrusion head is heated to melt the thermoplastic materials past their glass transition temperature. This way, the material can be deposited and afterward cures by cooling down again. Since they are comparably inexpensive, FDM printers generally find widespread applications even as desktop-scale printers in non-commercial use [11].

In the **foil-based** laminated object manufacturing (LOM), objects can be fabricated out of sheets from plastic or composite laminates. First, a new sheet of the material is placed on the build platform. The sheet is then bonded to either the base or the previous layer by a heated roller. An adhesive coating on one side of the layer is used for bonding. A laser traces the cross-sections of each layer and cross-hatches the excess area for later waste removal [11].

7.1.3 Comparison of AM technologies for manufacturing of packings for RPBs

It is important to note, that **it is not sufficient to focus on the printing resolution only, but also on the maximum piece size or the ability to position multiple structures relative to one another.** Last but not least the printing speed and availability of materials are crucial for rapid prototyping and evaluation [5]. The material selected should provide a sufficient chemical and mechanical resistance while also being stable at the temperature required for the application [5]. For example, the thermoplastic material used in FDM can lose its structural integrity when used in high-temperature applications such as distillation. When developing new packings, in some cases it might be beneficial to use a translucent printing material, as it allows an optical evaluation of the printing success. Additionally, a highly translucent and clear material provides the opportunity to investigate the flow pattern of the liquid inside the packing directly. It is, of course, necessary to have inspection windows or cameras to look inside the RPB for that. Due to the necessity to remove the remaining printing material from the inner spaces of a packing structure after the printing process, foil-based technologies are not recommended for most packings. Lastly, the surface properties of the selected material can have significant influence on the pressure drop and mass transfer of the packing [5]. A comparison of representative example-specifications from different manufacturers for the AM technologies mentioned is given in Tab. 7.1. It can be used to assess whether an AM technology might be suited for a desired application. If, for example, an equipment design department regularly needs to develop customized packings for different applications, DLP or SLA would be a good option. Both offer a relatively high printing speed, accuracy, and, depending on the desired application, a suitable photopolymer can be selected. Translucent polymers would also allow to evaluate the printing success and flow patterns within the various tailor-made inner packing structures. If, in another example, particularly long-lasting packings of high mechanical stability are required, metal might be the material of choice and the packings should rather be produced, e.g., in an SLS printer.

In general, DLP and SLA offer good resolutions and printing of small details. Inkjet 3D printing allows the use of a wide range of materials, but the removal of the essential support material for this printing technology can be a significant limitation, especially if the desired packing geometry contains a lot of less accessible cavities. The liquid-based SLA and the powder-based SLS have similar limitations in terms of printing precision. A major advantage of SLS is its applicability for processing functional polymers as well as metals. However, this comes at the expense of 5–10 times higher overall costs for equipment and printing material compared to SLA or DLP. Wire-based technologies like FDM offer the lowest printing costs but also a comparably low manufacturing precision for the manufacturing of packings for RPBs [5].

7.2 Applications

Gładyszewski and Skiborowski [5] have created the first additively manufactured packing for an RPB. It was a replica of a metal foam with a porosity ε of 92%, a specific surface area a_{pack} of 1,000 $m^2\,m^{-3}$, and an estimated pore diameter of 1.4 mm, as stated

Tab. 7.1: Specifications, materials, advantages, and disadvantages of various AM technologies, edited from [11].

Technology	Resolution	Maximum build-space [mm]	Printing speed	Material	Advantages	Disadvantages
Liquid based						
Stereo-lithography (SLA)	3D Systems: 50 μm (1.27 μm; by laser resolution)	3D Systems: 1,500 × 750 × 550	N/A	Photopolymers	1. Capable of printing complex geometry 2. Good accuracy	1. Requires support structures 2. Hazardous resins 3. Limited to curable materials 4. slow
Digital light processing (DLP)	15–150 μm	192x120x230	N/A	Photopolymers	1. Faster than SLA	1. Same as SLA 2. Small build size
Material jetting	3D systems: 750 × 750 × 890 dpi; 29 μm layers Stratasys: X/Y: 300 dpi; Z: 1600 dpi	3D systems: 517.78 × 380.75 × 294.39 Stratasys: 1000 × 800 × 500	3D systems: 5–15 mm/h	Photopolymers, metal, polyimide, etc.	1. Short build time 2. high resolution 3. pieces can be printed in range of colors	1. Limited mechanical properties 2. Relatively poor surface finish 3. Newly printed pieces require post-processing

(continued)

Tab. 7.1 (continued)

Technology	Resolution	Maximum build-space [mm]	Printing speed	Material	Advantages	Disadvantages
TPP	0.15 µm	100 × 100 × 3	100 µm/s	Photoresists	1. Nanometer resolution	1. Small build size 2. Expensive 3. Slow printing speed 4. Requires materials of high optical transparency
Powder based						
Binder jetting	Voxeljet: 200–600 dpi ExOne: X/Y: 0.0635–100 µm Z: 1 µm	Voxeljet: 4,000 × 2,000 × 1,000 ExOne: 2200 × 1200 × 700	Voxeljet: 12–36 mm/h ExOne: 20–400 L/h	Polymer, metals, alloys and ceramics	1. Short build time 2. Cheap 3. Does not employ heat 4. pieces can be printed in range of colors	1. Limited mechanical properties 2. Additional postprocessing can add significant time to the overall process 3. Not always suitable for structural pieces
Selective Laser Sintering (SLS)	3D systems: 80–150 µm EOS: Plastic: 60–180 µm Metal: 30–500 µm	3D systems: 550 × 550 × 750 EOS plastic: 700 × 380 × 560 EOS metals: 400 × 400 × 400	3D systems: 0.9–5 L/h EOS plastic: 7–48 mm/h	Thermoplastics, metal powders, ceramic powders	1. No need for post-curing (except for ceramic) 2. Do not require support structures	1. Porous surface 2. Poor surface finishes 3. May require long heating and cooling time

Powder based

	Build size	Z/layer	Build speed	Materials	Advantages	Disadvantages
Selective Laser Melting (SLM)	3D systems: 500 × 500 × 500 SLM: 500 × 280 × 325 Renishaw: 245 × 245 × 300	3D systems: 10–50 μm SLM: 20–200 μm Renishaw: 20–100 μm	3D systems: 55–80 cm3/h SLM: 20–105 cm3/h Renishaw: 5–20 cm3/h	Titanium alloys, cobalt chrome alloys, nickel-base alloys, tool steels, stainless steel, aluminum etc.		

Wire based

	Build size	Z/layer	Build speed	Materials	Advantages	Disadvantages
Fused Deposition Modeling (FDM)	Stratasys: 914 × 610 × 914	Stratasys: 127–330 μm	N/A	Thermoplastics, modeling clay, plasticine, metal clay, eutectic metals	1. Vast range of materials 2. Availability of desktop scale printer 3. Consistent throughput	1. Requires support structures 2. Reduced strength in the vertical direction 3. Grainy appearance 4. Slower than laser or print head but overall process speed is comparable with other technologies

Foil based

	Build size	Z/layer	Build speed	Materials	Advantages	Disadvantages
Laminated Object Manufacturing (LOM)	N/A	Z: 100 μm (paper); 150 μm (plastic)	N/A	metal foil, plastic film	1. Fast build 2. Relatively cheap	1. Poor strength 2. Wooden pieces may absorb moisture 3. Newly printed pieces require post-processing

by the manufacturer. Using digital light processing (DLP), it was possible to manufacture a packing with a porosity $\varepsilon = 91.2\%$ and specific surface area $a_{pack} = 1{,}136$ m^2 m^{-3}. With a deviation of 0.7% in the porosity and 3.7% in the specific surface area from the specifications of the used 3D model ($\varepsilon = 0.918\%$; $a_{pack} = 1{,}094$ m^2 m^{-3}), this was considered a successfully manufactured replica of the metal foam. Thus, the applicability of the liquid-based AM technology for accurately manufacturing even complex microscale geometrical packing structures like foams was proven. The slight deviations were caused by the deterioration of the printing tray, which holds the liquid resin and is traversed by the projected UV light. Every time the light traversed a part of the tray, that part became slightly foggier which resulted in increased scattering of the light and blurring of the borders of the projected image, only marginally increasing the pieces' size [5]. Since then, several applications of 3D printed packings in RPBs have been reported. The morphologies include multi-inlet [3] and anisotropic[2] wire mesh packings as well as different structured packings. The Zickzack packing and the spiral-packing as two examples for structured packings are presented in the following.

7.2.1 Zickzack packings

The first additively manufactured packing with a novel design was an implementation of the zigzag internal concept (see Chapter 5). Unlike in the common rotating zig-zag bed (RZB) design, in the newly developed Zickzack packing the concentric tray-like baffles, between which the fluid phases move in a winding path, are all mounted in a single block. To provide a constant available cross-sectional area for fluid flow throughout the packing, the positions and dimensions of the baffles are chosen as a function of the radius. The advantages of the Zickzack implementation are the comparably simple design of the rotor and high mechanical durability of the single-block packing [7]. Furthermore, the separation efficiency has proven to be comparable to RZBs while the pressure drop and power consumption were significantly reduced [1]. The concept of the single-block design of the Zickzack packing is shown in Fig. 7.1.

Compared to isotropic packings like metal foam or knit mesh packings, the ZickZack packing design is intended to provide a higher holdup and, thus, increased liquid residence time for an increased gas-liquid mass transfer [1]. An advantage lies in the redistribution of the liquid phase in the cavity of every baffle. While in isotropic packings an increased tangential maldistribution of the phases at the outer radii of the packing has been observed, especially at higher rotational speeds (see Chapter 3), the liquid-holdup in the baffles is intended to provide a uniform liquid distribution around the entire circumference. At lower rotational speeds a ring-shaped volume of high liquid holdup forms at the inner diameter of an isotropic foam packing, while the holdup is significantly lower in the remaining packing (see Chapter 3). This radial maldistribution should also be reduced by the baffles.

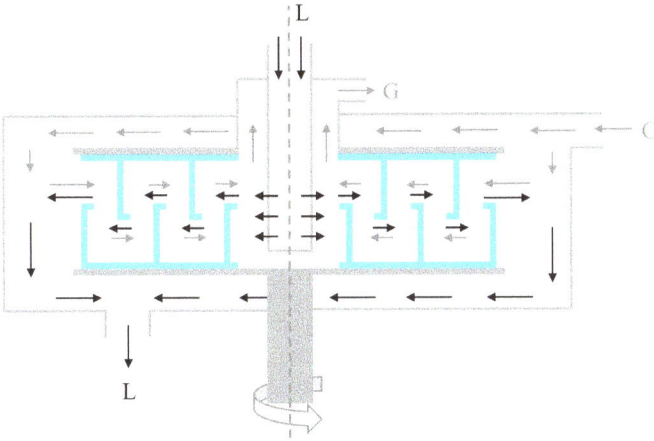

Fig. 7.1: Concept of the single-block design of the Zickzack packing.

Similarly, the baffles reduce the dependence of mass transfer on nozzle performance as the spraying behavior can affect the liquid distribution in the first top and bottom baffle only.

7.2.2 Spiral-packings

Another concept that provides constant cross-sectional areas for the fluid flow is the use of an Archimedean spiral as the shape of the channels inside the packing. This design is intended to provide a laminar and segregated film flow of the fluid phases. Thus, it enables a precise control of the interfacial area between the phases. Due to the structure of the spiral and the high rotational speeds of the rotor, the thickness of the liquid film and therefore the mass transfer resistance is reduced [4]. To adjust the interfacial area for the desired mass transfer between the fluid phases, the design parameters of the spiral-structure can be adjusted as well as multi-channel spiral structures can be used. The laminar liquid film flow with no disruptive obstacles like in mesh or foam packings is especially advantageous when gentle processing of liquid phases containing living microorganisms is required [4].

7.3 Modeling and design

For additive manufacturing of RPB packings, at first a CAD model has to be created according to the desired morphology. The underlying mathematical description for the modeling of the Zickzack – and spiral-structures are described in the following sections. It has to be noted that in many AM-technologies the need to remove residual printing material after a print can restrict the design of certain morphologies of a packing. For instance, within the work of Gładyszewski and Skiborowski [5] it was not possible to remove the residual resin from foam packings with smaller pores than the foam described in Section 7.2.

7.3.1 Modeling and design of Zickzack packings

Within the scope of the zigzag internal concept, several shapes and customized deployments of the tray-like baffles exist [13]. In the original Zickzack packing implementation, L-shaped baffles without perforation are used (see Fig. 7.2). To provide approximately uniform gas and liquid loads along the packing, the **equal area principle** is implemented in the deployment of the baffles according to eq. (7.1).

$$A_{\text{cross}} = \pi \left(R_{i+1,\text{inner}}^2 - R_{i,\text{outer}}^2 \right) = \text{const.} \tag{7.1}$$

In order to keep the cross-sectional area A_{cross} for fluid flow between the individual baffles constant, the distance between the outer radius of one baffle $R_{i,\text{outer}}$ and the inner radius of the next baffle $R_{i+1,\text{inner}}$ is decreasing over the radius of the packing (see Fig. 7.2).

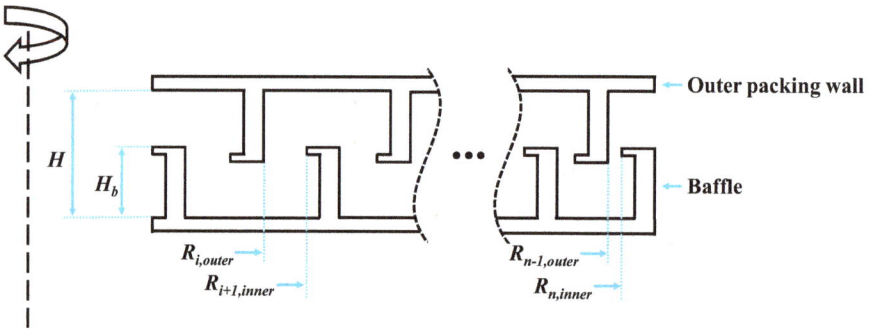

Fig. 7.2: Deployment of the baffles in the Zickzack packing according to the equal area principle.

In other implementations of the zigzag internal concept, the cross-sectional area between the individual baffles and the top and bottom wall of the channels can also be kept (equally) constant by changing the dimensions of the baffles. In that case there is also a decreasing distance over the radius between the individual baffles of height $H_{b,i}$ and the surrounding walls which are themselves separated from each other by the height H.

$$A_{cross,i,wall} = 2\pi R_{i,inner} \, (H - H_{b,i}) = const. \tag{7.2}$$

Depending on the resolution of the 3D printer used, the radially decreasing distance between individual baffles and between baffle and outer packing wall may lead to a constraint in the printable packing dimensions. Another phenomenon to consider is the curvature of the liquid in the pockets of the concentric tray-like baffles during operation. A distinct liquid curvature might lead to an increased pressure drop and risk of flooding in the packing since the liquid fills up the space between the baffles in the lower row of the packing internal. However, the liquid curvature can be assumed to be negligible under the condition that the rotation induced Bond number Bo_ω is higher than 20 [14]

$$Bo_\omega = \frac{\rho_1 \, D^3 \, \omega^2}{8 \, \sigma_{lg}} > 20 \tag{7.3}$$

For a given component system and operating conditions, the density of the liquid ρ_1 and the surface energy σ_{lg} are fixed and the minimum baffle- (and packing-)diameter D and the minimum rotational speed ω during operation are coupled with this criterion.

Furthermore, there are certain general design implications of this packing structure. Due to the baffles in the Zickzack packing, the cross-sectional area of the channels for fluid flow is often smaller than in isotropic packings like metal foams and knit meshes of the same outer dimensions (height and inner and outer diameter of the packing). Due to the consequently increased liquid loadings and F-factors, flooding of the packing can occur at operating conditions that are fine for isotropic packings. Thus, when designing a packing for a specific separation task, the operability of the packing with the design parameters chosen must be checked early in the development. In addition, the general approach of choosing the axial height of the packing according to the desired capacity of the RPB is also more difficult since the cross-sectional area between two individual baffles is not a function of the Zickzack packing's height.

In isotropic packings with increasing cross-sectional area along the packing radius, often the critical, minimum cross-sectional area, and thus maximum F-factor (at the inner diameter), is used to characterize the gas-load in the packing. In the Zickzack packing, the gas load is also characterized using the smallest cross-sectional area in the packing between the baffles. The purpose of the Zickzack packing is to keep this critical cross-sectional area approximately constant throughout the packing using the equal cross-sectional area principle. Thus, there is no single critical area with the highest F-factor but multiple places where flooding might occur. Furthermore, for a fixed uniform baffle height, it should be noted that as the inner diameter of the rotor and packing decreases, might not the cross-sectional area between the baffles, but the cross-sectional area between the first inner baffle and the outer packing wall defines the highest F-factor present in the packing.

If a certain cross-sectional area has been chosen for a specific separation task, it has to be checked that the cross-sectional area between the first baffle and the outer packing wall is larger than the cross-sectional area defined. The maximum outer radius of the Zickzack packing is then limited by the resolution of the printer. The maximum outer radius of a Zickzack packing is calculated by adding the thickness of the baffles including the weir to the maximum inner radius of the outermost baffle $R_{n, inner, max}$. From eq. (7.1) an expression can be derived for the maximum possible inner radius of the last baffle $R_{n, inner, max}$ as a function of the minimum printable clearance between two objects Cl_{min} in a printer, given in eq. (7.4).

$$R_{n, inner, max} = \frac{A_{cross}}{2 \pi Cl_{min}} + \frac{Cl_{min}}{2} \tag{7.4}$$

Lastly, supports need to be added into the Zickzack structure to ensure structural integrity. Vertical pillars from the baffles to the surrounding walls are used in the original Zickzack packing to connect the lower half of the packing to the upper half at a constant distance. To increase the mechanical stability further, especially for polymer-packings, horizontal pillars can also be added to connect the individual baffles [1].

7.3.2 Modeling and design of spiral-packings

The equation of the Archimedean spiral can be used to define the morphology of the spiral channels. It is given in polar coordinates in eq. (7.5). The design parameters of each spiral in the structure of the internal are the inner packing radius $R_{packing, inner}$ and the parameter for controlling the separation between the turns of the spiral b. φ is the rotation angle.

$$r(\varphi) = R_{packing, inner} + b \, \varphi \tag{7.5}$$

The distance between the turns of a spiral is constant at $2 \cdot \pi \cdot b$. If a packing is made up of a single spiral of thickness w_s, the constant cross-sectional area A_{cross} (**equal area principle**) for fluid flow can be calculated from the spiral parameter b and the height of the channels in the packing H_{ch} as given in eq. (7.6).

$$A_{cross} = H_{ch} (2 \pi b - w_s) \tag{7.6}$$

In the spiral-packing the minimum printable clearance of the printer sets a boundary for the spiral-walls in the packing. For a single spiral-packing the minimum separation between the turns of the spiral b_{min} can be calculated using eq. (7.7).

$$b_{min} = \frac{Cl_{min} + w_s}{2 \pi} \tag{7.7}$$

The rotational speed and the spiral baffle pitch angle (set by b and $R_{packing, inner}$) can be used to influence the thickness of the liquid film and therefore the diffusion resistance in the liquid phase. To increase the capacity of processed fluids in a spiral-packing, the height of the packing can be increased. Furthermore, several spiral structures can be superpositioned in a spiral-packing. Unlike the Zickzack packing, the width (and height) of the channels is constant over the packing's radius. Thus, the minimum clearance of the used printer Cl_{min} only restricts the design of the general morphology (spiral baffle pitch angle, number of spirals) but not the radial scalability of this kind of packing. In the spiral-packing, the surrounding walls of the packing and the spiral structure are inherently connected forming the channels in the packing. Thus, no internal support structures in the packing are needed to print this kind of internal as a single block packing.

7.4 Practical implications

As one complete packing can often be too large to print for most AM processes with a sufficient resolution, the CAD model in these cases has to be prepared and printed as a fraction of the complete packing [5]. Afterward these fractions have to be put together to form the complete packing. Ideally, the complete packing structure can be made up of multiple identical pieces, so that only one CAD-model needs to be prepared for all required pieces of the packing. Two example packings and its manufacturing process are presented in the following.

7.4.1 Manufacturing of an exemplary Zickzack packing

The manufacturing process of a Zickzack packing, as first done by Qammar et al. (see Chapter 6), is shown in Fig. 7.3. The model of the packing (see Fig. 7.3A) has been created using Rhinoceros 3D software and it was printed with a Formlabs Form 2 SLA-printer [1] (see Fig. 7.3B) with a minimum clearance of 0.5 mm according to the manufacturer [15]. The cross-sectional area between the baffles was kept constant throughout the packing. Since the packing was destined for distillation experiments, the (translucent) Formlabs High-Temp resin [16] was chosen. That resin also allows optical evaluation of the printing success of the inner structures. To support the packing structure, Qammar et al. used 168 vertical pillars and 72 radial pillars in the packing [1]. To fit the pieces in the available build space in the printer of 145 mm × 145 mm × 175 mm [17], the packing of Qammar et al. was fabricated from nine identical pieces. After printing, these pieces were freed from the external supports and from residual liquid photopolymer on the surfaces (see Fig. 7.3C), cured in a heated UV-chamber (see Fig. 7.3D) and bonded together with Technicoll

9464 2-component epoxy adhesive. That way, the full packing (see Fig. 7.3E) with inner diameter of 146 mm and outer diameter of 356 mm was manufactured. The height of the packing, including the top and bottom plate on which the baffles were mounted, was 10 mm. The pressure drop and mass transfer results of this packing in comparison to different wire mesh and metal foam packings are reported in Chapter 6.

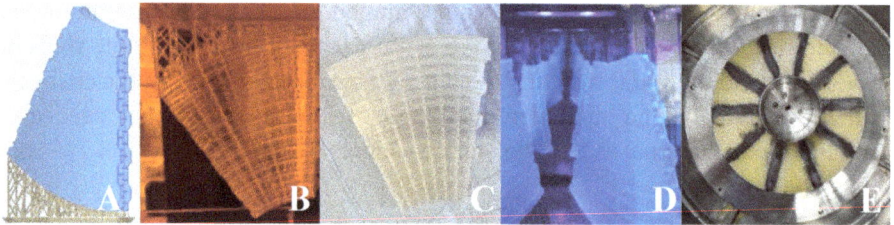

Fig. 7.3: A: CAD model of Zickzack packing piece including internal pillars for stabilization (blue) and external supports for the printing process (beige) in PreForm (Formlabs software); B: Packing piece connected to the built platform of the Formlabs printer by external supports after the printing process; C: Packing piece without external supports and freed from residual liquid photopolymer on its surfaces; D: Two packing pieces being cured in a heated UV-chamber (FormCure); E: Packing pieces glued together to one full packing ring on the lower rotor plate of an RPB.

7.4.2 Application of a 3D printed spiral-packing in aroma stripping

One example of a 3D printed spiral-packing is shown in Fig. 7.4. This packing was printed using Formlabs Standard Clear resin[18], also with a Formlabs Form 2 SLA-printer [17]. Since the cured resin is translucent, the inner spiral structures can be seen through the top- and bottom-plate of the packing. Four spiral walls in the packing form four channels with a width of 2.14 mm each [4]. Two identical pieces (87 mm × 20 mm × 174 mm) were printed and glued to form the packing. The packing has an inner diameter of 45 mm, outer diameter of 174 mm and a height of 20 mm[4].

Lukin et al. [4] used this packing to perform inline stripping experiments to recover aroma compounds from crude fermentation broth in an RPB. **In the spiral-packing the liquid phase containing living cells of Saccharomyces cerevisiae does not encounter disruptive obstacles as in metal foams or wire mesh packings.** Thus, the stripping of the aroma compounds could be performed without destroying the cells in the liquid[4].

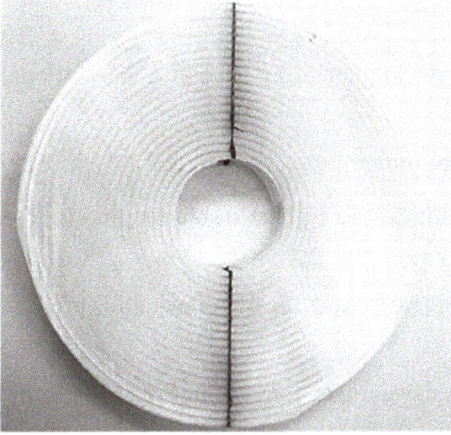

Fig. 7.4: 3D printed spiral-packing used in the experimental study [4].

7.5 Take-home message

– Additive manufacturing offers the possibility to quickly manufacture new types of packings for rotating packed beds. This will lead to an accelerated rate, in which rotating packed bed-technology is developed.
– Tailored packings can be designed for specific applications and tested quickly.
– Packing structures should be developed to provide a scalable packing design. In that regard, the aim of the packing design should be to enable homogeneous operating conditions throughout the entire packing.
– Each additive manufacturing technology offers individual advantages and disadvantages so that there is no perfect technology for every application.
– For choosing an additive manufacturing technology, it is important not to focus on the printing resolution only, but also on the maximum piece size or the ability to position multiple structures relative to one another. Last but not least the printing speed and availability of materials are crucial for quick and efficient packing design evaluation.
– An overview of the characteristics of different additive manufacturing-technologies is given in this chapter. However, specifications vary by manufacturer and important parameters such as the minimum printable clearance between objects should be considered when selecting a particular printer.
– Reported 3D printed packings for rotating packed beds in literature include a variety of morphologies like isotropic foam packings, multi-inlet and anisotropic wire mesh packings as well as structured Zickzack and spiral-packings.

7.6 Quiz

Question 1. What advantages does additive manufacturing have over conventional manufacturing processes?

Question 2. Name four categories of AM technology.

Question 3. Name the AM technology with (a) the highest resolution and (b) the lowest printing cost

Question 4. Which category of AM technologies allows the use of metals as printing material?

Question 5. Why is SLA often used for the development of packings for RPBs?

Question 6. What was the first additively manufactured packing for an RPB?

Question 7. What advantages does the Zickzack packing have over a) RZBs and b) isotropic metal foam or wire mesh packings?

Question 8. Which design-parameter is kept constant in the Zickzack and spiral-packings presented in order to provide more uniform operating conditions along the paths of the fluid phases through the packing?

Question 9. Why is the spiral-packing morphology especially suited for processing of fluids containing fragile objects like living cells of *S. cerevisiae*?

7.7 Exercises

For a counter-current gas-liquid operation the Zickzack packing is considered as the internal of a pilot-scale RPB in your laboratory. Packings with an inner diameter of 160 mm, a maximum outer diameter of 450 mm, and a height of 10 mm fit into the rotor of the RPB. The cross-sectional area between the baffles of this packing should be constant in order to provide approximately uniform operating conditions throughout the packing. The volume flow rate of the gas phase in your application is 5 $m^3\,h^{-1}$ and previous experiments have shown that an F-factor of 1.5 $Pa^{0.5}$ leads to optimal mass transfer performance in Zickzack packings for your application. The density of the gas phase can be assumed to be constant at 1.2 kg m^{-3}. You can use a printer with a maximum built space of 260 mm × 160 mm × 260 mm (X, Y, Z) and a minimum printable clearance between two printed objects of 0.5 mm.

Exercise 1. Determine the constant cross-sectional area between the individual baffles of the Zickzack packing.

Exercise 2. What is the maximum printable position of the last baffle of the packing for this cross-sectional area and the used printer? Is it possible to accurately print a packing with an outer diameter of 450 mm?

Exercise 3. Is the constant cross-sectional area between the individual baffles the smallest cross-sectional area for fluid flow in this packing, if the distance between the baffle and the wall is 3 mm for every baffle?

Exercise 4. How many (identical) pieces do you have to print so that the pieces fit in the printer?

7.8 Solutions

Solution to Question 1: Time and cost-effective prototyping of individualized packings

Solution to Question 2: Liquid-based, powder-based, wire-based, and foil-based AM technology

Solution to Question 3: (a) Two-photon polymerization (TPP); (b) Fused deposition modeling (FDM)

Solution to Question 4: Powder-based AM

Solution to Question 5: SLA offers the advantages of the ability to print complex geometries and good accuracy at relatively low cost compared to, for example, SLS. For the development of new packings it is beneficial that some photopolymers are translucent and in some cases allow an easy optical evaluation of the printing success of the inner packing structures.

Solution to Question 6: A replica of a metal foam packing with a porosity of 91.2% and a specific surface area of 1,136 $m^2\,m^{-3}$

Solution to Question 7: (a) compared to RZBs: simple design of the Rotor, high mechanical durability (single-block packing), comparable separation efficiency at a significantly reduced pressure drop and power consumption; (b) compared to isotropic packings: higher holdup/ residence time and redistribution of the liquid phase in the packing, less dependent on nozzle-performance, more uniform operating conditions throughout the packing

Solution to Question 8: The cross-sectional area

Solution to Question 9: No disruptive obstacles like in mesh or foam packings

Solution to Exercise 1:

$$A_{cross} = 1014.3 \, mm^2 \tag{7.8}$$

Solution to Exercise 2:

$$R_{n, inner, max} = 323.1 \, mm,$$
$$D_{n, inner, max} = 646.2 \, mm \tag{7.9}$$

It is possible to print a Zickzack packing with an outer diameter of 450 mm in this printer without fusing of baffles in the outer radii, blocking the channels for fluid flow.

Solution to Exercise 3: Besides the constant cross-sectional area between the baffles A_{cross}, the cross-sectional area between the first baffle and the top-plate $A_{cross, 1, wall}$ can also be the smallest cross-sectional area for fluid flow in the packing.

$$A_{cross, 1, wall} = 1508 \, mm^2 > A_{cross}, \tag{7.10}$$

Solution to Exercise 4: The packing can be printed as four identical pieces; if the pieces are positioned like in Fig. 7.5, all four pieces can be printed simultaneously, resulting in a minimum manufacturing time for the packing.

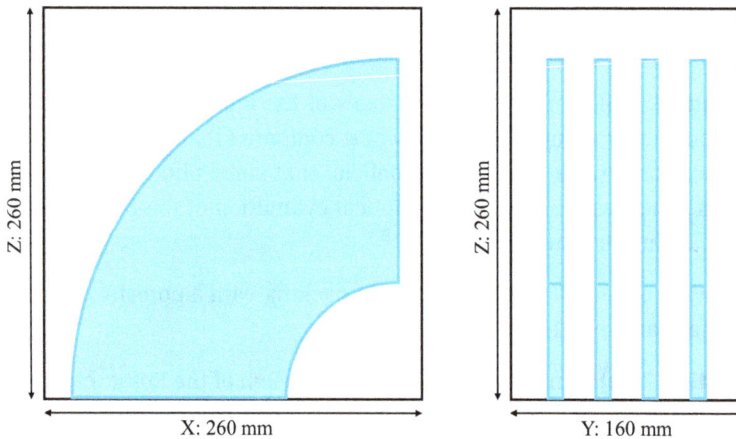

Fig. 7.5: Positioning of four identical pieces of a packing with an outer diameter of 450 mm and height of 10 mm.

List of symbols

Latin letters

A_{cross}	cross-sectional area	m^2
a_{pack}	specific geometric packing surface area per unit	$m^2\,m^{-3}$
b	spiral parameter for separation between turns	m
Bo_ω	rotation induced bond number	–
Cl_{min}	minimum printable clearance between two objects	m
D	diameter	m
H	height	m
R	radius	m
r	variable radius	m
w_s	width of one spiral	m

Greek letters

ε	porosity	–
ρ	density	$kg\,m^3$
φ	angle	rad
σ	surface energy	$N\,m^{-1}$
ω	angular velocity	$rad\,s^{-1}$

Subscripts

b	baffle
ch	channel
g	gas
l	liquid
packing	packing
wall	wall

List of abbreviations

RAM	additive manufacturing
CAD	computer-aided design
DLP	digital light processing
EBM	electron beam melting
FDM	fused deposition modeling
LOM	laminated object manufacturing
RPB	rotating packed bed
RZB	rotating zig-zag bed
SLA	stereolithography
SLM	selective laser melting

| SLS | selective laser sintering |
| TPP | two-photon polymerization |

References

[1] Qammar H, Gładyszewski K, Górak A, Skiborowski M. Towards the development of advanced packing design for distillation in rotating packed beds. Chem. Ing. Technik. 2019;91 (11):1663–73.

[2] Wen Z-N, Wu W, Luo Y, Zhang -L-L, Sun B-C, Chu G-W. Novel wire mesh packing with controllable cross-sectional area in a rotating packed bed: Mass transfer studies. Ind. Eng. Chem. Res. 2020;59(36):16043–51.

[3] Wu W, Luo Y, Chu G-W, Su M-J, Cai Y, Zou H-K, et al. Liquid flow behavior in a multiliquid-inlet rotating packed bed reactor with three-dimensional printed packing. Chem. Eng. J. 2020;386:121537.

[4] Lukin I, Wingartz I, Schembecker G. Application of rotating packed bed for in-line aroma stripping from cell slurry. J. Chem. Technol. Biotechnol. 2020;95(11):2834–41.

[5] Gładyszewski K, Skiborowski M. Additive manufacturing of packings for rotating packed beds. Chem. Eng. Process. Process Intensif. 2018;127:1–9.

[6] Wang GQ, Yuan XG, Yu KT. Review of mass-transfer correlations for packed columns*. Ind. Eng. Chem. Res. 2005;44(23):8715–29.

[7] Cortes Garcia GE, van der Schaaf J, Kiss AA. A review on process intensification in HiGee distillation. J. Chem. Technol. Biotechnol. 2017;92(6):1136–56.

[8] Neumann K, Gladyszewski K, Groß K, Qammar H, Wenzel D, Górak A, et al. A guide on the industrial application of rotating packed beds. Chem. Eng. Res. Des. 2018;134:443–62.

[9] Femmer T, Flack I, Wessling M. Additive manufacturing in fluid process engineering. Chem. Ing. Technik. 2016;88(5):535–52.

[10] Neukäufer J, Hanusch F, Kutscherauer M, Rehfeldt S, Klein H, Grützner T. Methodology for the development of additively manufactured packings in thermal separation technology. Chem. Eng. Technol. 2019;42(9):1970–77.

[11] Low Z-X, Chua YT, Ray BM, Mattia D, Metcalfe IS, Patterson DA. Perspective on 3D printing of separation membranes and comparison to related unconventional fabrication techniques. J. Memb. Sci. 2017;523:596–613.

[12] Formlabs. Guide to Stereolithography (SLA) 3D Printing. [April 28, 2021]; Available from: https://formlabs.com/blog/ultimate-guide-to-stereolithography-sla-3d-printing/.

[13] Wang GQ, Zhou ZJ, Li YM, Ji JB. Qualitative relationships between structure and performance of rotating zigzag bed in distillation. Chem. Eng. Process. Process Intensif. 2019;135:141–47.

[14] Lubarda VA. The shape of a liquid surface in a uniformly rotating cylinder in the presence of surface tension. Acta Mech. 2013;224(7):1365–82.

[15] Formlabs. Design specifications for 3D models (Form 2). [June 04, 2021]; Available from: https://support.formlabs.com/s/article/Design-Specs?language=en_US.

[16] Formlabs. Material data sheet – High temp resin. [June 02, 2021]; Available from: https://for mlabs-media.formlabs.com/datasheets/High_Temp_Technical.pdf.

[17] Formlabs. Form 2 Tech Specs. [June 02, 2021]; Available from: https://formlabs.com/de/3d-printers/form-2/tech-specs/.

[18] Formlabs. Material data sheet – Standard resin. [June 02, 2021]; Available from: https://for mlabs-media.formlabs.com/datasheets/Standard-DataSheet.pdf.

Markus Illner, Matthias Hilpert, Jens-Uwe Repke

8 Rotating packed beds in distillation: rate-based modeling for multicomponent systems

Despite the manifold potential for process intensification no general distribution of RPBs is perceivable. Especially for the European chemical industry to date no reports on industrial implementations of a TRL above 5 have been published. One of the major reasons for that is the need for investments with high probability of success, scalability, and low uncertainty. This is rather limited for new or less established technologies where little or no in-house experience exists. **Unlocking the potential of RPB or HIGEE processes hence largely depends on available knowledge on design specifications, respective design and scale-up guidelines, as well as the operation behavior regarding mass transfer efficiency, process dynamics, and its connection to the design of RPB machines.** Potential benefits and drawbacks must be clearly stated and should be quantifiable regarding CAPEX and OPEX for a desired application. Suitable models to describe separation processes using HIGEE are a fundamental basis to this and furthermore enable the integration of such process options into concepts and frameworks for process synthesis. In the ideal case, process models have reached such a maturity that experiments can be omitted for the design process. This is partially true for conventional distillation columns [1], but far off for RPBs. Major reasons for that are a severe lack of available experimental studies (less than 25 publications in English literature concerning distillation in RPBs are available) and an exclusive focus on binary systems for distillation operations. However, especially data on multicomponent separation operations are vital to verify developed descriptions or correlations of vapor and liquid-side mass transfer coefficients and their application for RPB modeling for multicomponent systems. Describing RPB operation and separation efficiencies for such systems broadens the scope of possible applications of RPBs since real-world engineering problems and industrial applications typically involve multicomponent separation tasks [2].

This chapter thus sketches a procedure for developing and validating a rate-based model for multicomponent distillation in RPBs. Special focus is laid on deriving vapor and liquid-side mass transfer correlations for a given RPB design and subsequent model validation.

As a starting point, fundamentals on mass transfer modeling and required experimental procedures for separation processes are discussed in Section 8.1. The status quo on experimental investigations and modeling of distillation processes

Markus Illner, Matthias Hilpert, Jens-Uwe Repke, Technische Universität Berlin

https://doi.org/10.1515/9783110724998-008

using RPBs is discussed in Section 8.2 and used to point out major obstacles and derive requirements as well as suitable procedures for model development. As an introduction to that, Chapters 1 and 3 present fundamental principles of centrifugally enhanced separations for a detailed analysis of hydrodynamic phenomena and their connection to the rotational speed. Based on Chapters 5 and 6, the status quo on applied equipment designs, experimental investigations of distillation using RPBs, as well as existing process models is outlined in Section 8.2. This is then used to point out major obstacles regarding modeling approaches for RPBs and derive requirements as well as suitable procedures for model development.

A rate-based stage model including a discretization scheme is presented in Section 8.3 followed by a comprehensive case study in Section 8.4. To gain industrially relevant insights, systematic experiments are conducted in a pilot-plant-scale RPB machine. The data are subsequently evaluated and used to derive mass transfer correlations for multicomponent distillation. Developed model structures are then validated for distillation application using the binary system ethanol/water and the ternary system ethanol/n-propanol/water.

This chapter is based on the dissertation by Matthias Hilpert carried out at the Process Dynamics and Operations Group at Technische Universität Berlin [3]. Furthermore, parts of the outlines have already been published in our own contributions [4, 5].

8.1 Fundamentals

To assess separation efficiency in RPBs via process models, required fundamentals on mass transfer in gas-liquid systems, as well as necessary experimental investigations are outlined. Furthermore, approaches for specifically modeling mass transfer in RPBs are outlined. This section builds on the extensive deliberations given in Chapters 1 and 3 regarding RPB operation, hydrodynamics, pressure drop, liquid holdup and residence time, as well as operation ranges. The following fundamentals are thus solely focused on aspects of mass transfer and the derivation of respective correlations applicable for RPBs.

8.1.1 Mass transfer modeling

The overarching principle to most separation processes is the generation of an interface between adjacent phases at which fluid properties abruptly change and the transport of heat or/and material across this interface [1]. Focusing on mass transfer, respective models are subdivided into equilibrium-based (see fundamentals in Book 1, Chapter 3) or rate-based models, which explicitly consider transfer of mass due to driving forces between adjacent phases. **Rate-based models are considered more suitable for RPBs, as they link equipment geometry, fluid dynamics, and operational parameters with separation performance, account for mass transfer interactions occurring in multicomponent systems, and generally provide a more realistic representation of the occurring phenomena [6].**

In general, the molar mass transfer rate N_i of component i in a binary mixture (with j) across the effective interfacial area A_e from phase L is [1]:

$$N_i = \underbrace{k_{i,j}^L A_e \rho_{Mix}^{L,mol} \left(x_i^I - x_i \right)}_{\text{diffusive}} + \underbrace{x_i N_{net}}_{\text{convective}} \tag{8.1}$$

The effective interfacial area A_e defines the area across which mass transfer between phases occurs. This is not necessarily the provided geometrical area by internals, such as structured packings or metal foams, as areas not wetted, areas covered in stagnant liquid, or additional area created by liquid droplets or ligaments not attached to the packing may occur [7]. Regarding this, the specific effective interfacial area a_e is defined and calculated with respect to the total packing volume:

$$a_e = \frac{A_e}{V_p} \tag{8.2}$$

For the description of mass transfer in gas-liquid contactors, typically film theory and penetration or surface-renewal theory are applied (see Chapter 4 of the first book of this series on reactive and membrane-assisted separations). Throughout this chapter, focus is laid on film theory extended to the two-film theory as resistances in series. Generally, film theory assumes that all mass transfer resistance is located in a thin film at the interface between adjacent phases. Only steady-state diffusive transfer normal to the film occurs while all gradients vanish for the bulk region due to turbulence. No accumulation of material or energy takes place in the film and mass transfer parallel to the surface is assumed negligible. Following these assumptions, a linear dependency between rate of mass transfer and molecular diffusivity can be derived. This however was proven insufficient in real-world experiments, as dependencies of $k \sim D^{0.5\ldots0.75}$ were found [1]. Hence, other theories, such as penetration theory or empirical approaches have been developed. The film theory was extended to the two-film model, in which the interface is considered to be of infinitesimal thickness without any resistance to mass transfer. At the interface, phase equilibrium is assumed. The two films around the interface then represent mass transfers resistances in series. These resistances and mass transfer coefficients are mathematically described by introducing an overall mass transfer coefficient that describes the mass transfer rate if all the resistance is assumed to be located in one phase. To introduce an overall vapor-side mass transfer coefficient $K_{i,j}^V$, a virtual vapor mole fraction y_i^* at the interface is defined which is in equilibrium with the bulk liquid x_i. The molar diffusional flux of component i can then be described as:

$$j_i = K_{i,j}^V \rho_{Mix}^{V,mol} \left(y_i - y_i^* \right) = k_{i,j}^V \rho_{Mix}^{V,mol} \left(y_i - y_i^I \right) = k_{i,j}^L \rho_{Mix}^{L,mol} \left(x_i^I - x_i \right) \tag{8.3}$$

Using m as the local slope of the equilibrium curve between both phases, the overall mass transfer resistance can be written as:

$$\frac{1}{K_{i,j}^V} = \frac{1}{k_{i,j}^V} + \frac{\rho_{Mix}^{V,mol}}{\rho_{Mix}^{L,mol}} \cdot \frac{1}{k_{i,j}^L} \left(\frac{y_i^I - y_i}{x_i^I - x_i} \right) = \frac{1}{k_{i,j}^V} + \frac{\rho_{Mix}^{V,mol}}{\rho_{Mix}^{L,mol}} \cdot \frac{m}{k_{i,j}^L} \tag{8.4}$$

The concept provided by (two-)film theory with assuming equilibrium at the interface and the summation of resistances in series also enables the usage of descriptions of mass transfer coefficients from other theories, such as penetration or surface-renewal theory. Those theories assume infinitesimal fluid elements (eddies) to move towards the interface and contacting the other phase for a defined exposure time t_e or residence time distribution before returning to the bulk phase and being replaced by other eddies at bulk phase conditions. For this time period unsteady diffusion normal to the interface occurs [8].

8.1.2 Mass transfer experiments

The rate-based modeling of multicomponent distillation processes requires knowledge of the individual gas and liquid-side mass transfer coefficients, as well as the effective interfacial area. These specific variables are accessible by chemical or physical absorption or distillation experiments. An in-depth review on this is given by Hegely et al. [7], who summarize suitable methods. Applicable test systems are stated in [9]. Most methods are limited to evaluating the volumetric mass transfer coefficient $(k \cdot a_e)$, averaged over packings or internals, as the effective area is typically difficult to assess. Hence, they are measured from inlet and outlet flow conditions, whereby end effects should be strictly avoided.

Liquid-side mass transfer coefficients

Liquid-side volumetric mass transfer coefficients are measured by ab- or desorption of gaseous compounds with very low solubilities, e.g., oxygen in water. Following eq. (8.4), gas-side resistance then becomes negligible for very high gradients m. In the case of high mass transfer coefficients or large packing heights, experimental uncertainty may increase due to saturation (absorption) or depletion (desorption) of the transferring component. Alternatively, chemical absorption can be applied. However, care is required regarding control of concentrations and maintaining the reaction regime required for determining $k^L \cdot a_e$.

Gas-side mass transfer coefficients

Gas-side volumetric mass transfer coefficients are accessible via chemical absorption encompassing very fast kinetics. Danckwerts [10] recommends an instantaneous irreversible second-order reaction, leading to a reaction plane at the gas-liquid interface and thus eliminating liquid-side resistance and any back pressure. Most often the chemical absorption of SO_2 from air in aqueous sodium hydroxide solution is used. Here, very low concentrations of SO_2 are still analytically accessible, allowing for significant absorption to occur in the apparatus. Gas-side volumetric mass transfer coefficients could also be determined by vaporization of a pure liquid, as then no liquid-side resistance exists. However, saturation of the gas phase should be avoided and thus only investigations on short bed lengths are possible. Physical absorption of components with high solubility at high liquid to gas ratios has been used but faces the disadvantage of the additional influence of liquid-side resistance. Hence, the liquid-side volumetric mass transfer coefficient has to be accounted for by using other correlations or experiments.

Specific effective interfacial area

The specific effective interfacial area is determined using chemical absorption systems of very fast kinetics, such as CO_2/air/sodium hydroxide solution [7]. Given the volumetric mass transfer coefficients, k^L and k^V can be calculated. However, uncertainties in a_e are propagated to the mass transfer coefficients by this method. Hence, correlations for k^L and k^V should only be used together with the corresponding a_e correlation from which they were determined [11].

Transferability of mass transfer coefficients from ab- or desorption to distillation

Despite widely assumed analogy and transferability between coefficients obtained from sorption processes and coefficients used to describe distillation [12, 13], several significant differences are visible and transferability is still a subject of ongoing research. Specifically, differences in surface tension should be accounted for, as wetting of packing material and internals is directly influenced. While absorption and desorption processes are typically conducted using aqueous systems, distillation is most often applied on organic mixtures [14]. Onda et al. and Coughlin et al. compared systems with different wetting behavior by the addition of surfactants or modification of the surfaces of packing material [15, 16]. Using surfactants, a decrease in $k^L \cdot a_e$ is visible, but mostly accounted to increased mass transfer resistance at the interface, as surfactant molecules likely aggregate at phase boundaries and interfaces. By increasing the hydrophobicity of packing materials, a reduced specific effective area is caused. However, this is partly compensated by increased surface renewal.

Additionally, larger temperature differences between absorption and distillation processes result in altered fluid properties, especially viscosity. By the addition of glycerol to aqueous systems Song et al. [11] and Delaloye et al. [17] found decreasing liquid-side mass transfer coefficients with increasing viscosity caused by a reduced liquid diffusivity and a postulated second influence of decreased liquid turbulence (proportional to $\left(\eta^L\right)^{-0.4\ldots-0.5}$). However, Delaloye et al. [17] concluded that uncertainties due to deviations in viscosity have only a moderate impact on distillation column design and are only of importance for absorption processes with viscous solvents.

Moreover, the general diffusion regime differs for absorption (unidirectional) and distillation (approximately equimolar counter-diffusion, net zero molar flow rate through A_e). This effect is assumed to be negligible in comparison to changes in fluid properties and composition along the packing [16]. To account for this, the application of discretized rigorous models is suggested, whereby discretization is performed over the length or height of the packing material. Apart from that, the derivation of mass transfer correlations from absorption and desorption experiments remains best-practice if no superior methods are available.

To avoid uncertainty in the transfer of correlations from sorption processes to distillation, methods for directly measuring liquid- and gas-side mass transfer coefficients have been derived. Prominent examples are profile methods using captured temperature or liquid and vapor composition profiles from distillation experiments combined with model simulations and parameter estimation to directly obtain volumetric vapor and liquid-side mass transfer coefficients [18–21]. However, profile-based methods are difficult to adapt to RPBs, since temperature and concentration profiles must be gathered from the rotating equipment using wireless transmitters [22]. Refer to Chapter 6 for a description of a rotor telemetry system to enable such temperature measurements along the radius of the packing. Alternatively, design options using rotor stator concepts are suitable to enable sampling of liquids and vapor along the radius of the RPB packing.

8.1.3 Mass transfer in RPB

As compared to conventional columns, liquid-side mass transfer is significantly enhanced in RPBs. High centrifugal acceleration and related shearing, impingement, and mixing of the liquid result in the formation of high specific effective surface areas and rapid surface renewal [23–25]. In contrast, gas-side mass transfer is not affected to the same extent as the difference in velocity of gas phase and packing or the liquid is negligible. However, as gas velocities could be significantly increased without causing flooding, gas-side mass transfer coefficients in single-block packing RPBs could still be increased compared to conventional packed columns [26]. Regarding the mass transfer efficiency of RPBs several general trends can be stated:

- increasing mass transfer efficiency with increasing rotational speed, decreasing liquid flow rate and increasing gas-to liquid ratio [26]
- a distinct maximum of the separation efficiency depending on rotational speed and gas to liquid ratio [27, 28]
- increasing effective interfacial area with increasing rotational speed, liquid flow rate and gas-to-liquid ratio [29–31]
- high voidage packings are more advantageous to apply in RPBs than extremely high surface area packings due to lower pressure drop but comparable mass transfer performance [26, 32]
- wire meshes provide high separation efficiency [32, 33]

End effects

The geometric design of RPBs poses additional characteristic features regarding mass transfer. End effects at the inner and outer radius of the packing have to be considered as previously described in Chapters 3 and 6 (see also Fig. 8.1).

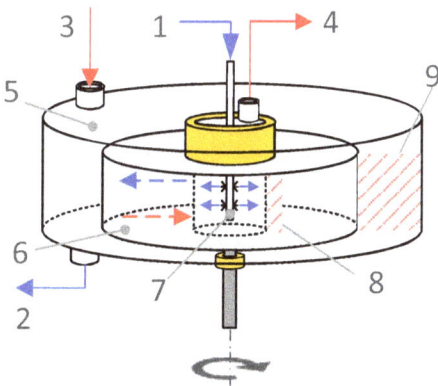

Fig. 8.1: Sketch of end effect zones in an RPB (1 – liquid inlet, 2 – liquid outlet, 3 – gas inlet, 4 – gas outlet, 5 – casing, 6 – packed rotor, 7 – liquid distributor, 8 – inner end effects zone, 9 – outer end effects zone). Image taken from [3].

At the inner radius of the packing, the large difference in radial velocity of packing and entering liquid results in increased shearing. Additionally, interactions of gas and liquid are largest in this section due to the highest gas stream velocities in the smallest cross-sectional area [34]. The outer end effect zone is set up by the casing at which additional interfacial area is formed by liquid exiting the packing and hitting the casing. Due to the rotation of the packing, it is rather difficult to separate contributions to mass transfer into packing and end effects. Additionally, mass transfer coefficients and effective surface area change over the packing radius due to changing centrifugal acceleration and velocities, formation of rivulets, or shadowing. Hence, averaged values for mass transfer coefficients for applied packings are usually stated in literature (see Tabs. 8.1 and 8.2). This of course complicates comparison of mass transfer correlations, as the geometric setup must be considered. Several approaches for separating end effects from mass transfer in packings have been tested, such as measuring liquid concentrations directly at the outer diameter of the packing or gas compositions inside of the casing [31, 35, 36]. Reddy et al. [37] used a gas shortcut over the casing to measure absorption of SO_2 only for the casing, obtaining similar results for $k^V \cdot a_e$ for this method and gas sampling in the casing for standard operation.

In general, end effects are desired as they significantly contribute to the total mass transfer. But they severely complicate the characterization of the mass transfer of the actual packing. To access this contribution, experiments with varying packed bed lengths with respect to inner and outer radii have been used [35, 38]. **The overall contribution of end effects differs significantly for different RPB designs and operation conditions with fractions between 1–40% of the total mass transfer being reported in literature [37, 39–41].**

Transferability of operation data

To account for geometric specifications an integrated centrifugal acceleration has been proposed as characteristic operational parameter instead of rotational speed (cf. Section 1.3.3). It is described as the energy input to an area of the rotor and is unique for a given design. However, this parameter has not yet found widespread application. Regarding mass transfer, volumetric coefficients, HTU, and HETP are used, whereby the latter two are defined using the radial length of the packing. **In analogy to HTU, the Area of a Transfer Unit ATU was proposed to account for the radial rotor geometry [35], and has repeatedly been employed. However, due to the changing mass transfer efficiency along the packing and large impact of end effects, comparison of RPBs of different sizes using these key performance indicators is still challenging.** Especially for RPBs with short radial length end effects contribute disproportionately to the separation efficiency and volumetric mass transfer coefficients may be up to an order of magnitude higher compared to larger RPBs [39, 41].

8.1.4 Modeling of mass transfer in RPB

Following Section 8.1.1, rate-based modeling is suggested for RPBs, as it links equipment geometry and operational parameters with separation performance and accounts for mass transfer interactions occurring in multicomponent systems. Consequently, mass and energy balances for the individual phases are deployed and linked by rate equations for mass and heat. Several approaches for the calculation of required mass transfer coefficients have already been developed and are summarized in this section. In general, these coefficients are used in discretized rate-based models [42, 43] or differential equations [44], which then are solved numerically. Hence, equation-oriented modeling is still dominant. RPB model implementations are usually not available for flowsheet simulators or limited to augmentation of rate-based models for columns [45].

Mass transfer coefficients derived from first principles models

As a rigorous approach to mass transfer coefficients, first principles modeling has been used for RPB modeling. Key therein are assumptions regarding the liquid flow patterns occurring, such as a falling film in conjunction with penetration theory under the influence of the centrifugal field [46], laminar film flow over randomly inclined plates to account for the random structure of packings coupled with penetration theory [47], or combinations of droplet and film flow [48]. For the latter, a surface renewal model was used together with hydrodynamic measurements regarding film thicknesses and droplet diameters. Based on this approach Luo et al. [31] modeled $k^L \cdot a_e$ for different sections of packed and blade RPBs incorporating results from hydrodynamic investigations, such as liquid holdup and droplet diameters in combination with surface renewal theory. Liu et al. [49] used this approach for a mesh-pin rotor RPB, employing a correlation for the liquid holdup based on performed X-Ray CT scans. Sectional $k^L \cdot a_e$ values were however calculated using geometrical mean radii of the individual rotor zones. The developed model was later adapted for an upscaled mesh-pin rotor by using droplet diameter correlations collected from the larger machine [50]. However, first principles modeling for RPB is usually applied for absorption and desorption processes, where gas-side resistance can be neglected [51, 52]. A theoretical derivation of $k^L \cdot a_e$ and $k^V \cdot a_e$ correlations for distillation is given by Li et al. [53] using a Counterflow Concentric Ring Rotating Bed (CRB), whereby a hydrodynamic fitting parameter based on distillation experiments is required to account for the rate of gas surface renewal.

Semi-empirical correlations

Semi-empirical correlations for mass transfer coefficients based on dimensionless numbers and groups are common practice for RPB modeling and usually applied for correlating volumetric coefficients (accessible via experiments). A generalized form is given in eq. (8.5), whereby variations in the variable x are visible to account for additional influences, e.g., surface tension [54–57].

$$Sh = c_1 Re^{c_2} Ga^{c_3} Fr^{c_4} We^{c_5} Sc^{c_6} \left(\frac{x}{x_{ref}}\right)^{c_7} \tag{8.5}$$

Due to the lack of information content in experimental data, exponents for liquid-side Schmidt numbers Sc^L are often set to 1/2, while 1/3 is used for Sc^V. The choice for Sc^L is in accordance with penetration theory $(k^L \sim (D^L)^{1/2})$ [58], while 1/3 for Sc^V is confirmed by experimental observations of Onda et al. $(k^V \sim (D^V)^{2/3})$ [16]. Additionally, the influence of centrifugal acceleration is captured by the Galilei number Ga. Note, that in RPB literature, Ga is often incorrectly termed as Grashof number, which is defined differently in heat and mass transfer, as it is used to describe free convection.

One of the first correlations has been set up by Singh et al. [35] for the overall liquid ATU based on the specific packing area as characteristic measure for Reynolds and Galilei number. Further developments then aimed for incorporating the influence of end effects by varying inner and outer radii and adding terms holding the volumes of RPB, eye, and outer cavity of the casing, as well as the influence of viscosity and packing material [27, 32, 38]. In general, rather small lab scale machines (radii 10–60 mm) were used. To account for changes in mass transfer efficiency over the radial length, several groups suggested correlations for $k^L \cdot a_e$ and $k^V \cdot a_e$ which are integrated using eq. (8.6) or employ Re and Ga based on the arithmetic mean of the radius of the packing [36, 37, 59].

$$ka_e = \frac{\int_{r_i}^{r_o} k_{loc} a_e r dr}{\int_{r_i}^{r_o} r dr} \tag{8.6}$$

In general, correlations for liquid-side volumetric mass transfer coefficients are more widely available than those for the gas side. Additionally, often only the overall gas-side mass transfer was captured since most experiments showed non-negligible liquid-side resistance to mass transfer. In this case, $k^V \cdot a_e$ is accessible via back-calculation from the overall gas-side coefficient $K^V \cdot a_e$ and available data or correlations for $k^L \cdot a_e$ [60].

Machine-learning-based correlations

Machine learning algorithms can be applied to predict the mass transfer behavior in case large and high value (information content) data sets are available. Same as for empirical correlations, variables to be correlated should be defined to narrow down suitable model structures. Exemplarily, artificial neuronal networks have been used to predict the overall volumetric mass transfer coefficient $K^V \cdot a_e$ for gas-

and liquid-side limited absorption and stripping processes in RPBs based on gas flow, liquid flow, and rotational speed from three different sets of experiments [61]. To increase the number of data points available for training, new points were generated by interpolation from experimental data. This way superior model structures and interpolation capabilities compared to empirical correlations can be achieved, while also facing prediction errors due to artificial and not validated training data. Support vector machines have been applied to connect the Sherwood number for CO_2-absorption in sodium hydroxide solution with dimensionless groups based on data from several sources [62]. Comparing the result with a fitted empirical correlation and use of an ANN with back-propagation on the same dataset, superior performance of the support vector machines was found.

Comparison of modeling approaches for mass transfer coefficients

Naturally, first principles models offer best extrapolation capabilities and the most realistic representation of mass transfer in RPBs, also including, e.g., end effects. However, they are relying on detailed knowledge of the fluid dynamics in RPBs. This is particularly difficult for gas-side mass transfer and usually several simplifying assumptions have to be made to enable mathematical modeling. These models often require additional hydrodynamic data, such as droplet sizes or holdup and are dependent on the according accuracy. This also limits generalization capabilities, as new measurements for variations in geometry may be necessary. In contrast, empirical correlations are often based on dimensional analysis [63] and correlate mass transfer coefficients with experimental data using dimensionless groups. Here, profound understanding of physical processes and causalities is not mandatory but can be lumped together in empirical structures. They are easy to implement and evaluate, but naturally cannot capture all influences and dependencies in their full extent and numbers. In combination with experimental error, obtained deviations for predicting mass transfer are typically in the range of $\pm 30\%$ [64]. Furthermore, empirical correlations are limited to operating conditions, fluid properties, and RPB design of the original experimental data and inherently do not allow for extrapolation. Applicability of correlations to other devices or geometries then significantly depends on the variables or phenomena included and is typically very limited.

Machine learning methods require a vast set of training data of high information content, which is then translated into an easy to evaluate black-box model. Besides high experimental effort and necessity to significantly vary operation conditions and design, physical causalities and influences are masked in the black-box model and are difficult to access for further research. Again, extrapolation capabilities are limited and even interpolation might be hampered for model domains with sparse experimental training data or high uncertainty.

8.2 Status quo: experimental studies and mass transfer correlations

To assess already available experimental data, as well as correlations for mass transfer coefficients, a brief outline of available literature is given. The focus here is on reusability of data and models. Hence, a closer look is taken at modeling approaches, while an overview over distillation studies along with additional information on geometrical configurations and applied component systems is given in Chapter 5 and collected in Tab. 5.2.

Distillation in RPBs

Taking a closer look at experimental studies on distillation in RPBs, a significant scarcity of literature is visible (see selection in Tab. 5.2). This poses a major deficiency, as such data are vital to verify the application of vapor and liquid-side mass transfer coefficient correlations for RPBs for modeling of industrially relevant multi-component mixtures. Additionally, key design or operation parameters are often enough not reported and actual concentration measurements are masked using *HETP*, *NTU*, or $K^V \cdot a_e$. These variables are partially displayed using simplified correlations, which are generally restricted to binary mixtures and hardly adaptable for correlations in the form of eq. (8.5) (see Chapters 5 and 6). Additionally, the applied designs of RPBs strongly differ in terms of dimensions, types (single-block packing, split packing, multi-rotor arrangements), and packing (wire mesh, metal foam, Zickzack). For a detailed overview on rotor configurations, design of packings, and their performance please refer to Chapter 5.

As a special characteristic for RPB operation in distillation, several authors report a distinct optimal rotational speed connected to a maximum in separation efficiency [33, 65–70]. This behavior likely results from the trade-off between an intensification of heat and mass transfer and the decrease in contact time with increasing centrifugal acceleration [33, 70]. Being able to measure, describe, and predict this maximum for a given process is crucial for improving RPB design and operation strategies. A first attempt to this was made by Li et al. [68], who investigated distillation of ethanol/water mixtures at pressure from vacuum to ambient. A process model was set up using the Aspen BatchSep module and mass transfer correlations of Chen et al. [32] for $k^L a_e$ (Tab. 8.1) and Chen [60] for $k^V a_e$ (Tab. 8.2). The maximum in separation efficiency observed in experiments is well predicted by the model. Unfortunately, reproduction of the data is not possible, as no throughputs are reported.

Mass transfer correlations for distillation

In addition to the overview on experimental data on distillation provided in Chapter 5, available correlations for the liquid- and gas-side mass transfer are outlined and discussed regarding underlying experimental data, RPB setups, and applicability.

Following Tab. 8.1, most publications apply semi-empirical correlations to describe liquid-side mass transfer, while mechanistic approaches are scarce. Singh et al. [71] present a first correlation for the ATU applied for the stripping of volatile organic compounds from water. Here, Reynolds and Galilei number were included using the specific geometric packing area as characteristic length, while excluding Sc^L due to lacking information content in the experimental data.

A series of correlations with subsequent improvements have been developed for water deaeration by the group of Chen: Initially, a fully empirical function was fitted to experimental data. Due to the lack of dimensionless numbers, transferability to other separation processes is not given [72]. In 2005, correlations based on the nowadays commonly used general form in eq. (8.5) were developed, accounting for the influence of bed geometries [27], shear behavior of applied component systems [38], and packing materials [32]. In the case of geometry, a variation of inner and outer radii was performed to investigate the influence of end effects. These were then included into the correlation via terms which incorporate volumes of RPB, eye, and outer cavity of the casing. Specifically for split packing RPBs in co- and counter-rotation Reddy et al. found correlations for local liquid-side mass transfer coefficients, which were rotor-averaged using eq. (8.6) [37]. In contrast, Shivhare et al. developed a correlation for $k^L a_e$ based on Re^L and Ga^L calculated at the arithmetic mean radius of the bed [59]. Additionally, correlations for structured stainless steel wire meshes were developed, whereby typically the square mesh opening size c and fiber thickness d are considered as characteristic parameters [73, 74]. In general, developed semi empirical correlations apply to the given case study in terms of RPB geometry, applied packing, and partially component system. This is especially critical, as most correlations (based on larger experimental data set) are developed for small-scale laboratory RPBs with radial bed lengths ranging from 10 to 50 mm. A predominant influence of end effects is hence expected. Transferability is thus often not given and appropriate choice of available correlations, validation experiments, or own developments based on experimental data should be considered.

Furthermore, all correlations encompass a proportionality of $k^L a_e \sim (a_c)^{const.}$ and hence cannot mathematically describe a local maximum in mass transfer efficiency as observed from experimental studies. A novel correlation to overcome this obstacle is thus presented later in this chapter (Section 8.4.2).

Referring to gas-side mass transfer coefficients, a short overview of selected correlations is given in Tab. 8.2. Chen et al. incorporated terms to account for RPB geometry and associated inner and outer end-effects at the eye and in the casing to enable scale-up capabilities [60]. However, again, rather small RPB dimensions

Tab. 8.1: Overview of semi-empirical correlations for volumetric liquid-side mass transfer coefficients in RPBs obtained from desorption of oxygen or VOCs from water. Packing Material: Stainless Steel (SS).

Author	RPB type, packing; dimensions $d_i/d_o/h_p$ (mm); No. of exp. points	Correlation	Comments
Singh [71]	single-block, SS wire mesh, metal foam; 254/457, 610, 762/127; ≈36	$$k^L a_e = \frac{\dot{V}^L}{h_p \cdot ATU}, ATU = \frac{337143.86}{a_p^2} \cdot \left(\frac{\rho^L \cdot \dot{v}^{L,avg}}{a_p \eta^L}\right)^{0.6} \cdot \left(\frac{\rho^{L2} a_c^{avg}}{\eta^{L2} a_p^3}\right)^{-0.15},$$ $$\dot{v}^{L,avg} = \frac{\dot{V}^L}{2\pi h_p} \frac{\ln(r_o/r_i)}{r_o - r_i}, \quad a_c^{avg} = \frac{r_o^2 - r_i^2}{2(r_o - r_i)} \omega^2$$	based on o-xylene, $d_o = 457$ mm rotor
Chen et al. [72]	single-block, SS metal sheets; 77/165/20, 46/118/20; 17	$$k^L a_e = 3.38 u_L(r_i)^{0.747} u_G(r_o)^{0.282} \omega^{0.311} 1.016^{\vartheta - 20}$$	not suitable for other component systems (no dimensionless groups)
Chen et al. [27]	single-block, SS wire mesh/ plastic beads; 20...100/60...120/20; ≈320	$$k^L a_e = 0.65 \cdot (Sc^L)^{0.5} \cdot (Re^L)^{0.17} \cdot (Ga'^L)^{0.3} \cdot (We^L)^{0.3} \cdot$$ $$\left(\frac{a_p \cdot D^L}{d_p}\right) \cdot \left(1 - 0.93 \cdot \frac{V_{out}}{V_t} - 1.13 \cdot \frac{V_{in}}{V_t}\right)^{-1}$$	influence of bed geometry
Chen et al. [38]	single-block, SS wire mesh; 20/120/20; ≈100	$$k^L a_e = 0.9 \cdot (Sc^L)^{0.5} \cdot (Re^L)^{0.24} \cdot (Ga'^L)^{0.29} \cdot (We^L)^{0.29} \cdot \left(\frac{a_p \cdot D^L}{d_p}\right)$$	varying viscosity, Newtonian and non-Newtonian fluids
Chen [32]	single-block, various packings; 20/120/20; ≈240(+320);	$$k^L a_e = 0.35 \cdot (Sc^L)^{0.5} \cdot (Re^L)^{0.17} \cdot (Ga'^L)^{0.3} \cdot (We^L)^{0.3}$$ $$\cdot \left(\frac{a_p \cdot D^L}{d'_p}\right) \cdot \left(1 - 0.93 \cdot \frac{V_{out}}{V_t} - 1.13 \cdot \frac{V_{in}}{V_t}\right)^{-1} \cdot \left(\frac{a_p}{a_b}\right)^{-0.5} \cdot \left(\frac{\sigma_{cr}}{\sigma_w}\right)^{0.14}$$	various internals

Reference	Packing	Correlation	Notes
Reddy et al. [37]	co-/counter-rotating split packing, Ni-Cr metal foam; 81/171/30;18;	$k^{L,loc} a_e = 0.152 \cdot (Re^L)^{0.569} \cdot (Ga^L)^{0.14} \cdot (Sc^L)^{0.5} \cdot \left(\dfrac{a_p \cdot D^L}{d_p}\right)$	local correlation for corotation shown
Lin and Jian [54]	blades, covered with SS wire mesh; 39/125/29.5;20;	$k^L a_e = 0.478 \cdot (Re^L)^{0.906} \cdot (Ga''^L)^{0.275} \cdot \left(\dfrac{a_p \cdot D^L}{d_p}\right)$	
Shivhare et al. [59]	single-block & co-/counter-rotating split packing, Ni-Cr metal foam; 81/121 … 171/30;36;	$k^L a_e = 2.59 \cdot (Sc^L)^{0.5} \cdot (Re^L)^{0.93} \cdot (Ga''^L)^{0.25} \cdot \left((r_o^2 - r_i^2) \cdot \dfrac{a_p}{d_p}\right)^{-1.01} \cdot \left(\dfrac{a_p \cdot D^L}{d_p}\right)$	correlation for single block shown
Chen et al. [73]	single-block, SS and PTFE wire meshes; 80/164/15; ≈ 36;	$k^L a_e = 25.237 \cdot (Re''^L)^{0.815} \cdot (We''^L)^{0.214} \cdot (Fr^L)^{-0.270} \cdot (\varphi)^{-0.154} \cdot \left(\dfrac{\sigma}{\sigma_{cr}}\right)^{0.936} \cdot \left(\dfrac{a_p \cdot D^L}{d_p}\right)$	correlations for structured wire meshes, for plain-woven packing shown, mesh form factor $\varphi = c^2/(c+d)^2$
Zhao et al. [75]	rotor-stator (static pins, perforated rotating rings); 70/190/65; > 45;	$k^L a_{e c_1} \cdot (Re''^L)^{c_2} \cdot (Re^V)^{c_3} \cdot (We''^L)^{c_4} \cdot (We^V)^{c_5} \cdot (Fr'^L)^{c_6} \cdot (Sc^L)^{c_7} \cdot (Sc^V)^{c_8} \cdot (Eu^L)^{c_9} \cdot (Eu^V)^{c_{10}} \cdot \left(\dfrac{D^L}{r_i^2}\right)$	atmospheric and vacuum
Wen et al. [74]	single-block, polyamide wire mesh; 40/80/16.7; ≈ 75;	$k^L a_e = 7.749 \cdot (Re''^L)^{1.171} \cdot (We''^L)^{0.086} \cdot (Fr^L)^{-0.234} \cdot (\varphi)^{-0.291} \cdot \left(\dfrac{\sigma}{\sigma_{cr}}\right)^{-0.851} \cdot \left(\dfrac{a_p \cdot D^L}{d_p}\right)$	correlations for structured wire meshes, mesh form factor $\varphi = c^2/(c+d)^2$

Tab. 8.2: Overview of selected semi-empirical correlations for volumetric gas-side mass transfer coefficients in RPBs. Packing Material: Stainless Steel (SS).

Author	RPB-Type, Packing; dimensions $d_i/d_o/h_p$ / mm; no. of exp. points	Correlation	Comments
Reddy et al. [37]	co-/counter-rotating split packing, Ni-Cr metal foam; 81/296/30;18;	$k^V a_e = 7.62 \cdot 10^{-3} \cdot \left(Re^V\right)^{0.852} \cdot \left(Ga''^V\right)^{0.194} \cdot$ $\left(Sc^V\right)^{1/3} \cdot \left(\dfrac{a_p \cdot D^V}{d_p}\right)$	Rectangular box as casing
Chen [60]	*single-block, various packings;* 4...9/8... 18/3; ≈ 240(+320);	$k^V a_e = 0.023 \cdot \left(Re^V\right)^{1.13} \cdot \left(Re^L\right)^{0.14} \cdot \left(Ga'^V\right)^{0.31} \cdot$ $\left(We^L\right)^{0.07} \cdot \left(\dfrac{a_p}{a_b}\right)^{1.4} \cdot \left(a_p\right)^2 \cdot \left(D^V\right) \cdot \left(1 - 0.9 \cdot \dfrac{V_o}{V_t}\right)^{-1}$	Data from RPBs of different sources

were used and hence the inner end-effects zone, i.e. the zone of increased mass transfer near the eye of the rotor, where liquid first contacts the rotating packing, contributes more strongly to the overall observed mass transfer compared to larger RPBs. Reddy et al. developed a correlation for $k^V a_e$ based on an RPB with a split packing design. It uses metal foam rings alternatingly fixed to both rotor plates and applied in a co-rotating rotor, simulating a single-block RPB [37]. Furthermore, an attempt was made to exclude casing contributions by directly measuring gas concentrations in the casing.

8.3 Discretized rate-based modeling of an RPB

The current section provides the formulation of a rate-based mass transfer model for the RPB used in the subsequently described case study in Section 8.4 and builds on our own contributions [4, 5]. The overarching aim regarding modeling is the steady-state representation of separation efficiency and mass transfer depending on geometry and operation conditions for an RPB for multicomponent distillation. According to Fig. 8.2, the RPB is modeled as a series of N_j discrete elements of the packing which each holds first principles balance equations for mass and energy for contacting liquid and vapor phases (using hydraulic equations as simplified momentum balances). Contacting phases are connected through mass and energy balances around the interface. The equilibrium constants are determined by the interface compositions and temperature. The model development is then based on the following assumptions:

- steady-state process
- bulk phases on each stage are ideally mixed

- mechanical equilibrium $P_j^L = P_j^V = P_j$ for vapor and liquid in each discrete element j
- negligible mass transfer resistance of the phase interface
- no accumulation of energy or material at the interface
- equilibrium of phases at the interface
- no chemical reactions
- vapor phases are treated as ideal due to desired operation at atmospheric pressure and type of applied components
- liquid phases are considered as non-ideal for equilibrium calculations and heat of mixing is neglected, since it amounts to less than 3% of enthalpy of vaporization for applied components [76–79]
- mass transfer coefficients are assumed equal for finite flux and low flux, hence for the calculation of mass transfer rates at high flux, the correction factor matrix is set equal to the identity matrix, as it has been found to be close to unity in distillation [8]
- negligible heat losses from the rotor to the environment as the rotor is surrounded by hot vapor in the casing.
- condenser (stage $j = 0$) and reboiler (stage $j = N_j + 1$) are considered as total heat exchangers and no separation occurs in these units; both are modeled as equilibrium stages

The model is implemented in the modeling environment MOSAICmodeling[1] and exported to AMPL. The resulting model, notation, and applied parameter sets are available on request in MOSAICmodeling and provided as full evaluations.[2] The algebraic equation system is then solved using the interior point solver IPOPT [80]. Parameters required for property and equilibrium models are taken from Aspen Plus V10 databanks PURE32 and VLE-IG [81].

8.3.1 MERSHQ-equations

According to the approach for rigorous rate-based column modeling, **m**aterial and **e**nergy balances, **r**ate, **s**ummation, **h**ydraulic, and e**q**uilibrium (MERSHQ) equations are set up at the interface for each discrete element or non-equilibrium stage [8].

---

1 mosaic-modeling.de.
2 ID 146,637 (ethanol/n-propanol), ID 146,649 (ethanol/water), ID 146,665 (ethanol/n-propanol /water).

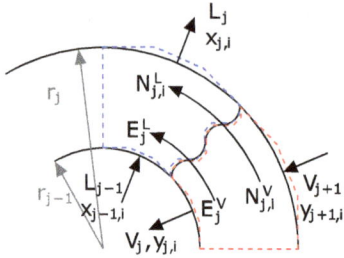

Fig. 8.2: Schematic representation of the jth non-equilibrium stage of an RPB for countercurrent gas-liquid mass transfer. Figure taken from [3].

M: Material balances

Generic component material balances are given for non-equilibrium stages j and component i according to Fig. 8.2:

$$\text{Liquid:} \qquad 0 = L_{j-1} \cdot x_{j-1,i} - L_j \cdot x_{j,i} + N_{j,i}^L \qquad (8.7)$$

$$\text{Vapor:} \qquad 0 = V_{j+1} \cdot y_{j+1,i} - V_j \cdot y_j - N_{j,i}^V \qquad (8.8)$$

$$\text{Interface:} \qquad 0 = N_{j,i}^L - N_{j,i}^V \qquad (8.9)$$

Liquid molar streams L and vapor molar streams V are denoted with index j for exiting streams. Mole fractions are denoted as $x_{j,i}$ and $y_{j,i}$, respectively. Mass transfer rates between liquid and vapor phase are given by $N_{j,i}$, while E_j is the rate of energy transfer.

E: Energy balances

Respective energy balances are given by:

$$\text{Liquid:} \qquad 0 = L_{j-1} \cdot h_{j-1}^L - L_j \cdot h_j^L + E_j^L \qquad (8.10)$$

$$\text{Vapor:} \qquad 0 = V_{j+1} \cdot h_{j+1}^V - V_j \cdot h_j^V - E_j^V \qquad (8.11)$$

$$\text{Interface:} \qquad 0 = E_j^L - E_j^V \qquad (8.12)$$

R: Transfer rate equations

Generic mass transfer rate equations are given by eqs. (8.13)–(8.14) applying the component index $i = 1, 2 \ldots, N_c - 1$ with N_c as the total number of components.

Liquid: $\qquad 0 = \left(N_j^L\right) - \rho_{Mix,j}^{L,mol} \cdot \left[k^L a_{e,j}\right] \cdot V_j^{discr} \cdot \left(x_j^I - x_j\right) - \sum_{i=1}^{NC} N_{j,i}^L \cdot \left(x_j\right)$ (8.13)

Vapor: $\qquad 0 = \left(N_j^V\right) - \rho_{Mix,j}^{V,mol} \cdot \left[k^V a_{e,j}\right] \cdot V_j^{discr} \cdot \left(y_j - y_j^I\right) - \sum_{i=1}^{NC} N_{j,i}^V \cdot \left(y_j\right)$ (8.14)

Superscript I indicates values at the liquid-vapor or liquid-gas interface. $\left(N_j^L\right)$ and $\left(N_j^V\right)$ display the vectors of elements of $N_{j,i}^L$ and $N_{j,i}^V$, while $\left[k^{L/V} a_{e,j}\right]$ are matrices of volumetric low-flux mass transfer coefficients (product of mass transfer coefficients and specific effective surface area). This implementation is chosen since correlations from literature usually aim for the volumetric coefficients and no specific effective surface areas are determined in the case study. (x_j), (y_j) and displayed differences with respective mole fractions at the interface are vectors with elements $x_{j,i}$ and $y_{j,i}$.

Generic energy transfer rate equations are given by eqs. (8.15)–(8.16):

Liquid: $\qquad 0 = E_j^L - \alpha^L a_{e,j} \cdot V_j^{discr} \cdot \left(T_j^I - T_j^L\right) - \sum_{i=1}^{NC} N_{j,i}^L \cdot \bar{h}_{j,i}^L$ (8.15)

Vapor: $\qquad 0 = E_j^V - \alpha^V a_{e,j} \cdot V_j^{discr} \cdot \left(T_j^V - T_j^I\right) - \sum_{i=1}^{NC} N_{j,i}^V \cdot \bar{h}_{j,i}^V$ (8.16)

$$0 = V_j^{discr} - \pi \cdot \left((r_j)^2 - (r_{j-1})^2\right) \cdot h_p$$ (8.17)

$\bar{h}$ denote the partial molar enthalpy and $\alpha\, a_{e,j}$ the volumetric heat transfer coefficients. The volume of a discrete element is given by eq. (8.17) and depends on the applied discretization scheme, which is further detailed in Section 8.3.4.

S: Summation equation

Summation equations are given for bulk and interface for vapor and liquid phase, respectively:

$$0 = 1 - \sum_{i=1}^{NC} x_{j,i}, \quad 0 = 1 - \sum_{i=1}^{NC} y_{j,i}$$ (8.18)

$$0 = 1 - \sum_{i=1}^{NC} x_{j,i}^I, \quad 0 = 1 - \sum_{i=1}^{NC} y_{j,i}^I$$ (8.19)

H: Hydraulic equation

As simplified momentum balances, pressures of adjacent non-equilibrium stages are related by eq. (8.20) with pressure drop ΔP_j calculated using the model of Neumann et al. [82] in eqs. (8.21)–(8.26). For the given application, $P_{j=1}$ is set to the pressure of the condenser.

$$0 = P_{j+1} - P_j - \Delta P_j \tag{8.20}$$

The model of Neumann et al. [82] was developed for an RPB of similar size and packing material, as used in the case study in Section 8.4. However, applicability for other model implementations and usage on experimental data or simulations should be validated.

$$0 = \Delta P_j - \zeta_{CH,j} - \Delta P_{f,stat,j} \tag{8.21}$$

$$0 = \zeta_{CH,j} - A_{CH} \cdot \frac{\rho_{Mix,j}^{V,mass} \cdot \omega^2}{2} \cdot \left((r_j)^2 - (r_{j-1})^2 \right) \tag{8.22}$$

$$0 = \Delta P_{f,stat,j} - \Psi \varphi_{0,j} \cdot (1 - \varphi) \cdot \frac{1-\epsilon}{\epsilon^3} \cdot \frac{\bar{F}_{int,j}^V}{d_p} \cdot (r_j - r_{j-1}) \tag{8.23}$$

$$0 = \bar{F}_{int,j}^V - \frac{V / \rho_{Mix,j}^{V,mol}}{2 \cdot \pi \cdot h_p \cdot (r_j - r_{j-1})} \cdot \ln\left(\frac{r_j}{r_{j-1}} \right) \cdot \sqrt{\rho_{Mix,j}^{V,mass}} \tag{8.24}$$

$$0 = \varphi_{0,j} - \frac{725.6}{\overline{Re}_{int,j}^V} - 3.203 \tag{8.25}$$

$$0 = \overline{Re}_{int,j}^V - \frac{d_p}{(1-\epsilon) \cdot \frac{\eta_{Mix,j}^V}{\rho_{Mix,j}^{V,mass}}} \cdot \frac{\bar{F}_{int,j}^V}{\sqrt{\rho_{Mix,j}^{V,mass}}} \tag{8.26}$$

It contains elements for centrifugal head ζ_{CH} (pressure difference across the rotor induced by rotation) and frictional pressure drop $\Delta P_{f,stat}$ of the gas flowing through the rotor, while neglecting the influence of liquid flow. ζ_{CH} is calculated using eq. (8.22) with A_{CH} as rotor specific adjustable parameter. $\Delta P_{f,stat}$ is calculated using eq. (8.23) with φ as the frictional form factor of the packing (see details on parameter fit in [82]).

Q: Interface equilibrium equation

Equilibrium is assumed at the interface with the vapor pressure $P_{j,i}^{LV}$ of the pure component and activity coefficients $\gamma_{j,i}$ calculated from the NRTL model.

$$0 = y_{j,i}^I \cdot P_j - x_{j,i}^I \cdot \gamma_{j,i} \cdot P_{j,i}^{LV} \qquad (8.27)$$

8.3.2 Calculation of mass transfer coefficients

Volumetric multicomponent mass transfer coefficients are calculated from RPB-specific binary mass transfer correlations by using the R-matrix in combination with the matrix of thermodynamic correction factors Γ as proposed by Taylor and Krishna [8]:

$$\left[ka_{e,j} \right] = \left[R_j \right]^{-1} \cdot \left[\Gamma_j \right] \qquad (8.28)$$

For calculating Γ, the elements of the R-matrix, and calculating binary mass transfer coefficients from (local) correlations, the linearized theory of Toor, Stewart, and Prober is applied [8]. Hence, arithmetic mean values of mole fractions and temperatures between interface and bulk are applied. Elements of R are then calculated by eqs. (8.29)–(8.30):

$$R_{j,m,m} = \frac{z_{j,m}}{ka_{e,j,m,NC}} + \sum_{\substack{k \neq m \\ k=1}}^{NC} \frac{z_{j,k}}{ka_{e,j,m,k}} \qquad (8.29)$$

$$R_{j,m,n} = -z_{j,m} \cdot \left(\frac{1}{ka_{e,j,m,n}} - \frac{1}{ka_{e,j,m,NC}} \right) \qquad (8.30)$$

$z_{j,m}$ and $z_{j,k}$ are thus the arithmetic mean molar phase compositions of component m or k on stage j between bulk and interface. $ka_{e,j,m,n}$ are the binary volumetric mass transfer coefficients for binary pairs $m - n$. The thermodynamic correction matrix $\left[\Gamma_j \right]$ is defined using eq. (8.31) with the Kronecker Delta given in eq. (8.32). The sum used in the partial derivative indicates constant mole fractions except the N_c'th to satisfy the summation equation. Again, activity coefficients are calculated using the NRTL model.

$$\Gamma_{j,m,n} = \delta_{m,n} + x_{j,m} \cdot \frac{\partial \ln \gamma_{j,m}}{\partial x_{j,n}} \bigg|_{T,P,\Sigma} \qquad (8.31)$$

$$\delta_{m,n} = \begin{cases} 1 \text{ if } m = n \\ 0 \text{ if } m \neq n \end{cases} \qquad (8.32)$$

The viscosity of the liquid mixture is calculated using the Generalized Corresponding States Method (GCSM) of Teja and Rice [83]. The method yields different results depending on the choice of the two reference fluids in multicomponent mixtures. Here, the reference fluids are chosen as the respective binary pairs for binary mixtures and ethanol/water for the ternary mixture used in the case study. Binary diffusion coefficients at infinite dilution are estimated using the Wilke-Chang and Wilke-Lee correlations for liquid and vapor phase, respectively [84]. Liquid diffusion coefficients at finite dilution are given by eq. (8.33), which is a geometrically consistent generalization of the Vignes equation [8].

$$D_{j,i,k}^L = \left(D_{j,i,k}^\infty\right)^{\left(1 + x_{j,k} - x_{j,i}\right)/2} \cdot \left(D_{j,k,i}^\infty\right)^{\left(1 + x_{j,i} - x_{j,k}\right)/2} \tag{8.33}$$

8.3.3 Calculation of volumetric heat transfer coefficients

Diffusivities are calculated as average values based on eq. (8.34) using arithmetic mean values of mole fractions and temperatures between interface and bulk. This approach is also used for averaging of mass transfer coefficients (eq. (8.35)).

$$\bar{D}_j = \frac{\sum_{m=1}^{NC-1} \sum_{n=m+1}^{NC} \left(x_{j,m} + \bar{\delta}\right)\left(x_{j,n} + \bar{\delta}\right) D_{j,m,n}}{\sum_{m=1}^{NC-1} \sum_{n=m+1}^{NC} \left(x_{j,m} + \bar{\delta}\right)\left(x_{j,n} + \bar{\delta}\right)} \tag{8.34}$$

Volumetric heat transfer coefficients for vapor and liquid phases are then estimated using the averaged volumetric mass transfer coefficients and diffusivities:

$$\alpha a_{e,j} = \rho_{Mix,j}^{Mass} \cdot c_{p,Mix,j}^{Mass} \cdot \overline{ka}_{e,j} \cdot \left(\frac{\lambda_{Mix,j}}{\rho_{Mix,j}^{Mass} \cdot c_{p,Mix,j}^{Mass} \cdot \bar{D}_j}\right)^{2/3} \tag{8.35}$$

$$St_\alpha Pr^{2/3} = StSc^{2/3} \tag{8.36}$$

In analogy to packed columns, eq. (8.35) is derived from the Chilton-Colburn-analogy (eq. (8.36)) [85]. It was found that uncertainties or errors introduced by using the Chilton-Colburn analogy have minor influence on obtained composition profiles in rate-based modeling of distillation columns.

8.3.4 Discretization scheme

As already stated, the volume of discrete elements strongly depends on the applied discretization scheme. Two approaches are reported in literature: the equidistant scheme (equal radial length of elements) and the equiareal discretization (equal volume of elements) by Sudhoff et al. [86]. The equiareal scheme also provides equal area per element, assuming constant packing height. However, radial length will decrease with increasing radius (see Fig. 8.3). The choice of discretization scheme and applied number of discrete elements depends on the specific application case regarding used RPB geometry, packing, operation conditions, and component systems. Sudhoff et al. claim that the equiareal approach results in lower required discrete elements [86].

This, however, should be checked with simplified simulation studies, where the number of discrete elements is increased until no significant changes in calculated outlet mass fractions occur.

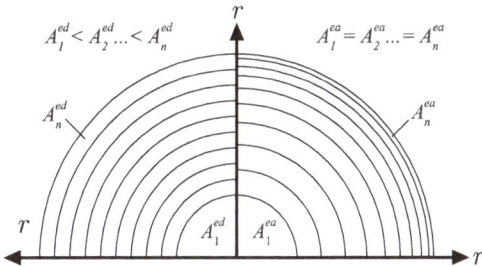

Fig. 8.3: Discretization approaches (left: equidistant (ed); right: equiareal (ea) discretization).

8.4 Case study: pilot-plant system for multi-component distillation

The status quo on RPB literature regarding mass transfer modeling revealed several obstacles especially in the development of respective correlations and their transferability for varying geometrical configurations. Furthermore, a major deficiency in industrially relevant applications is visible. The availability of suitable descriptive models however is a crucial step towards conceptual design of processes using RPBs in process synthesis, equipment design and scale-up, as well as unlocking the potential in flexible operation of these apparatuses. Due to the complexity of interactions between RPBs and packing design, operation conditions with resulting operation behavior and separation efficiency, generic and widely applicable models are currently out of reach.

To nevertheless enable the application of RPBs, it is desirable to have efficient modeling approaches at hand, which allow for the evaluation of specific RPB configurations and compare their performance to, e.g., other distillation equipment. One approach to this is outlined in the following case study. Based on the status quo, several goals are defined for this case study:
– Capture experimental data for multicomponent distillation as an industrially relevant case.
– Deploy pilot-scale RPB to reduce the influence of end effects and trigger industrial relevance.
– Cover large operation region and high loads with F-factors up to 5 $Pa^{0.5}$.

– Modeling and mass transfer correlations should focus on capturing the efficiency maximum depending on load and rotational speed as it is highly important for optimization of unit design and operation.

For the case study, a pilot-plant scale single-block RPB distillation process has been
built and operated. Based on the experimental data, a rate-based model for multicomponent distillation is developed and validated. Key to this model are suitable correlations for gas and liquid mass transfer coefficients, which are developed for the
given pilot-plant. To reduce experimental errors and uncertainty in evaluating specific
effective surface areas, volumetric mass transfer coefficients are used in accordance
with RPB literature. The ternary system ethanol/n-propanol/water is used as a test system since it is suitable for distillation at atmospheric pressures and provides sufficient
changes in composition given the expected numbers of theoretical stages (4 to 8). It
also offers analytical accessibility, literature data for comparison, and low toxicity.

The procedure for developing respective correlations is based on the resistances in
series model derived from the two-film theory (see Section 8.1.1) and sketched in
Fig. 8.4. **Without temperature and concentration profiles, which are hard to obtain experimentally for single-block RPBs, only $k^L \cdot a_e$ or $k^V \cdot a_e$ can be obtained
from distillation experiments (top and bottom composition measurements for
distillation at infinite reflux) [87].** Following Section 8.1.2, it is indicated that $k^L \cdot a_e$
is likely contributing significantly to the overall mass transfer resistance and cannot be
neglected for distillation operation. Hence, this case study captures both gas and liquid-side mass transfer resistances. **Ab- and desorption experiments are considered
to capture $k^L \cdot a_e$ in a first step.** Suitable test systems have to be used here in order to
assume $k^V \cdot a_e$ to be negligible. Following RPB literature (see also Tab. 8.1) and VDI
2761-2 [9] desorption of oxygen from water into a nitrogen gas stream is used. This system provides negligible vapor-side mass transfer resistance, is nonhazardous, and provides very good analytical access.

**The second step then applies binary distillation experiments to measure
the overall mass transfer coefficients $K^V \cdot a_e$, from which $k^V \cdot a_e$ values are back-
calculated in the third step.** This approach holds the advantage of minimizing experimental effort by eliminating separately required measurements of gas-side mass
transfer coefficients. The procedure is illustrated in Fig. 8.4.

Based on the collected experimental data, adequate correlations for mass transfer
coefficients can be developed or integrated from literature. Finally, the rate-based
model of the RPB is validated using data from ternary distillation experiments.

It is pointed out that the presented case study omits any (model-based) scale-
up studies or design optimization of RPBs. However, the presented approach is
thoroughly investigating model development for multicomponent distillation and
can be readily adapted for investigations on differently sized RPBs or modified designs. Based on this, the development of scale-up methodologies and generalized
models to assist process operation is attainable.

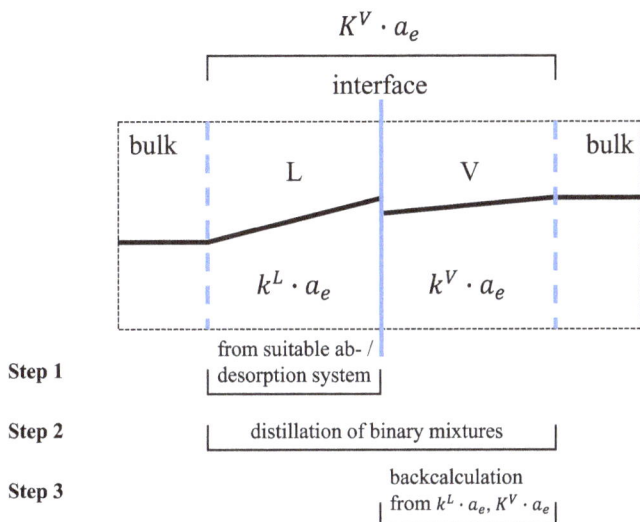

Fig. 8.4: Employed approach for obtaining volumetric mass transfer coefficients according to the resistances in series model based on the two-film theory.

8.4.1 Materials and pilot-plant setup

Applied chemicals and the pilot-plant setup are outlined in the following. The former refers to desorption, binary distillation, as well as ternary distillation experiments. The latter comprises the specific design of the employed RPB including packing and nozzle geometry as well as integration into a process for distillation experiments at total reflux and desorption experiments for measuring $k^L \cdot a_e$.

Applied chemicals

Applied substances are listed in Tab. 8.3.

Tab. 8.3: Applied substances for desorption or distillation experiments and analytics.

Substance	Purity	Supplier	Application
Ethanol	≥99.8%	Carl Roth GmbH + Co. KG	distillation
N-propanol	≥99.5%	Carl Roth GmbH + Co. KG	distillation
Water	deionized	–	distillation, desorption

Tab. 8.3 (continued)

Substance	Purity	Supplier	Application
2-propanol	≥99.8%	Carl Roth GmbH + Co. KG	analytics
Nitrogen	5.0	Linde	desorption

RPB design

The design of the RPB and the rotor setup are summarized in Tab. 8.4. For placing and adjustment of the liquid distribution system (two adjacent full jet nozzles with 3 orifices each) an inner radius of 60 mm is chosen. To avoid dominating influence of end effects and achieve industrial relevant sizing, an outer packing radius of 225 mm is chosen. Actual throughput is then determined by the bed height. For distillation experiments, a reboiler with 25 kW heat duty is employed. Using a bed height of 10 mm, F-factors of up to 5 $Pa^{0.5}$ can be realized at the eye of the RPB depending on the applied component system. Inner and outer support rings fix applied packing material on the rotor, whereby a large free cross-sectional area is ensured to avoid shadowing of the packing and increased end effects. NCX1116 Cr-Ni metal foam supplied by Recemat BV is used as packing material. It provides a specific geometric surface area of 1,000 m^2m^{-3}, a porosity of 0.92, and an average pore size of 1.4 mm [88]. Ring segments are used to realize a variable radial bed length. The casing of the RPB is designed to minimize liquid holdup and liquid residence time. The floor is beveled towards the outer diameter, where three liquid outlets are placed. Additionally, radial wipers are installed to direct liquid towards the outlets and prevent a churning of the liquid on the bottom plate of the casing. The gas inlet is located at the outer radius and provided through the top cover. Vapor flow is then driven by the pressure gradient and passes the packing before exiting the casing through the central eye of the rotor.

The RPB device is made from stainless steel and can be operated for distillation within the pressure range from 0.1 to 1.5 bar(a) and temperatures up to 150 °C. The rotational speed of the RPB is set by a frequency-controlled speed drive allowing for 0 to 2,800 min^{-1}.

Tab. 8.4: Rotating Packed Bed geometry, packing and liquid distributor specifications.

Geometry	Design specification
Casing – inner diameter/height	533 mm/105 mm
Packing – inner diameter/outer diameter/height	120 mm/450 mm/10 mm

Tab. 8.4 (continued)

Geometry	Design specification
Packing – inner support ring	10 vertical ribs, ≈ 90% free area
Packing – outer support	3 cylindrical disc spacers; diameter: 20 mm
Packing – type	Ni-Cr metal foam Recemat NCX1116 ($\epsilon = 0.92$, $a_P = 1{,}000\ \mathrm{m^2/m^3}$, average pore diameter 1.4 mm) [88]
Gas Flow – inlet/outlet diameter (both top)	63 mm
Liquid – diameter outlets (3 in bottom)	25 mm
Liquid – distributor	2 full jet nozzles with three orifices each 180° offset; inner diameter: 6 mm, diameter of orifices: 0.7 mm

Pilot-plant configuration and flow diagrams

Distillation and desorption are used as the two major operation modes. In both cases, specific peripheral equipment is used, whereas the general pilot-plant configuration remains unchanged (see Fig. 8.5). This enables fast switching between experimental procedures. Employed measurement devices and respective accuracies are listed in Tab. 8.5. The plant is fully automatized with Siemens PCS7, which allows for manipulation of operation modes and efficient online tracking of measurement data. Additionally, detailed HAZOP studies have been conducted to ensure safe operation of the setup.

Distillation operation for evaluation of separation efficiency is usually conducted at infinite reflux. The respective plant configuration for such an operation mode is given by the P&ID in Fig. 8.6. A 25 kW electrical reboiler is used for evaporating the liquid. The vapor then enters the RPB casing at the outer diameter, passes the packing due to pressure gradients, and exits the RPB at the eye. Here, a top seal prevents vapor from bypassing the rotor. Using a total condenser, the vapor is liquefied and subcooled to avoid cavitation in the reflux pump and collected in the reflux drum D-02. Using pump P-02 and controlled via the liquid mass flow meter FI-02 liquid is then fed via a reflux preheater to the spraying nozzles and evenly distributed on the inner radius of the RPB packing. For steady-state operation, the pump speed is regulated using level control on tank D-02. Liquid sprayed onto the packing is driven to the outside by centrifugal force, reaches the outer casing as ligaments and droplets, and is finally routed back to the reboiler. Several liquid sampling positions enable concentration measurements via offline GC analysis. Here, S-03 and S-06 are used for direct sampling at the edge of the packing and outlet tube, respectively. S-04 is used for sampling of condensed distillate, while S-02 samples vapor

Fig. 8.5: Pilot-plant setup including RPB (2) and peripheral equipment (1-condenser, 3-Drum D-01, 4-reboiler). Taken from [3].

Tab. 8.5: Sensors and measurement devices used within the pilot-plant and respective accuracies.

PID tag name	Measurement	Sensor type	Accuracy
TI01 – TI12	Temperature	Pt100	$0.15 + 0.002\ T(°C)$
PI01, PI02	Pressure	Piezoresistive	6 mbar
PDI01	Pressure difference	Piezoresistive	0.5 mbar
SI01	Rotational speed	Optical	n.a.
FI02	Liquid mass flow	Coriolis	0.15%
QI01	Dissolved oxygen	Optical	$|\Delta c_{rel}| = 4\ \%$
QI02	Dissolved oxygen	Optical	$|\Delta c_{rel}| = max\left(4\ \% + \dfrac{0.125\mu mol/L}{c(\mu mol/L)},\ 5\ \%\right)$
S-01 – SI06	Liquid sampling	GC-TCD	$|\Delta w_{rel}| = 4\ \%$

from the casing. Additionally, the plant is inertized using nitrogen fed to the reboiler via thermal mass flow controller MFC-01 and routed to atmosphere through magnet valve MV-01.

Desorption operation is carried out using the configuration shown in the P&ID in Fig. 8.7. A 220 L barrel (D-03) is used for saturating water with oxygen by gassing with oxygen. The liquid feed is then provided to the liquid distributor of the RPB using pump P-02, while measuring the feed flow rate (FI-02) and oxygen concentration QI-01

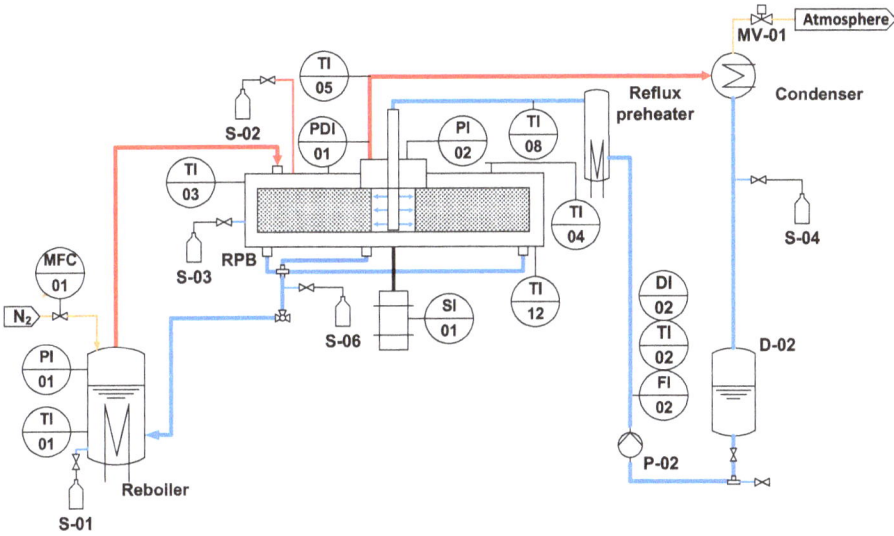

Fig. 8.6: Process flow diagram of the pilot-plant configuration used for distillation at infinite reflux. Taken from [3].

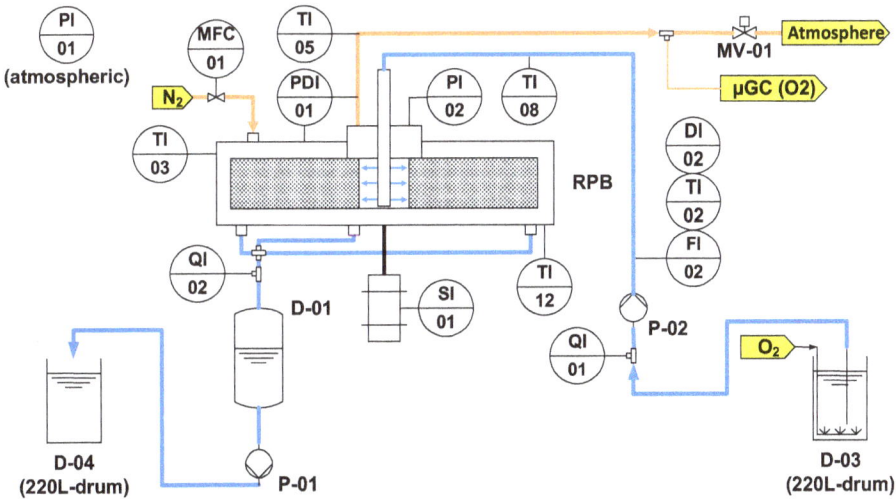

Fig. 8.7: Process flow diagram of the pilot-plant configuration used for oxygen desorption from water into a countercurrent nitrogen stream. Taken from [4].

(OXROB10 of Pyroscience GmbH). After oxygen desorption in the RPB, the water is collected in the casing and led to buffer tank D-01. For this outlet stream, again dissolved oxygen is measured using QI-02 (TROXROB10 of Pyroscience GmbH). Finally, the water is collected in D-04 using pump P-01 and level control on D-01.

Countercurrent operation of the RPB is then initialized using nitrogen supplied by gas cylinders to the outer part of the RPB casing. The gas stream passes the packing, exits the RPB at the eye, and is routed to the atmosphere. Nitrogen flow rate is controlled through thermal mass flow controller MFC01 and outlet oxygen concentrations are tracked online using a micro-GC (μGC).

Analytics

Analysis of liquid samples is carried out using an Agilent Technologies 6850 GC System with an HP-PLOT Q capillary column (30.0 m × 320 μm × 20.0 μm nominal) and a TC-Detector. *Iso*-propanol is used as solvent and helium as carrier gas. Calibration was done based on external standards.

Procedure of desorption experiments for oxygen-water

For preparation purpose, sensors for dissolved oxygen are calibrated using deionized water, which was initially gassed with nitrogen for ten minutes. To scavenge residual oxygen, sodium sulfite was dissolved to obtain a 30 g/L solution prepared in a covered flask. After 15 min oxygen sensors and a Pt100 were pierced through the cover and zero points were recorded upon reaching constant readings. A second calibration point was obtained by treating water with compressed air for 10 min within a flask and adding the oxygen and temperature sensors afterwards. Prior to plant operation, nozzle positions were adjusted to properly align jet impact points over the height of the packing. After pressure testing and ensuring a maximum pressure loss of typically <6 mbar/h at 1.5 bar(a), the system was inertized using pressure cycling of nitrogen and checking outlet oxygen concentration via the μGC.

Desorption experiments are carried out at a plant pressure of 40 to 100 mbar above ambient pressure, depending on the gas flow rate of nitrogen. Deionized water in barrel D-03 was saturated with oxygen by gassing pure oxygen from a gas container. In contrast to the recommendations in VDI 2761-2 [21], no pre-saturation of the feed N_2 stream with water was performed, because test runs showed that the influence on resulting liquid-side mass transfer coefficients is negligible.

Plant operation was then started by activating nitrogen and water feed to the plant. Different operating points were investigated by setting desired values of rotational speed, nitrogen flow rate, and water flow rate. For each operation point, steady state was assumed when constant measurements of flow rate, levels, pressures, and dissolved oxygen concentration were found. Sensor data required for the calculation of results were averaged over a period of ten minutes. Operating points were typically repeated twice on different days to demonstrate reproducibility. As hysteresis effects could not be observed in preliminary tests, no fixed procedure regarding setpoint changes and plant trajectories was applied.

Procedure of binary and ternary distillation

In general, distillation experiments are performed using the test system ethanol/*n*-propanol/water or its binary subsystems. Prior to plant operation, the nozzle positions were adjusted to properly align the jet impact points over the height of the packing. Pressure testing was conducted to ensure a maximum pressure loss below 6 mbar/h at 1.5 bar(a). Afterwards, the system was inertized using pressure cycling with nitrogen (purity > 99.8%).

Distillation operation was then started by applying desired setpoints for reboiler duty of liquid flow to the RPB via P-02. Reaching steady state took approximately 1 h, mainly due to the heat capacities of RPB, peripheral equipment, and piping. On setpoint change, the time of transient operation was mainly governed by the liquid hold-ups in the peripheral equipment (reboiler and reflux drum) and was approximately 30 min depending on the impact of setpoint changes. Main operation parameters are rotational speed, reboiler duty, and an estimated pre-heater temperature, which were varied to achieve several steady states. While operating the plant, a sub-cooling of the reflux fed to the RPB of approximately 0.5 K was targeted (based on bubble point calculations for distillate composition and pressure at the eye of the RPB). Required concentration data were obtained from sampling and GC measurements (ethanol/*n*-propanol and ethanol/*n*-propanol/water systems) or based on density and temperature measurements from the Coriolis flow meter (ethanol/water, using the correlation of Khattab et al. [89]).

After reaching the desired setpoint of the pre-heater, steady state is assumed on obtaining constant values for all measurements over a 30 min horizon. On reaching steady state, all relevant samples were taken using syringes pierced through septa. Samples were placed in a refrigerator until analysis on the same day. After another 10 min, a second set of samples was taken to verify steady-state conditions and a steady state was only accepted if concentration evaluations matched within the accuracy of the GC method. Respective sensor measurements were averaged over the time interval of sampling. To demonstrate reproducibility, each operating point used for model development was repeated once, usually on a different day. As hysteresis effects could not be observed in preliminary tests, no fixed procedure regarding set point changes and plant trajectories was applied.

8.4.2 Liquid-side mass transfer: experiments and correlation development

As outlined in Section 8.3.2, the mathematical description of liquid-side (and gas-side) mass transfer coefficients is a crucial part of rate-based modeling. Due to experimental effort and hence a limited set of experimental data, mechanistic or data-driven modeling approaches are out of scope and hence semi-empirical correlations are used for modeling. Prior to the testing and development of correlations, performed experimental studies on desorption processes within the RPB pilot-plant are evaluated with respect to the influence of the main operation parameters. Subsequently, available correlations from literature are tested on collected experimental data before developing a new correlation for appropriate description of observed separation efficiency.

Transport properties required for calculations within this section are calculated based on literature correlations for pure water. Since rather low concentrations of dissolved oxygen are expected, liquid diffusion coefficients are approximated using infinite dilution diffusion coefficients calculated with the Wilke-Chang method [90]. Density and viscosity of liquids are calculated using the PPDS equations and DIPPR 106 equation together with coefficients from VDI-Wärmeatlas 2013 [91]. Henry coefficients used in evaluations of the experimental results are calculated based on the fit of Garcia and Gordon for equilibrium dissolved oxygen concentration in water [92].

Results of desorption experiments

The evaluation of experimental data considers three main operation parameters influencing mass transfer: rotational speed, water flow rate, and nitrogen gas flow rate. Here, the water flow rate is characterized using the **liquid load at the inner cross-sectional area** of the rotor. The gas flow rate is then described using standard liters per minute (SLPM, at 0 °C and 1.01325 bar(a)) of nitrogen.

Resulting volumetric liquid-side mass transfer coefficients are calculated using eq. (8.37). It is derived by integrating the differential material balance of oxygen in the liquid phase and combining it with the overall material balance of oxygen in the rotor to calculate oxygen concentrations in the exiting gas stream. This equation is further based on several assumptions:

- process at steady state
- negligible change in liquid flow rate $\dot{V}_L$ with radial length
- vanishing gas-side resistance to mass transfer
- applicability of Henry's law and constant temperature over the rotor (i.e. constant Henry coefficients He)
- oxygen solely stems from the water feed to the RPB
- ideal gas phase due to low pressure
- $k^L a_e$ is averaged for the radial length and hence assumed constant

The resulting equation for countercurrent operation is [27]:

$$k^L a_e = \frac{\dot{V}_L}{h_p \pi \left(r_o^2 - r_i^2\right)} \frac{1}{1 - \frac{RT}{He}\frac{\dot{V}_L}{\dot{V}_G}} \ln\left(\frac{c_{L,in} - \frac{RT}{He}\left(c_{G,in} + \frac{\dot{V}_L}{\dot{V}_G}\left(c_{L,in} - c_{L,out}\right)\right)}{c_{L,out} - \frac{RT}{He}c_{G,in}}\right) \tag{8.37}$$

Here, the reference volume of the packed rotor is equal to the volume of the packing, while the reference volume for the empty rotor is chosen to be that of the empty rotor plates and hence both are assumed to be the same. With oxygen sensors placed at the inlet and outlet streams, end effects are attributed to the packing using eq. (8.37).

Experimental uncertainties reported and error bars shown in the following are derived from uncertainties of measurement devices, analytics, and repeat runs. Details can be found elsewhere [4].

The experimental results are then based on series of data for two different gas flow rates of nitrogen at 6.6 and 12.3 SLPM. The resulting mass transfer coefficients are displayed in Fig. 8.8 comparing empty and packed rotor. **As a general result and as already indicated from the status quo a distinct maximum in separation efficiency depending on rotor speed is visible for the packed rotor.** At very high centrifugal accelerations, the separation efficiency of the packed rotor approaches a lower limit, which is represented by the empty rotor. The maximum is more pronounced for higher liquid loads but is generally found for both investigated gas flow rates. **Higher gas flow rates move the maximum to higher rotor speeds at otherwise constant conditions.** For the lowest liquid flow rate of 4.8 m^3 m^{-2} h, no maximum is identified while also the difference in $k^L a_e$ for the two gas flow rates vanishes. With increasing liquid flow rate however, a significant improvement in $k^L a_e$ is found for the higher gas flow rate, starting at rotational speeds of 500 . . . 800 min^{-1}. The influence of the gas flow rate is also sporadically reported in literature and seems to depend on both RPB size and geometrical configuration [73, 72, 93]. As a reason, increased driving forces due to a higher dilution of oxygen in the gas phase could be assumed. This effect however is already accounted for in eq. (8.37). One should rather take a closer look at the specific effective surface area and mass transfer enhancement. The former was found to increase with higher gas flow rates [40, 41, 73], while the latter is known to be affected by increasing gas-liquid interaction [94].

Keeping rotational speed constant, this effect is more pronounced for higher liquid flow rates because higher liquid holdups are present and gas-liquid interaction is increased. However, if the rotational speed is too low, the liquid is not sufficiently dispersed to observe an influence of the gas flow rate. At higher gas flow rates, the improvement of mass transfer due to gas-liquid interaction is sufficiently high to counteract the reduction in residence time. Hence, the maximum in mass transfer efficiency is shifted to higher rotational speeds (see Fig. 8.8b). Further

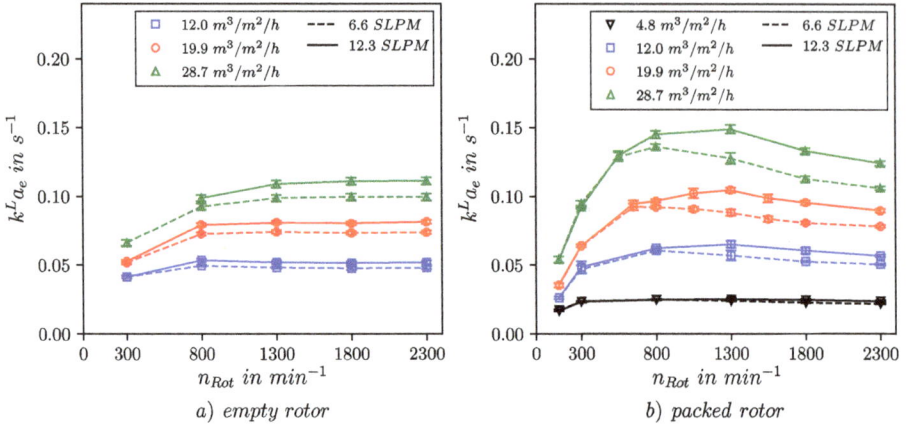

Fig. 8.8: Comparison of $k^L a_e$ values for countercurrent desorption using two nitrogen gas flow rates in the a) empty and b) packed rotor depending on rotational speed and liquid load at the eye. Values calculated from experimental data using eq. (8.34). Taken from [4].

increase of the gas flow rate however will likely yield entrainment and finally flooding, lowering the separation efficiency again.

Evaluation of literature correlations

To assess the applicability of available literature correlations for $k^L a_e$, simulation studies are compared with the collected experimental data. Of interest is the applicability of developed correlations (or their structure) for RPBs of different design and sizes compared to the originally used devices. Capabilities of capturing relevant operational phenomena, such as the outlined maxima in separation efficiency, are of special interest. Two correlations, already displayed in Tab. 8.1, are used. Chen [32] presents a correlation based on an extensive amount of data, including variations in fluid properties and packing geometry and material to account for end effects (correlation is a modification of Chen et al. [27]); Reddy et al. [37] obtained a correlation from an RPB employing metal foam packings. Since the investigation was conducted in a split-packing RPB, the correlation for co-rotation is used to describe the single-block packing applied in this case study. For the correlation developed by Chen et al. a design parameter for sphericity of the metal foam is required and set to $\psi = 0.75$ according to Gangosa Posodas [95], while a critical surface tension of the chromium-nickel material of 75 mN/m is used [96] along with the arithmetic mean radius $0.5(r_i + r_o)$. The comparison of simulated and experimentally obtained volumetric liquid-side mass transfer coefficients is shown in Fig. 8.9. Apparently, both correlations over-predict the experimental $k^L a_e$ values largely. This is likely due to the rather small geometries used for developing the correlations and hence to higher influence of end effects on rotor averaged $k^L a_e$. Additionally, the maximum in separation efficiency is not captured due to the inherent structure of these correlations, as centrifugal acceleration is only contained in one term raised to a constant power.

This is a general finding for correlations in Tab. 8.1, which are hence assumed not to be suitable to represent the observed operation behavior.

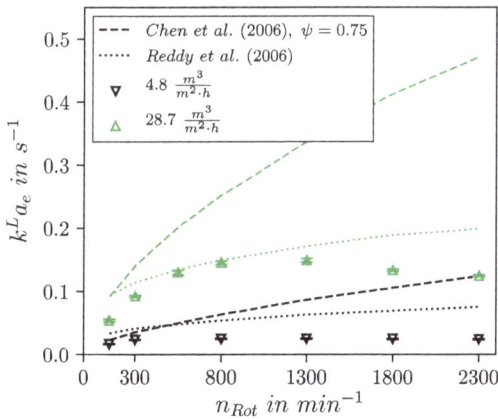

Fig. 8.9: Comparison of simulated values for $k^l a_e$ from literature correlations and from experimental data for countercurrent desorption at a nitrogen gas flow rate of 12.3 SLPM in the packed rotor at the highest and lowest investigated liquid loads at the eye.

Development of a new correlation for $k^l a_e$

Thus, it is necessary to develop a novel correlation, which is capable of predicting liquid-side volumetric mass transfer coefficients for the operated pilot-scale RPB and explicitly accounts for the maximum of separation efficiency.

The main novelty comprises the consideration of two additive terms representing the empty and the packed rotor, which both contribute to mass transfer. This is derived from the observation of converging $k^L a_e$ values for empty and packed rotor at high rotational speeds (see Fig. 8.8).

Reference lengths used for dimensionless numbers are the length of the liquid flow path $r_o - r_i$ for the empty rotor (as proposed by Munjal et al. [47]) and in case of the packed rotor the specific geometric surface area of the packing material (as proposed by Singh [71]). The general form of the correlation parts is aligned with the correlation by Reddy et al. [46]. Here, contributions of the liquid flow rate are represented by Re^L and centrifugal acceleration is represented through Ga^L. As described in Section 8.1.4, exponents of the Schmidt number are set to 0.5, as the collected experimental data do not contain significant variation in physical properties. **The maximum**

observed for the packed rotor is mathematically represented by introducing the factor $(Ga_L)c_6 \exp(c_7 \cdot Ga^L)$ in the packing term. The influence of gas flow rate is not implemented here due to a lack of data variation in this parameter and. Hence the correlation is build based on data for 12.3 SLPM. The influence of gas flow rate can be implemented by introducing $(Re^V)^c$ for the empty rotor term. This has already been shown by Hilpert and Repke and results in overall good agreement with experimental data of $\pm 30\%$ [4].

The final correlation is given in eq. (8.38) and aims at calculating local volumetric liquid-side mass transfer coefficients. Hence, eq. (8.6) is used to integrate those to obtain bed averaged values.

$$
k_{loc}^L a_e = \overbrace{c_1 \cdot \left(Re^{L,\,empty}\right)^{c_2} \cdot \left(Ga^{L,\,empty}\right)^{c_3} \cdot \frac{D^L}{\left(r_o - r_i\right)^2} \cdot \left(Sc^L\right)^{0.5}}^{empty\ rotor}
$$
$$
\underbrace{+ c_4 \cdot \left(Re^L\right)^{c_5} \cdot \left(Ga^L\right)^{c_6} \exp\left(c_7 \cdot Ga^L\right) \cdot \left(a_p\right)^2 \cdot D^L \cdot \left(Sc^L\right)^{0.5}}_{packed\ rotor} \tag{8.38}
$$

Parameter estimation has been performed, resulting in the final parameter set in eq. (8.39). A comparison of simulated $k^L a_e$ values and experimental data is given for the empty rotor in Fig. 8.10 and for the packed rotor in Fig. 8.11, whereby a parity plot is provided for each. Compared to the results from literature correlations, experimental data are predicted well using the novel correlation and deviations between simulation and experiment are well within 15% range. Furthermore, the maximum in separation efficiency is globally captured. However, despite these promising results, transferability of the developed correlation has yet to be proven for other RPB experiments using different sizes or geometries.

$$
k_{loc}^L a_e = 41.81 \cdot \left(Re^{L,\,empty}\right)^{0.8160} \cdot \left(Ga^{L,\,empty}\right)^{0.06993} \cdot \frac{D^L}{\left(r_o - r_i\right)^2} \cdot \left(Sc^L\right)^{0.5} \tag{8.39}
$$
$$
+ 1.955 \cdot 10^{-5} \cdot \left(Re^L\right)^{1.258} \cdot \left(Ga^L\right)^{0.8739} \exp\left(-4.963 \cdot 10^{-7} \cdot Ga^L\right)
$$
$$
\cdot \left(a_p\right)^2 \cdot D^L \cdot \left(Sc^L\right)^{0.5}
$$

8.4.3 Vapor-side mass transfer correlation

As a crucial part of a rate-based RPB model for describing distillation operations, vapor-side mass transfer is to be evaluated and incorporated into the model. Experimental data on binary and ternary distillation are hence evaluated to capture the influence of the main operation parameters on separation efficiency and enable the development of a suitable correlation for $k^V a_e$. Alongside, available correlations from literature are tested on collected experimental data for overall gas-side mass

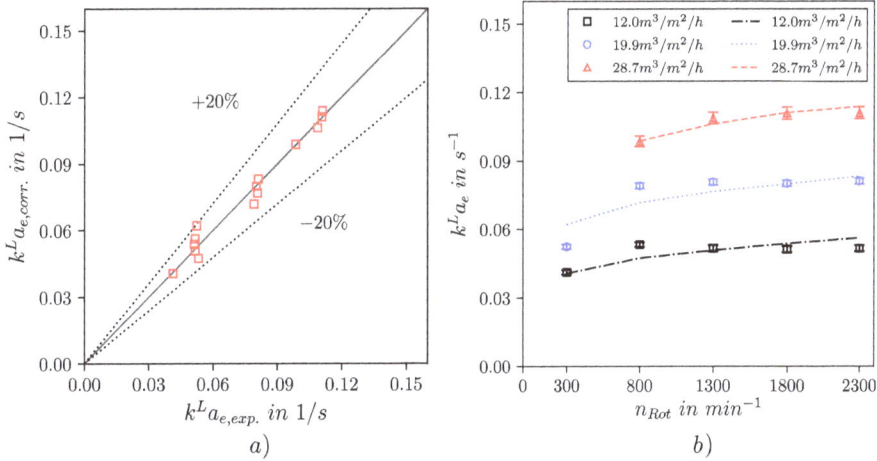

Fig. 8.10: Comparison of experimental (marker) and calculated data (lines) using the correlation for local $k^L a_e$ from eq. (8.39): (a) Parity plot, (b) $k^L a_e$ depending on rotational speed for countercurrent desorption with a nitrogen gas flow rate of 12.3 SLPM and different liquid loads at the eye in the empty rotor. Taken from [5].

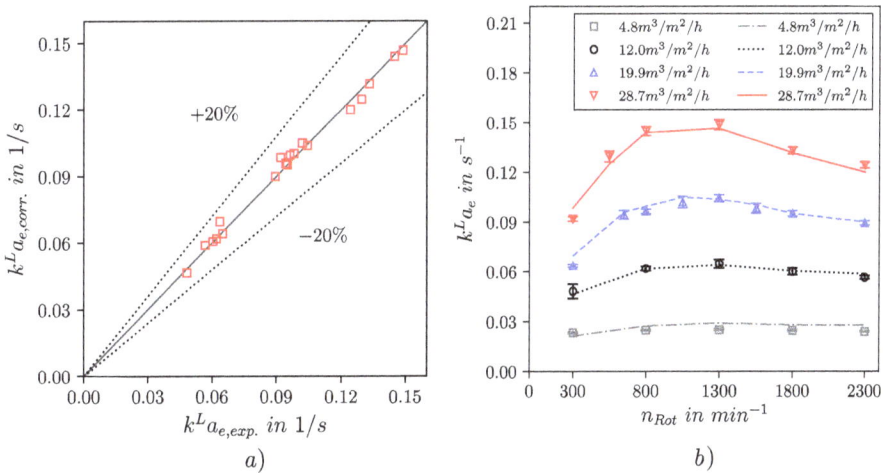

Fig. 8.11: Comparison of experimental (marker) and calculated data (lines) using the correlation for local $k^L a_e$ from eq. (8.39): (a) Parity plot, (b) $k^L a_e$ depending on rotational speed for countercurrent desorption with a nitrogen gas flow rate of 12.3 SLPM and different liquid loads at the eye in the packed rotor. Taken from [5].

transfer coefficients and in comparison to a newly developed correlation for appropriate description of observed separation efficiency.

Simulation data for $K^V a_e$ is based on the rate-based model for the RPB introduced in Section 8.3. All applications of the model imply a chosen number of discrete elements of 101. This was chosen based on a study on the relative change in simulated $K^V a_e$ values obtained for discrete element numbers from 30 to 200 compared to values obtained at 200 discrete elements. The study showed that an increase in the number of discrete elements beyond 101 resulted in insignificant changes of calculated separation efficiency as compared to experimental uncertainty.

Results of binary and ternary distillation experiments

The evaluation of experimental data considers three main operation parameters influencing mass transfer for distillation experiments at total reflux: rotational speed, F-factor, and liquid load. The latter two are defined using the cross section of the inner packing radius as a reference. Here, F-factor and liquid load are highest, whereas centrifugal acceleration is lowest.

 As experimental output on the separation efficiency, the overall volumetric gas-side mass transfer coefficients $K^V a_e$ are calculated using eq. (8.40) following the ATU_{og}-NTU_{og} concept of Singh [71] shown in eq. (8.41). This concept of Area per Transfer Units (with og: overall gas mass transfer resistance considered) was developed in analogy to the HTU-NTU concept for conventional columns and considers radial change of gas and liquid load, as well as centrifugal acceleration. Integration was conducted numerically using Romberg's method within the SciPy-package in Python. Vapor equilibrium compositions y^{eq} corresponding to y are obtained from equilibrium calculations based on the average bed pressure and assuming $x = y$ due to operation at total reflux. For calculating the total mass transfer resistance, integration boundaries are the concentrations measured over the RPB at sampling locations S-04 and S-06 (see Fig. 8.6). Hence, end effects are included in $K^V a_e$ and allocated to the rotor. Preliminary tests using variable liquid sampling positions (outer radius of rotor or casing), however, have shown that for the given RPB design the rotor generally accounts for more than 90% of mass transfer in the domain of the applied operation parameters. To enable comparability to literature data, also HETP values are provided based on the number of theoretical stages ($HETP = N_{th}/(r_o - r_i)$). N_{th} is obtained using the McCabe-Thiele short-cut method [97].

$$K^V a_e = \frac{\dot{V}_G}{h_p \pi (r_o^2 - r_i^2)} \int_{y_i}^{y_o} \frac{dy}{y^{eq} - y} \tag{8.40}$$

$$\pi (r_o^2 - r_i^2) = ATU_{og} NTU_{og} = \frac{\dot{V}_G}{h_p K^V a_e} \int_{y_i}^{y_o} \frac{dy}{y^{eq} - y} \tag{8.41}$$

For ternary distillation, NTU_{og} is of matrix form [8]. To nevertheless represent separation efficiency only by one single key variable, the number of theoretical stages required to obtain the observed separation performance for the key component ethanol is used (distillation line). It is calculated from equilibrium calculations assuming total reflux. Starting with the liquid outlet compositions, consecutive vapor-liquid equilibrium calculations are performed until the distillate mole fraction of ethanol is exceeded and N_{th} is obtained.

Binary distillation experiments are visualized in Figs. 8.12 and 8.13 for systems ethanol/n-propanol and ethanol/water. Overall gas-side mass transfer coefficients and numbers of transfer units are shown over rotational speed and the varied operation parameters F-factor and liquid load. Figure 8.12 additionally depicts a comparison of empty rotor and packed rotor performance for the ethanol/n-propanol system. For experiments without packing, $K^V a_e$ calculations use the empty rotor volume as geometric reference. For all investigated F-factors, empty rotor separation efficiency is globally well below obtained values for the packed rotor and shows a slight but monotonous increase with rotational speed. For NTU_{og}, a similar behavior is found, but values approach a distinct upper limit on higher rotational speed for all investigated F-factors.

For the packed rotor, a maximum in $K^V a_e$ over rotational speed is found, whose position is shifted to higher rotational speeds as gas and liquid loads increase. However, the maximum also significantly flattens with increasing load. This is also confirmed for the ethanol/water system in Fig. 8.12b, where an additional experiment at a higher F-factor of 4.9 $Pa^{0.5}$ was performed. In the case of low liquid flow rates, the $K^V a_e$ over rotational speed show an inflection point after the maximum using the packed rotor (Fig. 8.12a), 1.1 $Pa^{0.5}$). It is likely, that the $K^V a_e$ of the packed rotor converge to values close to the empty rotor observations for very high centrifugal accelerations, as the residence time of the liquid in the rotor decreases drastically. This behavior was also observed by Qammar et al. [67] for lower F-factors (0.35 $Pa^{0.5}$ and 0.65 $Pa^{0.5}$) using ethanol/water in an RPB of similar dimensions and a metal knit-mesh packing.

$K^V a_e$ values generally increase with increasing throughput, while NTU_{og} shows a diametral behavior before also reaching a maximum over rotational speed at which this dependency is inverted (see Fig. 8.13). For both binary mixtures, the maxima coincide with the behavior of $K^V a_e$ and comparable numbers of transfer units of ≈ 7 for ethanol/water and ≈ 8 for ethanol/n-propanol are found for F-factors of 2.1 $Pa^{0.5}$ and 3.0 $Pa^{0.5}$. For low liquid loads a comparably steep descent of NTU_{og} is visible after the maximum. This is caused by the rather low liquid holdups and reduced gas-liquid interaction at higher rotational speed, compared to higher liquid loads.

The achieved separation performance is equivalent to $N_{th}/HETP$ ranging from 1.9 to 9.7/17 to 87 mm for ethanol/n-propanol and 1.9 to 5.7/29 to 87 mm for ethanol/water. Respective ranges are well within typical results observed for distillation of binary mixtures (compare to Tab. 5.2).

The formation of the maximum in $K^V a_e$ and NTU_{og} with increasing rotational speed is discussed by several authors and attributed to two competing influences [33, 65, 70]: mass transfer intensification through increasing specific effective surface area and micro-mixing efficiency with increasing centrifugal

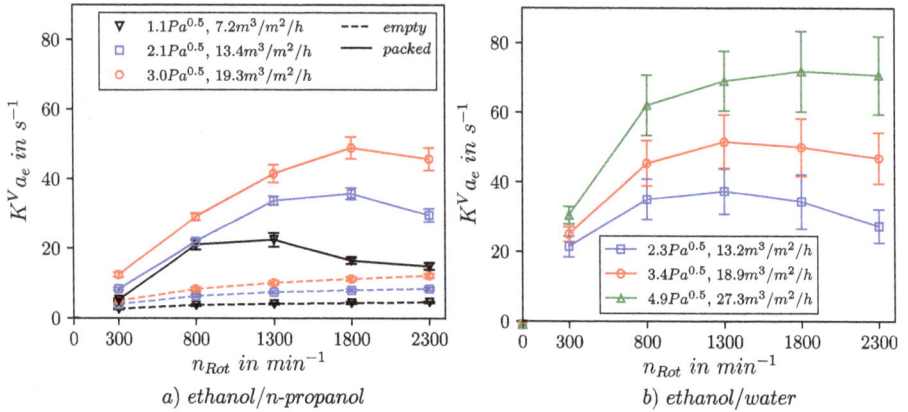

Fig. 8.12: Comparison of calculated $K^V a_e$ values for a) ethanol/n-propanol with packed and empty rotor and b) ethanol/water with packed rotor and both at varying F-factors and liquid flow rates. Taken from [5].

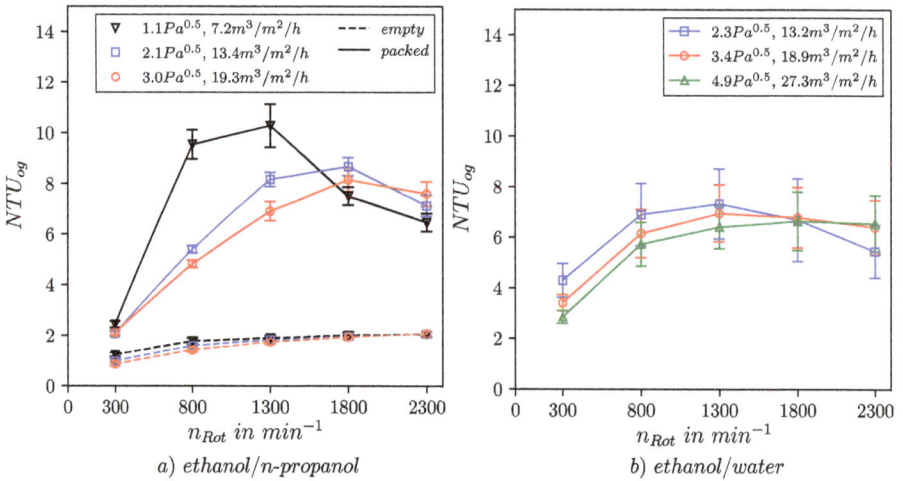

Fig. 8.13: Comparison of calculated NTU_{og} values for a) ethanol/n-propanol with packed and empty rotor and b) ethanol/water with packed rotor and both at varying F-factors and liquid flow rates. Taken from [3].

acceleration on the one hand, and the reduction of residence time of the liquid on the other hand. Additionally, maldistribution of liquid caused by shadowing and rivulet formation may increase at higher rotational speeds [98]. This, of course, does not apply for the empty rotor due to the lack of contact of gas and liquid compared to the porous packing. The location of the maximum is affected by the gas flow rate and the liquid load. On increasing gas flow rate, the specific effective surface area available for mass transfer increases, whereas for increasing liquid load the holdup in the packing increases while residence time decreases [40, 57, 99]. Larger liquid holdup however also causes intensified gas-liquid interaction and hence also intensified mass transfer. Thus, the specific rotational speed at which a shorter residence time equalizes mass transfer improvements is shifted to higher values if the throughput is increased.

Ternary distillation experiments are performed to provide a basis for validation of the multicomponent distillation model for RPBs and are discussed using Fig. 8.14 and Fig. 8.15. The system ethanol/n-propanol/water is used and exhibits two binary temperature-minimum azeotropes, between ethanol and water as well as n-propanol and water. Hence, a distillation boundary is formed, dividing the composition space in two distillation regions, which are both covered by experiments. Balance lines between top and bottoms compositions obtained from experiments at minimum and maximum separation efficiency for rotational speeds of 300 min^{-1} and 1,800 min^{-1} are visualized in Fig. 8.14. For the upper distillation region (etha-

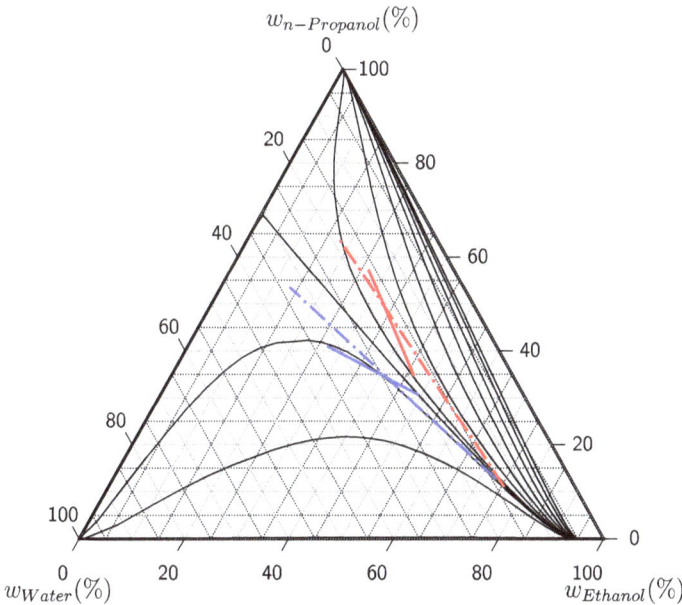

Fig. 8.14: Residue curve map of weight fractions of the ethanol/n-propanol/water system at 1.013 bar(a) with balance lines of experiments at 300 min^{-1} (full line) and 1,800 min^{-1} (dot line), at 3.5 Pa$^{0.5}$ and 18 to 19 m^3/m^{-2}h^{-1} with the packed rotor. Distillation boundaries (black lines) calculated at 1.013 bar using the NRTL property method in Aspen Plus V10. Taken from [5].

nol/n-propanol side), an F-factor of 3.5 Pa$^{0.5}$ and rotational speeds ranging from 300 to 1,800 min^{-1} in steps of 500 min^{-1} were applied. For the ethanol/water side, additionally throughputs of 1.4 Pa$^{0.5}$ and 2.5 Pa$^{0.5}$ and the additional rotational speed of 2,300 min^{-1} have been used. As illustrated in Fig. 8.15, general trends on the numbers of theoretical stages as measure for the separation efficiency follow

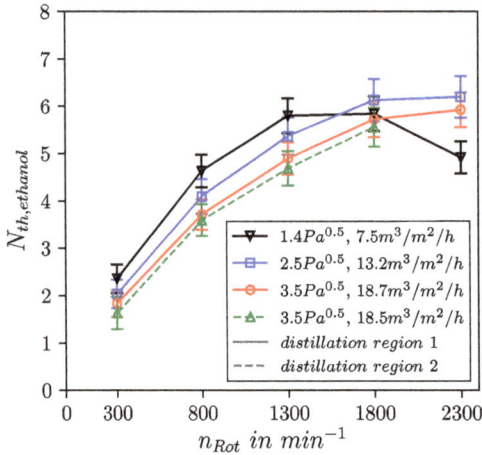

Fig. 8.15: Numbers of theoretical stages achieved in ternary distillation for the ethanol/
n-propanol/water system (calculated based on ethanol concentrations) with the packed rotor.
Taken from [5].

the same behavior as for the binary experiments. At low throughputs slightly higher
numbers of theoretical stages are achieved. A maximum can be observed, that is
shifted to higher centrifugal accelerations for higher throughputs, while appearing
less pronounced. For both distillation regions, a similar separation performance is
achieved due to the similar composition ranges covered.

Evaluation of literature correlations

To assess the viability and prediction capabilities of already published correlations
for mass transfer coefficients in RPBs for distillation operation, simulations based
on the rate-based RPB model can be applied. For predicting experimental results on
distillation operation, i.e. $K^V a_e$ values, correlations of $k^L a_e$ and $k^V a_e$ are required
for calculating the R-matrix eqs. (8.29) and (8.30) and mass transfer rate equations
eqs. (8.13) and (8.14).

    The comparison of experiments and model predictions is obtained by calculating
$K^V a_e$ twice: based on measured distillate and casing outlet concentration (experiment, $K^V a_{e, exp}$); and based on measured distillate compositions (S-04) and predicted
casing compositions calculated from the RPB model and additional pilot-plant
measurements (model prediction, $K^V a_{e, sim}$). The additional measured model in-
puts comprise distillate flow rate FI02, condenser pressure PI02, reflux temperature TI08, rotational speed SI01, and RPB as well as packing geometry (see P&ID
tags in Fig. 8.6).

As a test case, experimental data obtained for the thermodynamically near-ideal ethanol/n-propanol system are used together with two sets of correlations: Chen et al. [32]/Chen [60] and Reddy et al. [37] (listed and referenced in Tabs. 8.1 and 8.2). The results are shown in Fig. 8.16 comparing $K^V a_{e,sim}$ and $K^V a_{e,exp}$ in a parity plot and over rotational speed. For the Chen correlations [32, 60], separation efficiency in the RPB is generally overpredicted and maxima in separation efficiency cannot be captured. Although these correlations hold terms or adjustable parameters to account for RPB geometry as well as associated inner and outer end effects and have hence been used for several studies on distillation in RPBs [68, 100], applicability is not given for the presented case study. This is mainly due to the rather small RPB dimensions used by Chen et al. (see dimensions of RPB devices in Tabs. 8.1 and 8.2). Here, especially inner end effects near the eye of the rotor lead to increased mass transfer and contribute more strongly to the overall mass transfer efficiency, compared to larger-scale RPBs. Additionally, regions at greater radii in larger-scale RPBs are often not fully wetted due to shadowing by support structures at the inner radius and rivulet formation [98] (see CT scans in Chapter 3).

The second set of correlations by Reddy et al. [37] is based on a similarly sized RPB with a split packing design and metal foam rings alternatively fixed to both rotor plates and operation with co-rotating rotor plates simulating a single-block RPB. Reddy et al. attempted to exclude casing contributions from $k^V a_e$ by measuring gas concentrations directly in the casing and by conducting separate experiments with the innermost packing ring only (for separating casing contributions to $k^L a_e$). Consequently, the correlations of Reddy et al. enable better prediction of the mass transfer in the investigated RPB. The parity plot in Fig. 8.16 indicates that the magnitude of $K^V a_e$ values is met, although several points exhibit offsets well above 30%. This becomes apparent in Fig. 8.16b, as the maximum in separation efficiency observed in experimental data is also not captured by the set of correlations from Reddy et al. [37].

Development of a new correlation for $k^V a_e$

To this point, literature correlations show a deficiency in describing maxima in separation efficiency for binary distillation experiments. This is inherent to their mathematical structure, which does not account for any adverse effect of centrifugal acceleration on mass transfer coefficients. This is also the case for the other semi-empirical mass transfer correlations available in RPB literature (see Tabs. 8.1 and 8.2). It is thus mandatory to derive an adapted correlation of suitable mathematical structure. Starting point is the rate-based model of the RPB with implemented correlations for local gas- and liquid-side mass transfer coefficients. The model is used to simulate casing outlet concentrations of ethanol and accordingly simulated values of $K^V a_{e,sim}$ based on experimental distillate concentrations and the other state variables and parameters

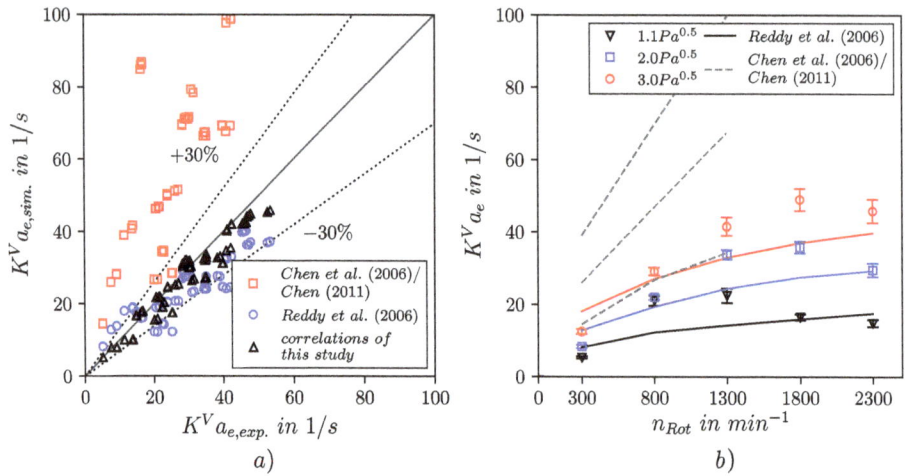

Fig. 8.16: Comparison of experimental and simulated data using literature correlations and eq. (8.43) for the binary ethanol/n-propanol system using a packed rotor. Correlations and references are given in Tab. 8.2. a) Parity plot for $K^V a_e$ and b) plot of $K^V a_e$ over rotational speed. Taken from [3].

(distillate flow rate, condenser pressure, reflux temperature, rotational speed, and RPB geometry). Using the experimental $K^V a_{e,exp}$, a parameter estimation problem deploying a weighted least squares formulation is set up to parameterize new formulations of correlations for local gas-side mass transfer coefficients. This approach is required, as rotor averaged vapor-side mass transfer coefficients obtained from two-film theory according to eq. (8.4) are not suitable due to changes in slope m of the equilibrium curve and in properties between inlet and outlet of the RPB. The correlation is developed based on distillation of the near-ideal binary system ethanol/n-propanol and applying eq. (8.39) as the correlation for local liquid-side mass transfer coefficients.

The adapted and improved correlation is structurally based on the formulation of Reddy et al. [37] shown in Tab. 8.2. Based on experimental observations and findings of Qammar et al. [67], it considers additive contributions from the empty rotor and the packed rotor. **Additionally, the packed rotor term is modified with the product $\left(Ga^V\right)^{c_6} \cdot \exp\left(c_7 \cdot Ga^V\right)$ to enable the representation of a maximum depending on rotational speed.** The reference lengths chosen for the dimensionless expressions are the length of the liquid flow path $r_o - r_i$ (empty rotor) and the specific geometric surface area a_P of the packing (packed rotor). As described in Section 8.1.4 exponents of the Schmidt number are set to 1/3, as the collected experimental data do not contain significant variation in physical properties. The resulting correlation is given by:

$$
k_{loc}{}^V a_e = \overbrace{c_1 \cdot \left(Re_{empty}^V\right)^{c_2} \cdot \left(Ga_{empty}^V\right)^{c_3} \frac{D^V}{(r_o - r_i)^2} \cdot \left(Sc^V\right)^{1/3}}^{\text{empty rotor}}
$$

$$
\underbrace{+ \; c_4 \cdot \left(Re^V\right)^{c_5} \cdot \left(Ga^V\right)^{c_6} \cdot \exp\left(c_7 \cdot Ga^V\right) \cdot \left(a_p\right)^2 \cdot \left(D^V\right) \cdot \left(Sc^V\right)^{1/3}}_{\text{packed rotor}}
$$

$$(8.42)$$

Fitted parameter values are obtained from a sequential two-step approach. First, parameters $c_1 - c_3$ of the empty rotor part are fitted based on respective data for the distillation of ethanol/n-propanol. Terms for the packed rotor are fixed to zero in eq. (8.42) and in the $k_{loc}{}^L a_e$ correlation eq. (8.39). For the second step, parameters $c_4 - c_7$ are estimated based on data for the packed rotor, while fixing the remaining parameters to their respective estimates. The resulting parameter set is given by:

$$
k_{loc}{}^V a_e = 0.02986 \cdot \left(Re_{empty}^V\right)^{0.8861} \cdot \left(Ga_{empty}^V\right)^{0.1954} \frac{D^V}{(r_o - r_i)^2} \cdot \left(Sc^V\right)^{1/3} +
$$

$$
2.6 \cdot 10^{-6} \cdot \left(Re^V\right)^{0.3431} \cdot \left(Ga^V\right)^{1.315} \cdot \exp\left(-1.0 \cdot 10^{-5} \cdot Ga^V\right) \cdot \left(a_p\right)^2 \cdot \left(D^V\right) \cdot \left(Sc^V\right)^{1/3}
$$

$$(8.43)$$

Results of the parameter estimation are furthermore presented in parity plots for simulated and experimental overall gas-side mass transfer coefficients in Fig. 8.17. The parity plot displays all respective experiments used for parameter estimation and additional repeat runs, while Fig. 8.17b) shows the uncertainty-weighted averages of experimental data. Data for both the empty and the packed rotor are well within ±30% range of deviation. More importantly, Fig. 8.17b) shows that the performance of the adapted correlation is well capable of predicting the separation performance and especially the maximum in separation efficiency. It is hence considered more suitable compared to literature correlations when applied for the investigated RPB. This could also be demonstrated for the simulation of experiments for F-factors of 0.8 Pa$^{0.5}$ and 3.9 Pa$^{0.5}$ with the ethanol/n-propanol system, which were not used for parameter estimation.

However, it has to be remarked that by using the back-calculation approach for obtaining $k^V a_e$ shown in Fig. 8.4 error propagation from the $k^L a_e$ into the $k^V a_e$ correlation is faced. In detail, the correlation for $k^L a_e$ in eq. (8.39) lacks the influence of the gas flow rate. Despite this being commonly done for conventional columns, results from performed desorption experiments for the applied RPB indicate a relevant influence (see Section 8.4.2). With the back-calculation approach, this influence is then incorporated into the correlation for $k^V a_e$ (eq. (8.43)). Hence, both correlations should always be used together, and applicability must be checked with respect to the desired usage.

Nevertheless, uncertainty reduction in the RPB model predictions is attainable by varying RPB geometries and operation ranges. Using different devices with significantly different sizing, especially the influence of end effects could be incorporated. Using different component systems, also physical properties of the substances and the mixtures could be varied (relevant for refining the exponent on vapor Schmidt number) to obtain higher information content for future parameter estimation.

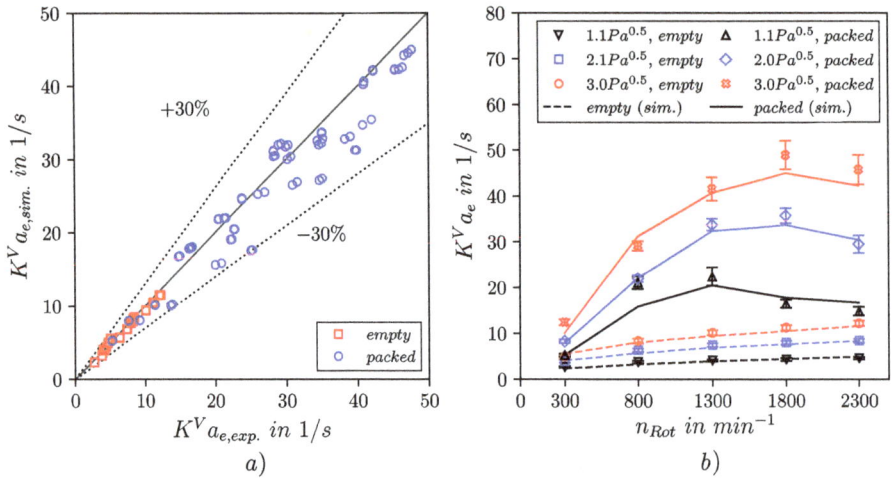

Fig. 8.17: Comparison of experimental (marker) and calculated data (lines) using estimated parameters of the developed $k_{loc}{}^V a_e$ correlation (eq. (8.43)) for the binary ethanol/n-propanol system: a) Parity plot and b) Plot of $K^V a_e$ over rotational speed both for empty and packed rotor. Taken from [5].

8.4.4 Model validation: application to distillation

For validating the rate-based RPB model and obtained new correlations, the thermodynamically non-ideal binary system ethanol/water and the ternary system ethanol/n-propanol/water are used. The respective experimental data were not used for parameter estimation. Model validation for the ternary system is considered the main result of this case study, as it was aimed at developing a suitable model for describing multicomponent distillation in RPBs based on data obtained from binary distillation.

In a first step, results for the binary system are given for experimentally obtained and simulated values for the overall gas-side mass transfer coefficient. Figure 8.18 provides a parity plot and $K^V a_e$ over rotational speed. In general, experimental data are captured within a ± 30% range of deviation. Dominating trends, such as the maximum in separation efficiency, especially its location and shift with rotational speed, are predicted. However, larger offsets are visible for the lowest rotational speed of 300 min^{-1}.

The validation for the multicomponent system is depicted in Fig. 8.19. Again, the discretized rate-based model is used to obtain simulation results for given

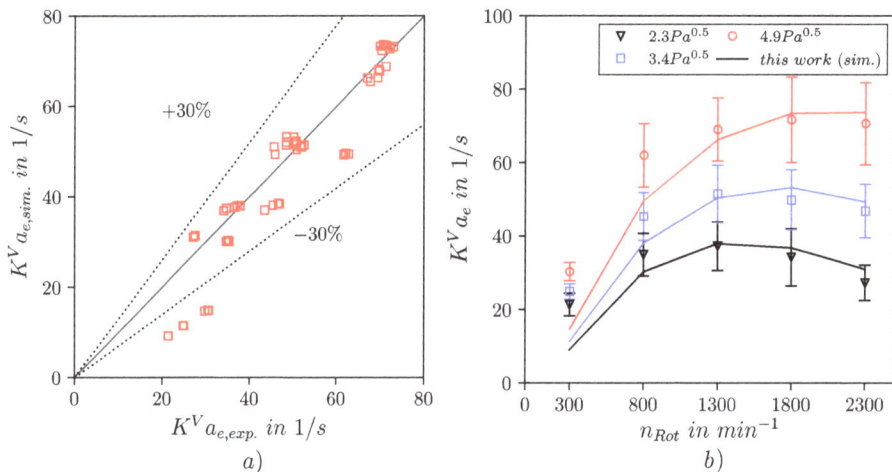

Fig. 8.18: Comparison of experimental (marker) and calculated data (lines) using the rate-based stage model with estimated parameters of the developed $k_{loc}{}^{V}a_e$ correlation for the binary ethanol/water system: a) Parity plot and b) Plot of $K^V a_e$ over rotational speed in the packed rotor. Taken from [5].

distillate composition and other experimental controls. Based on the simulation, $N_{th,ethanol}$ is obtained as key performance indicator for the separation efficiency. Following Fig. 8.19 b) the successful implementation of the adapted mass transfer correlations is stated, since the given experimental data are matched within 30% range of deviation between experiment and simulation. For higher rotational speeds and decreased flow rates, slight overprediction of the separation performance is visible and indicates potential for further improvement of the correlations, as already outlined.

This case study presents the first overall experimental results on ternary distillation at infinite reflux in an RPB and complements this with the development of new correlations capable of describing the specific mass transfer behavior of a larger scale RPB and specifically the distinct maximum in separation efficiency over rotational speed. A novel correlation for $k_{loc}{}^{V}a_e$ is embedded in a rate-based stage model of the RPB and completed by an additional new correlation for liquid-side mass transfer coefficients stemming from desorption experiments. The combination of desorption ($k_{loc}{}^{L}a_e$) and binary distillation experiments ($K^V a_e$) to develop and parameterize respective correlations for liquid- and gas-side mass transfer coefficients also applicable for ternary distillation has been found feasible. Respective models have been validated within 30% range of error for reproducing measured $K^V a_e$ and numbers of theoretical stages for the ternary system ethanol/n-propanol/water. Overall, it can be concluded that multicomponent distillation in an RPB can be accurately modeled with the derived rate-based model and suggested correlations for gas and liquid-side mass transfer coefficients.

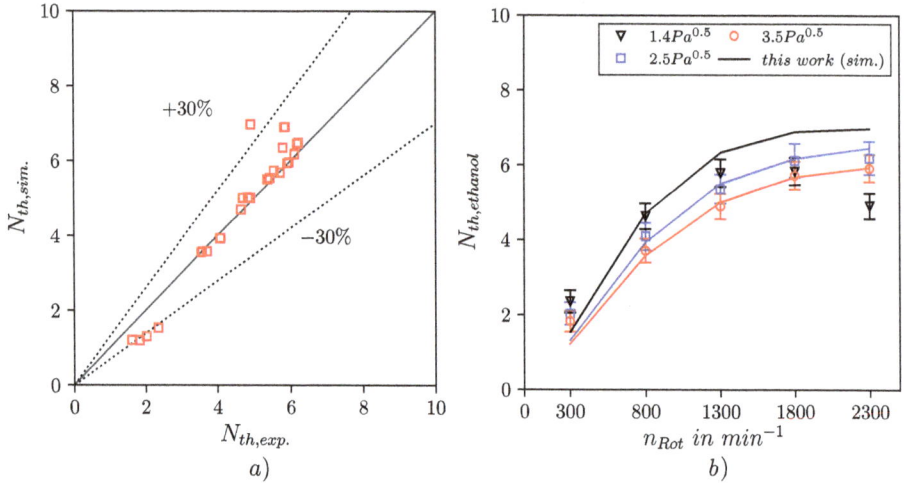

Fig. 8.19: Comparison of experimental (marker) and calculated data (lines) using the rate-based stage model with estimated parameters of the developed $k_{loc}{}^{V}a_e$ correlation for the ternary ethanol/n-propanol/water system: a) Parity plot and b) plot of $N_{th,ethanol}$ over rotational speed in the packed rotor. Taken from [5].

8.5 Take-home messages

- The complexity of fluid dynamics and mass transfer phenomena demand for a rate-based stage modeling approach for RPBs. This way, a physically more accurate description (influence of design, operation conditions, and mass transfer interactions) is achieved, as compared to equilibrium-based modeling.
- Liquid-side mass transfer cannot be neglected for describing distillation processes using RPBs and has to be considered, e.g., based on desorption experiments.
- Mass transfer hence could be described using the resistances in series approach derived from two-film theory.
- Experimental data and mass transfer correlations for distillation in RPBs is most often based on lab-scale devices and binary systems.
- For larger scale RPB, a distinct maximum in separation efficiency over rotational speed is found.
- This maximum is not captured by available literature correlations. By modification and augmentation of the Galilei number and separate handling of empty and packed rotor suitable correlations for $k_{loc}{}^{L}a_e$ and $k_{loc}{}^{V}a_e$ are derived.
- Due to the lack of concentration and temperature profiles, rotor averaged volumetric mass transfer coefficients are used and correlated.
- To describe and model mass transfer in a ternary system for the distillation at infinite reflux, desorption experiments and distillation on respective binary systems can be used as a first approximation.

– $k_{loc}{}^V a_e$ for distillation cannot be measured directly. It is obtained by simulating the rate-based stage model with a given liquid-side mass transfer correlation and calculated overall gas-side mass transfer coefficients obtained from distillation experiments.
– End effects represent a significant contribution to the total mass transfer in an RPB, especially for lab-scale devices.

8.6 Quiz

Question 1. Describe the difference in equilibrium-based and rate-based modeling. Which approach is suitable for RPB modeling and why?

Question 2. Name three approaches or theories to describe mass transfer in gas-liquid contactors.

Question 3. Sketch the concentration profile at a gas-liquid interface for the resistance in series model based on two-film theory. Mark contributions to mass transfer in the respective phases.

Question 4. What requirement must hold for an experimental test system to directly access local gas or liquid-side mass transfer coefficients? Describe a methodological approach to capture both for modeling distillation in RPBs.

Question 5. Describe two approaches to characterize the contribution of end effects to the total mass transfer.

Question 6. Radial discretization can be applied to derive an algebraic model for steady-state distillation operation of an RPB. What are possible discretization schemes and their respective advantages? How could the number of required discrete elements be obtained?

Question 7. Sketch the balance region of the jth non-equilibrium stage of an RPB for countercurrent gas-liquid mass transfer. Insert and name all occurring streams and phases.

Question 8. Which dominating effects likely cause the specific maximum of separation efficiency over rotational speed for RPBs?

Question 9. What is the Ga number, which physical phenomena does it represent, and how is it defined in the context of rotating equipment?

Question 10. Name three reasons that hinder the transferability of mass transfer correlations between different RPBs.

List of symbols

Latin letters

Symbol	Description	Unit
A_{CH}	RPB configuration dependent parameter to calculate the centrifugal head	—
ATU_{OL}	Area of an overall transfer unit based on the liquid phase	m^2
ATU_{og}	Area of an overall transfer unit based on the gas phase	m^2
a_e	Specific effective interfacial area	$m^2\,m^{-3}$
a_p	Specific surface area of packing	$m^2\,m^{-3}$
c	Correlation factor	—
c_p	Specific heat capacity at constant pressure	$J\,mol^{-1}\,K^{-1}$
D	Binary diffusion coefficient	$m^2\,s^{-1}$
d	Diameter	m
d_p	Spherical equivalent diameter of packing	m
d'_p	Spherical equivalent diameter of packing	m
E	Energy transfer flow rate	$J\,s^{-1}$
$F-factor$	Gas capacity factor	$Pa^{0.5}$
$\bar{F}^V_{int}$	Average value of the integrated gas capacity factor	$Pa^{0.5}$
h	Specific enthalpy	$J\,mol^{-1}$
h_p	Axial height of packing	m
$\bar{h}$	Partial molar enthalpy	$J\,mol^{-1}$
j	diffusional flux of component i	$mol\,m^{-3}\,s^{-1}$
Ka_e	Overall volumetric mass transfer coefficient	s^{-1}
$K^y a_e$	Overall gas-side volumetric mass transfer coefficient	$mol\,m^{-3}\,s^{-1}$
ka_e	Volumetric mass transfer coefficient	s^{-1}
L	Liquid flow rate	$mol\,s^{-1}$
M	Molar mass	—
m	Local slope of equilibrium curve	—
N	Mass transfer flow rate	$mol\,s^{-1}$
NTU_{og}	Number of overall transfer units based on the gas phase	—
n_{Rot}	Rotational speed	min^{-1}
P	Pressure	Pa
R	Inverted volumetric mass transfer coefficient	s
r	Radius	m
p^{LV}	Vapor pressure	Pa
$\bar{Re}^V_{int}$	Average value of the integrated vapor Reynolds number	—
T	Temperature	K
$\bar{u}$	Average flow velocity	$m\,s^{-1}$
V	Volume	m^3
V	Vapor flow rate	$mol\,s^{-1}$
$\dot{V}_L$	Volumetric liquid flow rate	$m^3\,s^{-1}$
$\dot{V}_G$	Volumetric gas flow rate	$m^3\,s^{-1}$
$v^{L,avg}$	Average liquid volume flux	$m^3\,m^{-2}\,s^{-1}$
w	Mass fraction	—
x	Liquid mole fraction	—
y	Vapor mole fraction	—
z	Mole fraction	—

Greek letters

α	Heat transfer coefficient	W m^{-2} K^{-1}
αa_e	Volumetric heat transfer coefficient	W m^{-3} K^{-1}
Γ	Thermodynamic factor	—
γ	Activity coefficient	—
ΔP	Pressure drop	Pa
$\Delta P_{f,stat}$	Pressure drop contribution due to friction in the packed bed without rotation	Pa
Δw_{rel}	Relative uncertainty of measured mass fraction	—
δ	Kronecker delta	—
$\bar{\delta}$	Numerical stability parameter for averaging	10^{-4}
ϵ	Porosity of packing	—
η	Dynamic viscosity	Pa s
λ	Thermal conductivity	W m^{-1} K^{-1}
ω	Angular velocity	s^{-1}
Ψ_0	Resistance coefficient for single-phase flow for classical, non-perforated packing elements	—
ψ^{Sphere}	Sphericity, 0.75 for metal foam	—
φ	Form factor of dry packing	—
ρ	Density	kg m^{-3}
σ	Surface tension	N m^{-1}
σ_{cr}	Critical surface tension of metal foam packing	0.075 N m^{-1}
σ_w	Surface tension of water at 298.15 K	0.072 N m^{-1}
ϑ	Celsius temperature	°C
ζ_{CH}	Pressure drop contribution of centrifugal head	Pa

Indices

i, j, k	Index for component (1 . . . NC)
j	Index for stage (1 . . . NJ)
k	Index for experimental point. Max NE
m, n	Index for component (1 . . . $NR = NC - 1$)

Subscripts

$\sum$	Constant mole fractions of all species except the nth
b	Bead
$casing$	Casing
CH	Centrifugal head
cr	Critical
e	Effective
$empty$	Empty rotor
exp	Experimental
f	Friction

G	Gas
i	Inner
int	Integrated
max	Maximum
min	Minimum
Mix	Mixture
o	Outer
P	Constant pressure
p	Packing
rel	Relative
sim	Simulation
$stat$	Static
T	Constant temperature
t	Total

Superscripts

∞	Infinite dilution
$discr$	Discrete element
eq	Equilibrium
I	Interface
L	Liquid phase
LV	Saturation
$mass$	Mass
mol	Molar
V	Vapor phase
$-$	Average

Dimensionless numbers

Eu^L	$\left(P(2\pi r h_p)^2 \, \rho_{Mix}^{L,Mass} / (LM_{Mix}^L)^2 \right)$
Eu^V	$\left(P(2\pi r h_p)^2 \, \rho_{Mix}^{V,Mass} / (VM_{Mix}^V)^2 \right)$
Fr^L	$\left((LM_{Mix}^L)^2 / \left((2\pi r h_p)^2 \, a_p \, r\omega^2 \, \rho_{Mix}^{L,Mass} \, d_p \right) \right)$
Fr'^L	$\left((LM_{Mix}^L)^2 / \left((2\pi r h_p)^2 \, a_p \, r\omega^2 \, \rho_{Mix}^{L,Mass} \, r_i \right) \right)$
Ga^L	$\left(\left((\rho_{Mix}^{L,mass})^2 \, r\omega^2 \right) / \left((\eta_{Mix}^L)^2 \, (a_p)^3 \right) \right)$
Ga^V	$\left(\left((\rho_{Mix}^{V,mass})^2 \, r\omega^2 \right) / \left((\eta_{Mix}^V)^2 \, (a_p)^3 \right) \right)$
Ga_{empty}^L	$\left(\left((\rho_{Mix}^{L,mass})^2 \, (r_o - r_i)^3 \, r\omega^2 \right) / (\eta_{Mix}^L)^2 \right)$
Ga_{empty}^V	$\left(\left((\rho_{Mix}^{V,mass})^2 \, (r_o - r_i)^3 \, r\omega^2 \right) / (\eta_{Mix}^V)^2 \right)$
Ga''^L	$\left(\left((\rho_{Mix}^{L,mass})^2 \, d_p^3 \, r\omega^2 \right) / (\eta_{Mix}^L)^2 \right)$
Ga''^V	$\left(\left((\rho_{Mix}^{V,mass})^2 \, d_p^3 \, r\omega^2 \right) / (\eta_{Mix}^V)^2 \right)$
Ga'^L	$\left(\left((\rho_{Mix}^{L,mass})^2 \, (d_p')^3 \, r\omega^2 \right) / (\eta_{Mix}^L)^2 \right)$

Ga'^V	$\left((\rho_{Mix}^{V,mass})^2 \left(d_p'\right)^3 r\omega^2 \right)/(\eta_{Mix}^V)^2$
Pr	$\left(c_{p,Mix}^{Mass}\, \eta_{Mix} \right)/\lambda_{Mix}$
Re^L	$(LM_{Mix}^L)/(2\pi rh_p\, a_p\, \eta_{Mix}^L)$
Re^V	$(VM_{Mix}^V)/(2\pi rh_p\, a_p\, \eta_{Mix}^V)$
Re_{empty}^L	$(LM_{Mix}^L\,(r_o-r_i))/(2\pi rh_p\, \eta_{Mix}^L)$
Re_{empty}^V	$(VM_{Mix}^V\,(r_o-r_i))/(2\pi rh_p\, \eta_{Mix}^V)$
Re'^L	$(LM_{Mix}^L\, d_p)/(2\pi rh_p\, \eta_{Mix}^L)$
Re'^V	$(LM_{Mix}^V\, r_i)/(2\pi rh_p\, \eta_{Mix}^V)$
Re''^L	$(LM_{Mix}^L\, r_i)/(2\pi rh_p\, \eta_{Mix}^L)$
Sc^L	$\eta_{Mix}^L/(\rho_{Mix}^{L,mass}\, D^L)$
Sc^V	$\eta_{Mix}^V/(\rho_{Mix}^{V,mass}\, D^V)$
Sh	$(kh)/D$
St	$\bar{k}/\bar{u}$
St_α	$\alpha/\left(\rho_{Mix}^{mass}\, c_{p,Mix}^{Mass}\, \bar{u} \right)$
We^L	$(LM_{Mix}^L)^2/\left((2\pi rh_p)^2\, a_p\, \rho_{Mix}^{L,mass}\, \sigma_{Mix} \right)$
We^V	$\left((VM_{Mix}^L)^2\, r_i\right)/\left((2\pi rh_p)^2\, \rho_{Mix}^{V,mass}\, \sigma_{Mix}^L \right)$
We'^L	$\left((LM_{Mix}^L)^2\, d_p\right)/\left((2\pi rh_p)^2\, \rho_{Mix}^{L,mass}\, \sigma_{Mix}^L \right)$
We''^L	$\left((LM_{Mix}^L)^2\, r_i\right)/\left((2\pi rh_p)^2\, \rho_{Mix}^{L,mass}\, \sigma_{Mix}^L \right)$

List of abbreviations

ANN	Artificial neuronal network
ATU	Area of a transfer unit
Cl_2	Chlorine
CO_2	Carbon dioxide
CRB	Concentric ring rotating bed
GC	Gas chromatography
H_2S	Hydrogen sulfide
HETP	Height equivalent to one theoretical plate
HTU	Height of a transfer unit
NRTL	Non-random two liquids
PPDS	Physical properties data services
RPB	Rotating packed bed
RZB	Rotating zig-zag bed
SLPM	Standard norm liters per minute
TCD	Thermal conductivity detector
TRL	Technology readiness level
TSCC	Two-stage countercurrent
VOC	Volatile organic compound

References

[1]   Henley EJ, Seader JD, Roper DK. Separation process principles. 3rd edition. Hoboken, NJ: Wiley; 2011.

[2]   Agarwal L, Pavani V, Rao DP, Kaistha N. Process intensification in HiGee absorption and distillation: Design procedure and applications. Ind. Eng. Chem. Res. 2010;49(20):10046–58.

[3]   Hilpert M. Development of a Rate-Based Model for Multi-Component Distillation in a Rotating Packed Bed based on Mass Transfer Experiments. PhD thesis. Berlin; 2022.

[4]   Hilpert M, Repke J-U. Experimental investigation and correlation of liquid-side mass transfer in pilot-scale rotating packed beds. Ind. Eng. Chem. Res. 2021;60(14):5251–63.

[5]   Hilpert M, Calvillo Aranda GU, Repke J-U. Experimental analysis and rate-based stage modeling of multicomponent distillation in a Rotating Packed Bed. Chem. Eng. Process. 2022;171:108651.

[6]   Kenig EY, Blagov S. Chapter 10 – Modeling of distillation processes. In: Gorak A, Sorensen E, eds. Distillation: Fundamentals and principles: Elsevier; 2014, 383–436.

[7]   Hegely L, Roesler J, Alix P, Rouzineau D, Meyer M. Absorption methods for the determination of mass transfer parameters of packing internals: A literature review. AIChE J. 2017;63 (8):3246–75.

[8]   Taylor R, Krishna R. Multicomponent mass transfer. New York: Wiley; 1993.

[9]   Verein Deutscher Ingenieure. Wet separators: Waste gas cleaning by adsorption (scrubbers); ICS 13.040.40(3679 Part 2). Düsseldorf: Beuth Verlag GmbH; 2014.

[10]   Danckwerts PV. Gas-liquid reactions. New York: McGraw-Hill Book Co; 1970.

[11]   Song D, Seibert AF, Rochelle GT. Mass transfer parameters for packings: Effect of viscosity. Ind. Eng. Chem. Res. 2018;57(2):718–29.

[12]   Deed DW, Schutz PW, Drew TB. Comparison of rectification and desorption in packed columns. Ind. Eng. Chem. 1947;39(6):766–74.

[13]   Ovejero G, van Grieken R, Rodriguez L, Valverde JL. The use of gas absorption correlations for mass transfer coefficients in distillation processes. Int. J. Heat Mass Transf. 1992;35 (11):2963–68.

[14]   Yoshida F, Koyanagi T. Mass transfer and effective interfacial areas in packed columns. AIChE J. 1962;8(3):309–16.

[15]   Coughlin RW. Effect of liquid-packing surface interaction on gas absorption and flooding in a packed column. AIChE J. 1969;15(5):654–59.

[16]   Onda K, Takeuchi H, Okumoto Y. Mass transfer coefficients between gas and liquid phases in packed columns. J. Chem. Eng. Jpn. 1968;1(1):56–62.

[17]   Delaloye MM, von SU, Xiao-ping L. The influence of viscosity on the liquid-phase mass transfer resistance in packed columns. Chem. Eng. J. 1991;47(1):51–61.

[18]   Rejl JF, Linek V, Moucha T, Prokopová E, Valenz L, Hovorka F. Vapour- and liquid-side volumetric mass transfer coefficients measured in distillation column. Comparison with data calculated from absorption correlations. Chem. Eng. Sci. 2006;61(18):6096–108.

[19]   Rejl JF, Valenz L, Linek V. "Profile method" for the measurement of kLa and kVa in distillation columns. Validation of rate-based distillation models using concentration profiles measured along the column. Ind. Eng. Chem. Res. 2010;49(9):4383–98.

[20]   Linek V, Moucha T, Prokopová E, Rejl JF. Simultaneous determination of vapour- and liquid-side volumetric mass transfer coefficients in distillation column. Chem. Eng. Res. Des. 2005;83(8):979–86.

[21]   Blum S. Eignung Des Stoffaustauschmodells für die Beschreibung der Vakuumrektifikation oleochemischer Grundstoffe: Zugl.: Diss. Düsseldorf: VDI-Verl; 1995.

[22] Hampel U, Schubert M, Döß A, Sohr J, Vishwakarma V, Repke J-U, et al. Recent advances in experimental techniques for flow and mass transfer analyses in thermal separation systems. Chem. Ing. Tech. 2020;92(7):926–48.

[23] Guo K, Guo F, Feng Y, Chen J-F, Zheng C, Gardner NC. Synchronous visual and RTD study on liquid flow in rotating packed-bed contactor. Chem. Eng. Sci. 2000;55(9):1699–706.

[24] Neumann K, Hunold S, de Beer M, Skiborowski M, Gorak A. Mass transfer studies in a pilot scale RPB with different packing diameters. Ind. Eng. Chem. Res. 2018;57(6):2258–66.

[25] Rao DP, Bhowal A, Goswami PS. Process intensification in rotating packed beds (HIGEE): An appraisal. Ind. Eng. Chem. Res. 2004;43(4):1150–62.

[26] Trent DL. Chemical Processing in High-Gravity Fields. In: Re-engineering the chemical processing plant. Stankiewicz AI, Moulijn JA, eds. New York: M. Dekker; 2004, 29–60.

[27] Chen Y-S, Lin -C-C, Liu H-S. Mass transfer in a rotating packed bed with various radii of the bed. Ind. Eng. Chem. Res. 2005;44(20):7868–75.

[28] Liu T, Wang D, Wang W, Liang Q, Shao L. Study on the efficient production of ozone water by a rotating packed bed. Ind. Eng. Chem. Res. 2019;58(17):7227–32.

[29] Leimbrink M, Neumann K, Kupitz K, Gorak A, Skiborowski M. Enzyme accelerated carbon capture in different contacting equipment – a comparative study. Energy Proc. 2017;114:795–812.

[30] Neumann K. Operating characteristics of rotating packed beds. München: Dr. Hut; 2018.

[31] Luo Y, Chu G-W, Zou H-K, Wang F, Xiang Y, Shao L, et al. Mass transfer studies in a rotating packed bed with novel rotors: Chemisorption of CO2. Ind. Eng. Chem. Res. 2012;51 (26):9164–72.

[32] Chen Y-S, Lin F-Y, Lin -C-C, Tai CY-D, Liu H-S. Packing characteristics for mass transfer in a rotating packed bed. Ind. Eng. Chem. Res. 2006;45(20):6846–53.

[33] Qammar H, Gladyszewski K, Gorak A, Skiborowski M. Towards the development of advanced packing design for distillation in rotating packed beds. Chem. Ing. Tech. 2019;91 (11):1663–73.

[34] Chu G-W, Luo Y, Xing Z-Y, Sang L, Zou H-K, Shao L, et al. Mass-transfer studies in a novel multiliquid-inlet rotating packed bed. Ind. Eng. Chem. Res. 2014;53(48):18580–84.

[35] Singh SP, Wilson JH, Counce RM, Lucero AJ, Reed GD, Ashworth RA, et al. Removal of volatile organic compounds from groundwater using a rotary air stripper. Ind. Eng. Chem. Res. 1992;31(2):574–80.

[36] Rajan S, Kumar M, Ansari MJ, Rao DP, Kaistha N. Limiting gas liquid flows and mass transfer in a novel rotating packed bed (HiGee). Ind. Eng. Chem. Res. 2011;50(2):986–97.

[37] Reddy KJ, Gupta A, Rao DP, Rama OP. Process intensification in a HIGEE with split packing. Ind. Eng. Chem. Res. 2006;45(12):4270–77.

[38] Chen Y-S, Lin -C-C, Liu H-S. Mass transfer in a rotating packed bed with viscous newtonian and non-newtonian fluids. Ind. Eng. Chem. Res. 2005;44(4):1043–51.

[39] Chu G-W, Sang L, Du X-K, Luo Y, Zou H-K, Chen J-F. Studies of CO2 absorption and effective interfacial area in a two-stage rotating packed bed with nickel foam packing. Chem. Eng. Process. 2015;90:34–40.

[40] Luo Y, Luo J-Z, Chu G-W, Zhao Z-Q, Arowo M, Chen J-F. Investigation of effective interfacial area in a rotating packed bed with structured stainless steel wire mesh packing. Chem. Eng. Sci. 2017;170:347–54.

[41] Yang K, Chu G, Zou H-K, Sun B, Shao L, Chen J-F. Determination of the effective interfacial area in rotating packed bed. Chem. Eng. J. 2011;168(3):1377–82.

[42] Sudhoff D, Leimbrink M, Schleinitz M, Gorak A, Lutze P. Modelling, design and flexibility analysis of rotating packed beds for distillation. Chem. Eng. Res. Des. 2015;94:72–89.

[43] Gudena K, Rangaiah GP, Samavedham L. Modeling and optimization of reactive higee stripper-membrane process for methyl lactate hydrolysis. Ind. Eng. Chem. Res. 2013;52 (23):7795–802.

[44] Yi F, Zou H-K, Chu G-W, Shao L, Chen J-F. Modeling and experimental studies on absorption of CO2 by Benfield solution in rotating packed bed. Chem. Eng. J. 2009;145(3):377–84.

[45] Gudena K, Rangaiah GP, Lakshminarayanan S. Modeling and analysis of solid catalyzed reactive HiGee stripping. Chem. Eng. Sci. 2012;80:242–52.

[46] Tung -H-H, Mah RSH. Modeling liquid mass transfer in HiGee separation process. Chem. Eng. Commun. 1985;39(1–6):147–53.

[47] Munjal S, Duduković MP, Ramachandran P. Mass-transfer in rotating packed beds--I. Development of gas-liquid and liquid-solid mass-transfer correlations. Chem. Eng. Sci. 1989;44(10):2245–56.

[48] Guo F, Zheng C, Guo K, Feng Y, Gardner NC. Hydrodynamics and mass transfer in cross-flow rotating packed bed. Chem. Eng. Sci. 1997;52(21–22):3853–59.

[49] Liu W, Chu G-W, Luo Y, Liu Y-Z, Meng F-Y, Sun B-C, et al. Mass transfer in a rotating packed bed reactor with a mesh-pin rotor: Modeling and experimental studies. Chem. Eng. J. 2019;369:600–10.

[50] Liu W, Luo Y, Li Y-B, Chu G-W. Scale-up of a rotating packed bed reactor with a mesh-pin rotor: (II) Mass transfer and application. Ind. Eng. Chem. Res. 2020;59(11):5124–32.

[51] Qian Z, Xu L, Cao H, Guo K. Modeling study on absorption of CO2 by aqueous solutions of N-Methyldiethanolamine in rotating packed bed. Ind. Eng. Chem. Res. 2009;48(20):9261–67.

[52] Li W, Wu W, Zou H-K, Chu G, Shao L, Chen J-F. A mass transfer model for devolatilization of highly viscous media in rotating packed bed. Chin. J. Chem. Eng. 2010;18(2):194–201.

[53] Li Y, Li X, Wang Y, Chen Y, Ji J, Yu Y, et al. Distillation in a counterflow concentric-ring rotating bed. Ind. Eng. Chem. Res. 2014;53(12):4821–37.

[54] Lin -C-C, Jian G-S. Characteristics of a rotating packed bed equipped with blade packings. Sep. Purif. Technol. 2007;54(1):51–60.

[55] Jiao WZ, Liu YZ, Qi GS. Gas pressure drop and mass transfer characteristics in a cross-flow rotating packed bed with porous plate packing. Ind. Eng. Chem. Res. 2010;49(8):3732–40.

[56] Zhang JLB, Li Q, Zhang X. CO2 Capture from Coal Fired Power Plant and EOR in Shengli oilfield of Sinopec China. Abu Dhabi; 2011.

[57] Ma C, Su M-J, Luo Y, Chu G-W, Sun B-C, Chen J-F. Wetting behavior of the stainless steel wire mesh with Al2O3 coatings and mass transfer intensification in a rotating packed bed. Ind. Eng. Chem. Res. 2019;59(3):1374–82.

[58] Bird RB, Stewart WE, Lightfoot EN. Transport phenomena. 2nd edition. New York: Wiley; 2002.

[59] Shivhare MK, Rao DP, Kaistha N. Mass transfer studies on split-packing and single-block packing rotating packed beds. Chem. Eng. Process. 2013;71:115–24.

[60] Chen Y-S. Correlations of mass transfer coefficients in a rotating packed bed. Ind. Eng. Chem. Res. 2011;50(3):1778–85.

[61] Saha D. Prediction of mass transfer coefficient in rotating bed contactor (Higee) using artificial neural network. Heat Mass Transf. 2009;45(4):451–57.

[62] Zhao B, Su Y, Tao W. Mass transfer performance of CO2 capture in rotating packed bed: Dimensionless modeling and intelligent prediction. Appl. Energy. 2014;136:132–42.

[63] Buckingham E. On physically similar systems; illustrations of the use of dimensional equations. Phys. Rev. 1914;4(4):345–76.

[64] Cussler EL. Diffusion: Mass transfer in fluid systems. 2nd edition. Cambridge: Cambridge Univ. Press; 2000.

[65] Luo Y, Chu G, Sang L, Zou H-K, Xiang Y, Chen J-F. A two-stage blade-packing rotating packed bed for intensification of continuous distillation. Chin. J. Chem. Eng. 2016;24(1):109–15.

[66] Luo Y, Chu G-W, Zou H-K, Xiang Y, Shao L, Chen J-F. Characteristics of a two-stage counter-current rotating packed bed for continuous distillation. Chem. Eng. Process. 2012;52:55–62.

[67] Qammar H, Hecht F, Skiborowski M, Gorak A. Experimental investigation and design of rotating packed beds for distillation. Chem. Eng. Trans. 2018;69:655–60.

[68] Li W, Song B, Li X, Liu Y. Modelling of vacuum distillation in a rotating packed bed by Aspen. Appl. Therm. Eng. 2017;117:322–29.

[69] Chu G-W, Gao X, Luo Y, Zou H-K, Shao L, Chen J-F. Distillation studies in a two-stage counter-current rotating packed bed. Sep. Purif. Technol. 2013;102:62–66.

[70] Nascimento JVS, Ravagnani TMK, Pereira JAFR. Experimental study of a rotating packed bed distillation column. Braz. J. Chem. Eng. 2009;26(1):219–26.

[71] Singh SP. Air stripping of volatile organic compounds from groundwater: An evaluation of a centrifugal vapor-liquid contactor. Dissertation. Knoxville, Tennessee; 1989.

[72] Chen YH, Chang CY, Su WL, Chen CC, Chiu CY, Yu YH, et al. Modeling ozone contacting process in a rotating packed bed. Ind. Eng. Chem. Res. 2004;43(1):228–36.

[73] Chen Q-Y, Chu G-W, Luo Y, Sang L, Zhang -L-L, Zou H-K, et al. Polytetrafluoroethylene wire mesh packing in a rotating packed bed: Mass-transfer studies. Ind. Eng. Chem. Res. 2016;55 (44):11606–13.

[74] Wen Z-N, Wu W, Luo Y, Zhang -L-L, Sun B-C, Chu G-W. Novel wire mesh packing with controllable cross-sectional area in a rotating packed bed: Mass transfer studies. Ind. Eng. Chem. Res. 2020;59(36):16043–51.

[75] Zhao Z, Wang J, Sun B, Arowo M, Shao L. Mass transfer study of water deoxygenation in a rotor-stator reactor based on principal component regression method. Chem. Eng. Res. Des. 2018;132:677–85.

[76] Larkin J. Thermodynamic properties of aqueous non-electrolyte mixtures I. Excess enthalpy for water + ethanol at 298.15 to 383.15 K. J. Chem. Thermodyn. 1975;7(2):137–48.

[77] Majer V, Svoboda V, Hynek V, Pick J. A new calorimeter for determination of the temperature dependence of heat of vaporization. Collect. Czech. Chem. Commun. 1978;43(5):1313–24.

[78] Ramalho RS, Ruel M. Heats of mixing for binary systems: N -Alkanes + n -Alcohols and n -Alcohols. Can. J. Chem. Eng. 1968;46(6):456–61.

[79] Counsell JF, Fenwick JO, Lees EB. Thermodynamic properties of organic oxygen compounds 24. Vapour heat capacities and enthalpies of vaporization of ethanol, 2-methylpropan-1-ol, and pentan-1-ol. J. Chem. Thermodyn. 1970;2(3):367–72.

[80] Wächter A, Biegler LT. On the implementation of an interior-point filter line-search algorithm for large-scale nonlinear programming. Math. Program. 2006;106(1):25–57.

[81] Aspen plus V10 (36.0.0.249). USA: Aspen Technology Inc; 2017.

[82] Neumann K, Hunold S, Skiborowski M, Gorak A. Dry pressure drop in rotating packed beds – systematic experimental studies. Ind. Eng. Chem. Res. 2017 56(43):12395–405.

[83] Teja AS, Rice P. Generalized corresponding states method for the viscosities of liquid mixtures. Ind. Eng. Chem. 1981;20(1):77–81.

[84] Poling BE, Prausnitz JM, O'Connell JP. The properties of gases and liquids. 5th edition. New York, NY: McGraw-Hill; 2001.

[85] Krishnamurthy R, Taylor R. A nonequilibrium stage model of multicomponent separation processes. Part I: Model description and method of solution. AIChE J. 1985;31(3):449–56.

[86] Sudhoff D, Neumann K, Lutze P. An integrated design method for rotating packed beds for distillation. In: 24th European symposium on computer aided process engineering. Elsevier; 2014, 1303–08.

[87] Pelkonen S. Multicomponent mass transfer in packed distillation columns: Zugl.: Dortmund, Univ., Diss., 1997. Aachen: Shaker; 1997.

[88] BV R. Data sheet nickel-chromium – extra foam. Geldermalsen, The Netherlands; 2014; Available from: https://www.recemat.nl/wp-content/uploads/2020/08/Datasheet_NCX.pdf. [March 02, 2021].

[89] Khattab IS, Bandarkar F, Fakhree MAA, Jouyban A. Density, viscosity, and surface tension of water+ethanol mixtures from 293 to 323K. Korean J. Chem. Eng. 2012;29(6):812–17.

[90] Perry RH, Green DW. Perry's chemical engineers' handbook. 8th edition. New York, NY: McGraw-Hill; 2008.

[91] Verein Deutscher Ingenieure. VDI heat atlas. 2nd edition. Berlin: Springer-Verlag Berlin Heidelberg; 2010.

[92] Garcia HE, Gordon LI. Oxygen solubility in seawater: Better fitting equations. Limnol. Oceanogr. 1992;37(6):1307–12.

[93] de Mello D, Marcus V, Huang H-M. System and methods for deaerating seawater using a rotating packed bed device; World Patent WO 2018/013350 A1 2018.

[94] Kapoustina V, Guffart J, Hien A, Strischakov J, Medina I, Rädle M, et al. Influence of a single microstructure on local mass transfer in liquid film flows. Chem. Eng. Res. Des. 2019;146:352–62.

[95] Gangoso Posadas I. Espumas Metálicas de Níquel. Caracterización y comportamiento en servicio. Bachelor Thesis. Valladolid: Escuela de Ingenería Mecánica; 2018.

[96] Kelleher T, Fair JR. Distillation studies in a high-gravity contactor. Ind. Eng. Chem. Res. 1996;35(12):4646–55.

[97] Krishna R, Martinez HF, Sreedhar R, Standart GL. Murphree point efficiencies in multicomponent systems. Trans. Inst. Chem. Eng. 1977;55:178–83.

[98] Burns JR, Ramshaw C. Process intensification: Visual study of liquid maldistribution in rotating packed beds. Chem. Eng. Sci. 1996;51(8):1347–52.

[99] Burns JR, Jamil JN, Ramshaw C. Process intensification: Operating characteristics of rotating packed beds -- determination of liquid hold-up for a high-voidage structured packing. Chem. Eng. Sci. 2000;55(13):2401–15.

[100] Gudena K, Rangaiah GP, Lakshminarayanan S. Optimal design of a rotating packed bed for VOC stripping from contaminated groundwater. Ind. Eng. Chem. Res. 2012;51(2):835–47.

Dawid Zawadzki, Maciej Jaskulski, Michał Blatkiewicz

9 Modeling of two-phase flows in rotating packed beds

9.1 Motivation

A flow can be defined as ability of a substance to continually deform under shear stress or external forces [1]. Substances that are capable of flowing are collectively referred to as fluids. Although in common speech fluids are synonymous with liquids, the technical term includes liquids, gases, plasmas, and even flowing heterogeneous mixtures, such as emulsions, suspensions, and foams. The part of physics concerned with fluids is called fluid mechanics. It can be divided into two basic branches: fluid statics, which relates to fluids at rest, and fluid dynamics, which relates to fluids in motion [2]. Therefore, fluid dynamics is, by definition, the study of flows.

Fluids and their flows are ubiquitous in nature and technology, and therefore are of interest in both life sciences and engineering. Currents in rivers and oceans, moving masses of air, viscous creep of magma, blood circulation in human body – these are just a few examples of naturally occurring, widely investigated flows. Technological topics concerned with fluid flows include aerodynamics of airplanes, wide varieties of compressors and fans, ventilations, dryers, and heat exchangers [2].

In many technological tasks, modeling of fluid flows can be substantially simplified. For example, while dealing with heat exchangers, we only need information about general aspects of the flows, as they can be described with widely investigated correlations, allowing to determine their behavior in macro-scale. In such applications, all the occurring phenomena are simplified to a set of differential equations, which can be solved for different values determined by process parameters. Unfortunately, such approach is not always possible. There are many cases where a flow cannot be solved analytically with general approximations, either when the issue is too complex, or when formulation of necessary equations would require too much time and experimental effort. There is, however, a method which can qualitatively and quantitatively predict flow behavior in micro-scale in a reliable and universal manner. This method is called computational fluid dynamics (CFD) [3].

Dawid Zawadzki, Maciej Jaskulski, Michał Blatkiewicz, Łódź University of Technology, Faculty of Process and Environmental Engineering, 90-924, Łódź, Poland

https://doi.org/10.1515/9783110724998-009

9.1.1 What is CFD?

Simply put, CFD is a method of calculation of flow dynamics with the use of numerical methods applied to discrete elements of fluids [4]. At its core it utilizes conservation equations, which include continuity equations, momentum equations (also known as Navier-Stokes equations), and energy equations. Together, they describe the elements of fluid in a particular frame of reference. In the CFD method that frame of reference, usually referred to as the domain, is divided into very large number of elements (cells), forming what is commonly known as a mesh. All cells within the domain interact with other adjacent cells or the limits of the domain specified by its boundaries. Thus, for each cell, which is treated as a finite element of fluid, conservation equations can be solved numerically, with faces serving as boundaries. Since this chapter is focused mainly on hydrodynamics within the RPB equipment, described flows are treated as isothermal; therefore, energy equations are not taken into account.

The main advantage of CFD is its universality. In most engineering tasks, analytical calculations rely on process-specific correlation equations, which are true only for particular problems that have already been investigated under specified conditions, like in the case of equilibrium-based and rate-based modeling approaches (see Chapter 4 FIRST BOOK). For CFD, regardless of the problem at hand, the underlying mathematics remain the same, as every fluid flow is determined by its continuity principle and forces described by the Navier-Stokes equations. Due to this fact, CFD may serve not only as a predictive modeling tool, which determines the impact of parameter shifts on the system's performance, but also as an exploratory tool, showing the behavior of systems that the researcher knows very little of [5].

The largest drawback of CFD is that it usually requires large computational power. At each simulation's iteration, modified conservation equations are calculated for up to tens of millions of cells within a single domain, which results in very long computation times, even with the use of high-performance processors [6]. The number of cells can be decreased, but the resolution of the mesh is a crucial factor in not only the eventual quality of the results, but even the simulation's internal stability. Since CFD works within the basic principles of fluid dynamics, theoretically its results may be as reliable as experimental data. However, due to the fact that cells are not differential objects, but elements of finite size, every simulation is still an approximation of real fluid motion, and prone to errors resulting from a variety of internal sources, such as approximations assumed by particular models. Therefore, CFD can rarely serve as a strong validation tool, and usually requires at least some experimental data to back it up.

9.1.2 CFD for rotating packed bed simulation

Compared to more traditional phase contactors, such as columns, rotating packed bed contactors pose a lot of difficulties in terms of process investigation and modeling. The main issue with RPBs is the fact that phase contact occurs mainly within the rotating internal. It significantly impedes insight, as many of the traditional sensors cannot be used in such conditions, although some advanced methods can partially circumvent this problem. Computer tomography has been used to measure liquid holdup in the rotating internal (see Chapter 3), while temperature telemetry can help estimate concentration profiles in steady state RPB distillation (see Chapter 6).

More importantly, rotation of the annular internal prevents uniformity of the flow across the domain, as centrifugal force increases with the radius [7]. As the centrifugal force acting on elements of fluid changes with the distance of the moving fluid, it not only gradually affects the process' operating variables, but also may also completely change the nature of the flow, which would require switching between various hydrodynamic models to properly describe mass transfer processes occurring within the RPB apparatus (described in detail in Chapter 3 of this book). CFD can be an extremely useful tool for RPB investigation and modeling, as it can simulate hydrodynamics of a variety of RPB equipment types with all sorts of internals and phase types, and under any set conditions. CFD simulations can be carried out in both steady and transient state, allowing the investigator to determine dry and wet pressure drops, liquid holdups and interfacial areas, determine flow patterns, end zone bounds, and many more. However, investigation of different phenomena requires uses of different CFD tools, which vary in terms of complexity. For example, dry pressure drop estimations require simulation of only one phase, which significantly decreases the computational load of calculations, compared to tasks including multi-phase interactions (see Section 9.3 – multiphase flows).

It may also serve as a heuristic tool for tentative identification of efficiency of novel types of internals before they are prototyped and examined experimentally. Although CFD simulations are focused mostly on hydrodynamics, they can be extended to include heat and mass transfer, which may make CFD software a complete tool for simulations of whole chemical processes such as absorption or distillation [8].

9.2 Introduction

9.2.1 Continuity equation

As stated above, each CFD simulation is based on solving the conservation equations. Although they are set for an infinitesimally small element of fluid, they are approximately true for elements of fluid with finite dimensions. The basic and most important one is the mass continuity equation [9]. For a simplified, incompressible flow it can be defined as follows:

$$\frac{\partial u}{\partial x} + \frac{\partial v}{\partial y} + \frac{\partial w}{\partial z} = 0 \tag{9.1}$$

where u, v, and w are velocity components in the x, y, and z direction, respectively.

The equation states that the flow satisfies the conservation of mass law, assuming that there are no mass sources or sinks within the control volume. For a compressible flow (where density is not constant), the mass continuity equation is defined as follows:

$$\frac{\partial \rho}{\partial t} + \frac{\partial(\rho u)}{\partial x} + \frac{\partial(\rho v)}{\partial y} + \frac{\partial(\rho w)}{\partial z} = \frac{d\rho}{dt} + \nabla(\rho \boldsymbol{U}) = 0 \tag{9.2}$$

where ρ is fluid density and $\boldsymbol{U}$ is the velocity vector.

9.2.2 Navier-Stokes momentum equations

As the continuity equation is the expression of mass conservation law for the flow, the Navier-Stokes equations describe the momentum conservation law for fluid flows [10]. The equations can be written in different ways, but the basic form for three-dimensional flow in direction x, y, and z is as follows:

$$\rho\left(\frac{\partial u}{\partial t} + u\frac{\partial u}{\partial x} + v\frac{\partial u}{\partial y} + w\frac{\partial u}{\partial z}\right) = \rho g_x - \frac{\partial p}{\partial x} + \mu\left(\frac{\partial^2 u}{\partial x^2} + \frac{\partial^2 u}{\partial y^2} + \frac{\partial^2 u}{\partial z^2}\right) \tag{9.3}$$

$$\rho\left(\frac{\partial v}{\partial t} + u\frac{\partial v}{\partial x} + v\frac{\partial v}{\partial y} + w\frac{\partial v}{\partial z}\right) = \rho g_y - \frac{\partial p}{\partial y} + \mu\left(\frac{\partial^2 v}{\partial x^2} + \frac{\partial^2 v}{\partial y^2} + \frac{\partial^2 v}{\partial z^2}\right)$$

$$\rho\left(\frac{\partial w}{\partial t} + u\frac{\partial w}{\partial x} + v\frac{\partial w}{\partial y} + w\frac{\partial w}{\partial z}\right) = \rho g_z - \frac{\partial p}{\partial z} + \mu\left(\frac{\partial^2 w}{\partial x^2} + \frac{\partial^2 w}{\partial y^2} + \frac{\partial^2 w}{\partial z^2}\right)$$

where μ is dynamic viscosity, and p is pressure.

The Navier-Stokes equations are an extension of Newton's second law of dynamics, which states that the mass multiplied by acceleration is equal to the sum of forces acting on a body, in relation to the volume of the element:

$$\frac{m \cdot \boldsymbol{a}}{dxdydz} = \frac{\sum \boldsymbol{F}}{dxdydz} \tag{9.4}$$

where m is mass, $\boldsymbol{a}$ is acceleration vector, and $\boldsymbol{F}$ is force vector.

There are three groups of forces acting on an element of fluid: external forces, body forces, and viscosity forces; therefore, the Navier-Stokes equation can be simply stated as follows:

$$\rho \frac{dU}{dt} = \frac{Fg + Fb + Fv}{dV} \tag{9.5}$$

where **Fg** is the sum of external forces, **Fb** is the body forces, and **Fv** is forces due to viscosity.

Combining partial differential equations stated in (9.3) results in overall Navier-Stokes equation for incompressible flows:

$$\rho \frac{dU}{dt} = -\nabla p + \mu \nabla^2 U + \rho g \tag{9.6}$$

which is sometimes presented otherwise as:

$$\frac{\partial U}{\partial t} + U \cdot \nabla U = -\frac{\nabla p}{\rho} + \nu \nabla^2 U + g \tag{9.7}$$

where ν is kinematic viscosity.

Upon taking phase compressibility into account, the overall Navier-Stokes equation is defined as follows:

$$\frac{\partial \rho U}{\partial t} + \nabla(\rho UU) = -\nabla p + \nabla\left(\mu\left((\nabla U) + (\nabla U)^T\right)\right) - \nabla\left(\frac{2}{3}\mu(\nabla U)\right) + \rho g \tag{9.8}$$

Together, eqs. (9.2) and (9.8) describe a continuous, Newtonian, isothermal flow of fluid. CFD solvers may calculate flows, which are non-Newtonian or non-isothermal, upon introducing viscosity expressions and energy equations, respectively.

9.2.3 Wall functions

One of the most common assumptions in fluid dynamics is the no-slip wall condition [11]. It states that at the wall boundary condition the tangential velocity of fluid is equal to the wall's tangential velocity, and gradually increases for the elements of fluid further away from the wall. This phenomenon is schematically shown in Fig. 9.1. The standard approach to capture this effect would be to gradually decrease the size of cells approaching the wall, however this often comes at the cost of higher computational loads. Experimental studies regarding tangential flow velocities in straight canals resulted in formulation of a profile between the dimensionless normal distance to the wall ($y+$) and dimensionless tangential velocity ($U+$), which is true for any fluid flow [12].

Due to the fact that different regions are described by different approximation functions (i.e., linear for the low $y+$ region, and logarithmic for the high $y+$ region) here are two main approaches in modeling the wall effect:

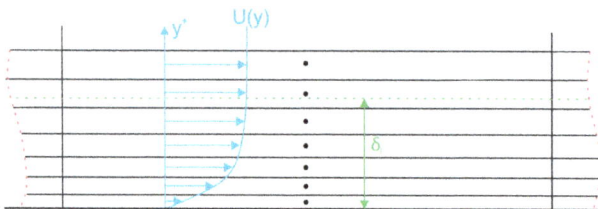

Fig. 9.1: Linear velocity gradient in the vicinity of the no-slip wall condition.

a) picking the approximation function dependent on the cell size – using appropriate functions for $y+$ values lower than 5 or higher than 30 accordingly, since in the buffer region between 5 and 30 both functions are burdened with large errors for both functions,
b) using the Spalding blended wall function [13], which is in excellent agreement with the experimental curve for $y+$ values up to 300.

Most CFD tools automatically fit the wall functions according to the characteristics of the mesh. For thermal wall problems, analogous approach is undertaken with the wall functions between dimensionless distance y^* and dimensionless temperature T^*.

9.2.4 Turbulence

Turbulences are chaotic results of local changes in pressure and velocity in fluids at motion, and they occur when local excess of kinetic energy cannot be suppressed by the fluid's viscosity. Due to their chaotic nature, turbulences are highly irregular, and as a result, it is very difficult to predict their characteristics. Although CFD tools technically are able to directly simulate turbulent velocities with the approach called Direct Numerical Simulation (DNS) [14], it is not usually advised due to extremely high spatial and temporal resolutions required for that purpose. To overcome that issue, Reynolds-Averaged Navier-Stokes (RANS) method can be used for very reliable approximation of turbulent phenomena [15]. RANS introduces an apparent stress expression to the momentum equation. The expression is defined as divergence of the averaged product of two fluctuating velocity components:

$$-\nabla\left(\rho\overline{U'U'}\right) \tag{9.9}$$

Since the apparent stress component is an unknown, it needs to be calculated with a separate equation. This is usually done by using the Boussinesq hypothesis, which assumes that the turbulent stresses are related to average velocity gradients analogously to the same case for the laminar stresses [16]. In other words, turbulences affect fluids in such a way that they behave as if they have higher viscosity. The Boussinesq hypothesis equation is as follows:

$$-\nabla\left(\overline{\rho U' U'}\right) = \mu_t\left((\nabla U) + (\nabla U)^T\right) - \frac{2}{3}\rho k I - \frac{2}{3}(\nabla U)I \qquad (9.10)$$

where k is turbulent kinetic energy
where the unknown is the apparent turbulent viscosity μ_t. Several models can be employed for the calculation of the μ_t parameter, including the k-ε, k-ω, and shear stress transport model.

k-Epsilon model

The k-epsilon turbulence model is an extension of the oldest turbulence viscosity model, which was based on the maximum mixing length of eddies occurring within the flow [17]. In the mixing length model the turbulent viscosity was calculated with simple algebraic function of the distance from the wall. This approach was, however, oversimplified for most practical applications, as turbulence is not a static characteristic, but diffuses and changes within the flow. Therefore, an improved model was suggested, in which the turbulent viscosity is based on turbulent kinetic energy k and kinetic energy dissipation rate ε, calculated with their corresponding transport equations:

$$\underbrace{\frac{\partial(\rho k)}{\partial t}}_{time\ term} + \underbrace{\nabla(\rho U k)}_{convective\ term} = \underbrace{\nabla\left(\left(\mu + \frac{\mu_t}{\sigma_k}\right)\nabla k\right)}_{diffusion\ term} - \underbrace{\rho\varepsilon}_{disspation\ term} + \underbrace{P_k + P_b + S_k}_{sources\ and\ sin\ ks} \qquad (9.11)$$

$$\underbrace{\frac{\partial(\rho\varepsilon)}{\partial t}}_{time\ term} + \underbrace{\nabla(\rho U \varepsilon)}_{convective\ term} = \underbrace{\nabla\left(\left(\mu + \frac{\mu_t}{\sigma_k}\right)\nabla\varepsilon\right)}_{diffusion\ term} - \underbrace{C_2\rho\frac{\varepsilon^2}{k}}_{dissipation\ term} + \underbrace{C_1\frac{\varepsilon}{k}(P_k + C_3 P_b) + S_\varepsilon}_{sources\ and\ sin\ ks} \qquad (9.12)$$

where ε is turbulence dissipation rate, P are source terms, S are sink terms, and C are empirical constants.

The viscosity parameter is then calculated with the following equation:

$$\mu_t = \rho C_\mu \frac{k^2}{\varepsilon} \qquad (9.13)$$

where C_μ is a constant with value specified by the chosen form of the k-epsilon model.

One of the important characteristics of the k-epsilon model is that C_1, C_2, C_3, and C_μ coefficients are not quite constant in the viscous sub-layer of the flow;

therefore, respective damping functions are introduced to account for turbulence damping in the vicinity of the no-slip wall condition.

k-Omega model

Although k-epsilon remains the most popular turbulent viscosity model, it has been proven ineffective for boundary layers with adverse pressure gradients, especially with the occurrence of supersonic shocks, as the damping functions are correct for flat boundary layers without adverse pressure gradients [18]. The k-omega model is, in principle, very similar to the k-epsilon model, as it is also based on solving the transport equations for turbulent kinetic energy and its dissipation rate, but it relies on the ω factor, which stands for the specific turbulence dissipation rate:

$$\omega = \frac{\varepsilon}{C_\mu k} \tag{9.14}$$

The transport equation for turbulent kinetic energy remains the same as for the k-epsilon mode (eq. (9.11)), while the transport equation for specific turbulence dissipation rate is as follows:

$$\underbrace{\frac{\partial(\rho\omega)}{\partial t}}_{time\ term} + \underbrace{\nabla(\rho U\omega)}_{convective\ term} = \underbrace{\nabla\left(\left(\mu + \frac{\mu_t}{\sigma_k}\right)\nabla\omega\right)}_{diffusion\ term} - \underbrace{\beta\rho\omega^2}_{dissipation\ term} + \underbrace{P_k}_{sources\ and\ \sin ks} \tag{9.15}$$

where β and σ_k are closure coefficients, and ω can be converted to ε with the use of eq. (9.14).

Although k-epsilon and k-omega are built on the same principles, the k-omega model relies on different expressions of empirical coefficients which do not account for viscous sub-layer damping effect.

Menter's shear stress transport (k-omega SST) model

The downside of the standard k-omega model is that its results rely heavily on the characteristics of freestream turbulence conditions, which means that small changes to the boundary condition may cause high inaccuracies for turbulent viscosity values, depending on the initial freestream condition. This effect may be a result of the lack of cross-diffusion term or imperfections in parameter value estimation [19]. At the same time, this susceptibility is not the case for the k-epsilon model.

Because of the limitations of both aforementioned models, there is a general rule that the k-omega model tends to be more accurate in the vicinity of the no-slip wall boundary condition, while the k-epsilon is more reliable in the regions away from the wall. Therefore, a combined approach was suggested, in regions close to the wall (low $y+$ values) the k-omega model is used, while the k-epsilon model is used in regions away from the wall (high $y+$ values), and in the region between them the blended approach, known as k-omega SST model, is undertaken [19].

As was stated previously, the dissipation rate ε and the specific dissipation rate ω can be mutually calculated with the eq. (9.14). So, when the expression for ε, calculated with its corresponding transport equation, is converted to ω, it can be compared to the analogous expression for ω in the k-omega model, which shows an additional term:

$$2\frac{\rho\sigma_{\omega2}}{\omega}\nabla k:\nabla\omega \qquad (9.16)$$

This term exists in the k-epsilon model, but does not exist in the k-omega model. The SST approach considers a blending function F_1 which provides smooth transition between the two models in the buffer region:

$$(1-F_1)\cdot2\frac{\rho\sigma_{\omega2}}{\omega}\nabla k:\nabla\omega \qquad (9.17)$$

where the value of F_1 equals 1 for the k-omega zone, and 0 for the k-epsilon zone. There are several expressions for the blending functions, all of which are empirical.

Other turbulence models

Besides DNS and RANS, other turbulence models, such as Large Eddy Simulation (LES) or Detached Eddy Simulation (DES) exist [20, 21], but those are applicable to special cases. Additionally, there is the Reynolds Stress Model (RSM), which relies on direct calculation of stress components instead of relying on eddy viscosities, but it is almost as computationally expensive as the DNS method. Thus, in small-scale simulations, such as the case of RPBs, the RANS method is usually recommended.

9.2.5 Courant number

Courant number (Co), also known as Courant-Friedrichs-Levy (CFL) number, is an essential criterial number in CFD simulations [22]. It is a measure of distance traveled by the flow in a specified discrete time step to total distance specified by discrete length and originates from the CFL condition, which determines the maximum time step allowing for convergence of numerical solutions. In CFD terms, it is velocity multiplied by time step, divided by the length of the cell. For one-dimensional flow, it may be simply written as:

$$Co = \frac{u\Delta t}{\Delta x} \tag{9.18}$$

Although the simplest definition of Co is applicable only to one-dimensional flow, it can be extended to multidimensional flows in complex geometries. The overall expression for the length of the cell in three-dimensional geometry is defined as the ratio of its volume to the sum of the areas of its faces:

$$\Delta x = \frac{V}{\sum_f A_f} \tag{9.19}$$

Since velocity of the flow is also a three-dimensional vector, the Co takes into account the velocity component normal to the face through which the fluid is flowing. Additionally, due to the fact that flows need to satisfy the continuity condition, the expression uses half the magnitude of the component normal to the face as magnitude operator and adds absolute values of positive and negative flux, effectively doubling the incoming flow. Thus, overall expression for three-dimensional flow through any type of geometry is defined as:

$$Co = \frac{1}{2}\frac{\sum_f |U_f \hat{n}_f| A_f}{V} \Delta t \tag{9.20}$$

where $\hat{n}$ is the normal component unit vector.

The Courant number can take any values between zero and infinity, but the important boundary value is one, as Co value of one means that within a certain time step, the flow travels across the whole cell. Choosing the proper Co is important to the simulation's stability, as at large values the solution may diverge, but too low values significantly extend calculation times.

Since velocities and mesh elements differ along the domain, Co also takes different values at spatial and temporal scales within the simulation. CFD solvers offer two different approaches to specifying the time step:
–   Fixed, where the time step has an arbitrarily set value, and Co is calculated without any boundaries at each iteration,
–   Adjustable, where only the initial time step is set along with maximum value of Co. When the calculated Co anywhere in the domain exceeds Co_{max}, the time step is lowered for the next iteration.

Adjustable time step is usually recommended, as it prevents the calculation's divergence.

9.3 Multiphase flows

9.3.1 Multiphase models

Any flow within a control volume can be described with one of two approaches:
- Lagrangian, where the reference point travels along the particular elements of fluid, and its properties can be determined by tracking its movement,
- Eulerian, where the reference frame is fixed in space, and properties of fluid flow are determined by scalar and vector fields within the control volume.

In the Eulerian approach, elements of fluid are described by independent spatial and temporal coordinates, while in the Lagrangian approach their movement is described as a function of time. Since individual tracking of each element of fluid in a control volume is very difficult in majority of practical applications, the Eulerian approach is commonly used in CFD simulations. However, the Lagrangian approach finds its uses in determination of streamlines of some of the fluid elements, as well as in discrete phase modeling.

Many CFD applications for engineering purposes are concerned with single-phase flows, mainly in the field of aerodynamics. RPB simulations sometimes rely on monophasic simulations, for example to determine dry pressure drops within the apparatus. However, since all practical RPB applications are concerned with multiphase mass transfer, they operate under multiphase flow regimes. Therefore, the contents of this section will focus on multiphase flow modeling.

Multiphase models can be essentially divided by the continuity of the simulated phases. A flow can be considered continuous when one phase occupies the whole cell. When the fraction of the phase within the cell is less than one, the phase is considered dispersed. Additionally, multiphase flows may be simulated with a special assumption that dispersed phases exist in the domain as discrete points, and by lacking volumes can be tracked only by Lagrangian approach. According to such criteria, continuous-continuous models, continuous-dispersed models, and discrete models can be distinguished (Fig. 9.2).

The continuous-continuous models are used in cases where clear distinction between contacting phases is of importance, and where the interface is of particular interest for the researcher. Such an assumption requires simulation of separate phases of specified volumes. Applications of continuous-continuous models include most notably free surface flows, but also gas bubble motions through liquids, sloshing, filling of tanks, and many other flow cases in which the interface needs to be tracked in steady or transient state. In the field of RPB simulations, continuous-continuous approach is of great use, since such values as liquid holdup and phase contact area are of particular interest for RPB efficiency evaluation and process modeling. However, an important limiting factor of the continuous-continuous models in such applications is mesh resolution, as fine liquid droplets, resulting

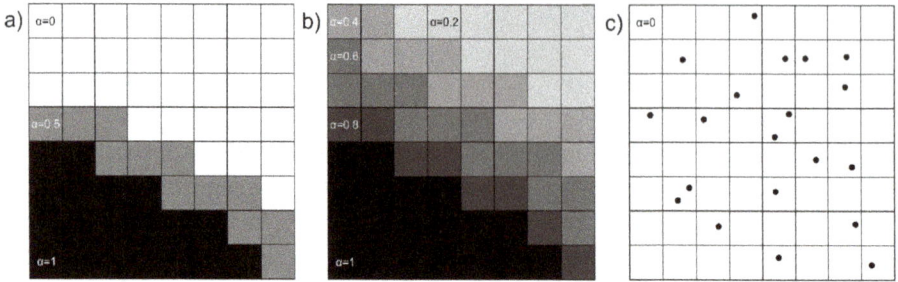

Fig. 9.2: Schematic representation of types of multiphase models: a) continuous-continuous, b) continuous-dispersed, c) discrete phase.

from high centrifugal forces, are often significantly smaller than volumes of cells, even when adaptive mesh refinement tools are used (see Section 9.4.3). Continuous-continuous models include the Eulerian model, the Volume Of Fluid (VOF) model, and the Eulerian Wall Film (EWF) model.

Although interface tracking is an important factor in many CFD simulations, continuous-continuous models pose several drawbacks. As mentioned before, their feasibility is limited by mesh resolution. Additionally, due to necessity of calculating separate continuity equations and, in the case of Eulerian model, separate momentum equations for each phase, the computational load may be relatively high compared to the other models. Continuous-dispersed models are therefore used in cases where the interfacial effects are not of particular interest, or where very high phase dispersion occurs and tracking of the interface requires too much computational power for practical purposes. Simulations using continuous-dispersed models do not offer sharp interfaces, but rather smooth fields of density of the particular phase. Typical systems simulated with continuous-dispersed models include emulsions, suspensions, sprays, and more. In the field of RPB studies mixture models do not serve much purpose, but they may be useful for initial simulations and estimations of parameters. Continuous-dispersed models include the mixture model and particular cases of the Eulerian model.

The Eulerian approach requires generation of a computational mesh for calculations, which covers the area occupied by a given phase. Thus, to simulate the movement of very small portions of fluid, such as several-micron size droplets typical to RPB processes, the computational mesh would have to be at least the size of the droplet diameter. But to accurately describe the interfacial contact surfaces, the mesh cells would need to be much smaller than the dimension of the simulated droplet. Such small mesh cells make simulations impossible, because they exceed the limitations of turbulence models, as very small mesh cells enforce calculation of micro-eddies, which should be averaged by RANS models. The use of mesh sizes too large may in turn cause missing of a large number of droplets, which results in poor mapping of the surface area of the interfacial contact. Thus, CFD calculations of multiphase flows using the Euler-Euler method in RPB units may result in inconsistencies with experimental measurements. The discrete models are a special type of multiphase models in which one or more phases exist in the domain in the form of particles dispersed so finely that their volumes can be effectively approximated to zero. Therefore, phase elements do not exist as total or partial continua, but as geometrical points scattered across the domain, tracked individually by Lagrangian approach. Since in such multiphase models at least one phase must be continuous, it is sometimes referred to as the Eulerian-Lagrangian approach.

9.3.2 Eulerian-Lagrangian multiphase models

Discrete Phase Model (DPM)

In the Lagrangian approach, the tracked points become sources of mass, momentum and heat exchange between both phases, and they are introduced into the Navier-Stokes momentum equations as external sources. The current position and trajectory of the droplets' movement is calculated from the balance of forces acting on the particle in a given direction of Cartesian coordinates:

$$\frac{du_p}{dt} = \underbrace{F_D(u - u_p)}_{\text{drag force}} + \underbrace{\frac{g(\rho_p - \rho)}{\rho_p}}_{\text{gravity force}} + \underbrace{F_x}_{\text{additional forces}} \tag{9.21}$$

where g is gravitational acceleration vector.
where the hydrodynamic drag force is calculated by the following equation:

$$F_D = \frac{18\mu}{\rho_p d_p^2} \frac{f_D Re}{24}. \tag{9.22}$$

where d_p is particle diameter, f_D is drag coefficient and Re is the Reynolds number.

The drag coefficient (C_D) depends on the shape, speed of the particle and the properties of the medium in which it is moving.

Additional forces (F_x) such as the so-called virtual mass forces, arise due to the pressure gradient in the fluid or the lift forces which act on the moving droplet, changing its flow trajectory. When a droplet moves in a rotating medium, as is the case in RPB devices, additional centrifugal forces act on the particle (see Section 9.4.1).

In the discrete phase model, droplets can undergo mass transfer processes, which may also in turn affect the energy state of the liquid due to the heat of change, e.g. evaporation/condensation, adsorption, heat of chemical reactions or liquid boiling. The possibility of defining multi-component droplets allows for precise calculation of multi-component evaporation processes using the available thermodynamic models describing the liquid-vapor equilibrium, such as NRTL or UNIFAQ. Due to the omission of mass and heat transport processes inside the droplet, which is one of the simplifications in DPM, it allows averaging of droplet parameters, which greatly facilitates the mathematical modeling process.

As droplets move through the domain, an important phenomenon of breakup and coalescence of the droplets may occur as a result of mutual collisions. The zero value of droplet volume in DPM makes it difficult to model such interactions, so the O'Rourke approach [23] is commonly used. It is a stochastic model that allows estimating the number of impacts between particles without the need to actually track the points of intersection between their trajectories. Such approach significantly reduces the calculation time of DPM simulations. If the spots of particle collisions are indicated, their results are subsequently determined. The basic criterion determining the result of a collision is the Weber number, the critical value of which is determined experimentally for a given material. Depending on the value of the Weber number, particles may coalesce or bounce. The O'Rourke algorithm assumes that a collision between particles can only occur if they are in the same continuous phase cell. This assumption makes the O'Rourke model highly dependent on mesh density and can only be used for dense meshes in which cell sizes are comparable to particle streams. Additionally, if the time step is too large, it can lead to uncontrollable skips of particles and omissions of potential collision sites. More accurate calculations of the coalescence and breakup processes are possible with the use of Euler-Euler multiphase models (see Section 9.3.3).

Droplets may interact not only with each other, but also with domain boundaries. Depending on the local conditions, different results may occur [24] (Fig. 9.3):
- reflection, where the droplet bounces against the surface and its momentum are restituted with a coefficient defined as the ratio of the rebound drop velocity to the pre-bounce velocity,
- trap, which occurs when the droplet stops at the point of impact upon hitting the wall. Although trapped droplets do not affect the flow of the continuous phase, the simulation acknowledges the position of impact,
- escape, which happens where the droplet reaches a boundary that is not a solid wall. This condition is most often assigned to surfaces that are outlets from the domain. After reaching the boundary condition, the calculations of the droplets' trajectories are terminated,
- Lagrangian Wall Film (LWF), where constitutive droplets join a collective surface. The LWF model allows to simulate the movement of a thin liquid film on the walls of the computational domain.

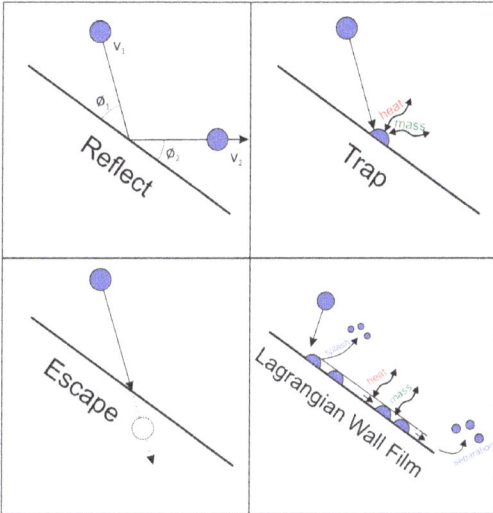

Fig. 9.3: Types of interactions between discrete fluid element and domain boundary.

Lagrangian wall film

The formation of a flowing liquid film formed by droplets impacting the surface is a complex phenomenon. Due to a large number of process parameters (diameter and velocity of droplets, impact angle, etc.), physicochemical properties of the liquid (viscosity, density and surface tension) or surface properties (roughness, wettability or slope), simulating the droplet behavior during the impingement requires complex mathematical models.

Two models are usually used to describe the formation of a Lagrangian liquid film. The first model, by Stanton and Rutland [25], uses wall temperature and value of Weber number as a criterion for determining the result of a droplet collision. The other, more complex, approach is the Kuhnke model [26], which takes into account the state of the wall at the moment of collision. Not only does it consider the properties of the droplet and the impact conditions, but also whether the wall is already wet or dry at the moment of the impact.

The runoff of the droplets forming the wall film is calculated from the relationship:

$$\rho h \frac{dU_p}{dt} = \tau_g \hat{m} + \tau_w + F + \rho h (g - a_w) \tag{9.23}$$

where τ_w and τ_g are stresses acting on the flowing liquid by the wall and gravity respectively, and h is the film thickness. τ_w has a significant impact on fluid movement in simulations with moving boundary conditions. It is described by the equation:

$$\tau_w = -2\frac{\mu_l}{h}\left(\boldsymbol{u_p} - \boldsymbol{u_w}\right).$$ (9.24)

Like DPM particles, the plots of particles flowing as surface film may undergo physicochemical interactions with the continuous phase as well as the wall itself.

The liquid film may detach from the surface. Such a situation is possible when the stresses on the edges of the film exceed the critical value, overcoming the force holding the film against the wall. Critical values of parameters at which the film peels off can be defined by specifying a critical Weber number value or by defining the angle of curvature at which the film will shear off the wall. The detached liquid is again traced in the domain with the discrete phase model. The process of forming droplets from the film is neglected.

9.3.3 Eulerian-Eulerian multiphase models

As was stated above, continuous-continuous models are applicable to cases in which a sharp interface between phases is important. Such an assumption requires simulating separate phases of specified volumes, which in turn implies that each of the phases must be treated with the Eulerian approach. This is why continuous models are based on Euler-Euler multiphase approach.

Eulerian multiphase model

The basic Euler-Euler model is called the Eulerian multiphase model. It is also the most complex, as it solves n momentum equations, one of each for every phase in the given system, and $n-1$ continuity equations, as they have to fulfill the summation condition (eq. (9.26)). At its basics it only assumes that phases share the same pressure. Eulerian model calculations result in separate velocity fields for each of the phases.

The Eulerian model encompasses continuous-continuous phase interactions as well as continuous-dispersed interactions, depending on the specified assumptions. The Volume Of Fluid model and the mixture model, described further in the section are simplifications of the Eulerian model for continuous-continuous systems and continuous-dispersed systems respectively. Regardless of the assumption, the Eulerian model allows multiple phases to exist within a single cell. To account for this, phase volume fraction of each phase α_q, is considered:

$$V_q = \int \alpha_q dV$$ (9.25)

where all phase volume fractions sum up to unity:

$$\sum_{q-1}^{n} \alpha_q = 1 \qquad (9.26)$$

wherever a cell is occupied by a single phase, its respective value of α is equal to one, while the others' is equal to zero. With thusly defined volume fraction, modified continuity and momentum equations are introduced:

$$\frac{\partial(\alpha_q\rho_q)}{\partial t} + \nabla\left(\alpha_q\rho_q\boldsymbol{U}_q\right) = \sum_{p=1}^{n}\left(\dot{m}_{pq} - \dot{m}_{qp}\right) \qquad (9.27)$$

$$\frac{\partial(\alpha_q\rho_q\boldsymbol{U}_q)}{\partial t} + \nabla\left(\alpha_q\rho_q\boldsymbol{U}_q\boldsymbol{U}_q\right) = -\alpha_q\nabla p + \nabla\tau_q + \alpha_q\rho_q g + \sum_{p=1}^{n}\left(\boldsymbol{D_{pq}} + \dot{m}_{pq}\boldsymbol{U}_{pq} - \dot{m}_{qp}\boldsymbol{U}_{pq}\right) + \boldsymbol{F}$$
$$(9.28)$$

where $\dot{m}$ is mass flux and $\boldsymbol{D}$ is interfacial drag.

These equations account for phase volume fraction, as well as mass and momentum transfer between phases in contact. An important thing to note is that in the momentum equation the element $\boldsymbol{D_{pq}}$ stands for interfacial drag, which is the predominant mechanism of momentum transfer due to difference in phase velocities in cell centroids for phase mixtures. The drag function is dependent on drag coefficient, which can be calculated with several available models.

Since the set of continuity and momentum equations does not distinguish between continuous-continuous and continuous-dispersed phase interactions by itself, an additional condition must be introduced. interface capturing schemes, described further in this section, allow the model to construct sharp interfaces between phases, thus separating regions which now consist of pure, continuous phases outside the interfacial region. The Eulerian model, as well as many other multiphase models, may be also used for fluid-solid interactions (e.g., granular flows), but this issue is not within the scope of RPB simulations, and therefore will not be covered in this chapter. Despite very high accuracy, the complexity of the Eulerian model hinders its application for RPB systems, as very intensive micro-mixing and developed interfaces reduce simulation stability even at very fine meshes.

Volume Of Fluid (VOF) model

The Volume Of Fluid (VOF) method is similar to the Eulerian model, but simplified in order to apply specifically to continuous-continuous flows [27]. Instead of separate momentum equations for each phase, only a single momentum equation is solved for averaged values of fluid properties. The volume fraction equation is analogous to the Eulerian model.

In the VOF method, volume fractions of each secondary phase is calculated with the discrete-time volume fraction equation, which may be formulated in one of two ways, implicit:

$$\frac{\alpha_q^{n+1}\rho_q^{n+1} - \alpha_q^n\rho_q^n}{\Delta t}V + \sum_f \left(\rho_q^{n+1}U_f^{n+1}\alpha_{q,f}^{n+1}\right) = V\sum_{p=1}^n \left(\dot{m}_{pq} - \dot{m}_{qp}\right) \qquad (9.29)$$

or explicit:

$$\frac{\alpha_q^{n+1}\rho_q^{n+1} - \alpha_q^n\rho_q^n}{\Delta t}V + \sum_f \left(\rho_q U_f \alpha_{q,f}\right) = V\sum_{p=1}^n \left(\dot{m}_{pq} - \dot{m}_{qp}\right). \qquad (9.30)$$

It can be noticed that the only difference between the two formulations is in the second expression on the left side of the equation. In the implicit formulation, the volume fraction at the current (n+1) time step depends on the quantities at the same time step, the equation must be solved in an iterative manner. The explicit formulation relies on the quantities of the previous time step, and therefore does not rely on the iterative solution. The time step can be modified with set values of the Courant number. The implicit time formulation is the only of the two which allows for steady-state simulations, as the explicit formulation is strictly time-dependent.

Interface capturing schemes

As was stated before, the main purpose of continuous approach is to determine the interface between the phases in question. There are several methods of the so-called interface capturing, which allow for interpolation of the boundary between the phases in the regions where phase volume fraction values are less than one and larger than zero. There are two general methods of interpolation, determining the interface by mass fluxes within the boundary cells. One, in which the slope between the phases is approximated by piecewise-linear approach, is called the Geometric Reconstruction Scheme [28]. The other approach is called the Donor-Acceptor Scheme [27], which divides the cells close to the interface into donors and acceptors, depending on whether they are the source or the receiver of the phase due to the flux.

Mixture model

Another model that simplifies the full Eulerian approach is the mixture model. This method is used to model only continuous-dispersed interactions, as it is not capable of tracking sharp interfaces. The mixture model uses common expression for momentum and mass continuity solved for the mixture with averaged values of velocity. This means that the phases may interpenetrate freely and that the values of α_p and α_q may vary between 0 and 1 at any point in the control volume.

A distinctive feature of the mixture model is that it allows the dispersed phases to move at different velocities with the use of relative (slip) velocity, which is the difference between the velocities of primary and secondary phase.

$$U_{pq} = U_p - U_q \tag{9.31}$$

The so-called drift velocity is connected to the slip velocity with the following expression:

$$U_{dr,p} = U_{pq} - \sum_{k=1}^{n} c_k U_{kq} \tag{9.32}$$

where c_k is mass fraction of the phase k.

Based on the assumption that local equilibria between phases are reached over short distances, the slip velocity can be formulated with an algebraic function:

$$U_{pq} = \frac{\tau_p}{f_D} \frac{\left(\rho_p - \rho_m\right)}{\rho_m} \boldsymbol{a} - \frac{\eta_t}{0.75} \left(\frac{\nabla\alpha_p}{\alpha_p} - \frac{\nabla\alpha_q}{\alpha_q}\right) \tag{9.33}$$

where τ_p is particle relaxation time, η is turbulent diffusivity coefficient, and **a** is the acceleration calculated for the mixture:

$$\boldsymbol{a} = \boldsymbol{g} - \boldsymbol{U}_m \cdot \nabla \boldsymbol{U}_m - \frac{\partial \boldsymbol{U}_m}{\partial t} \tag{9.34}$$

The second expression on the right side of eq. (9.33) is a diffusion term due to dispersion in turbulent flows, where η_t is turbulent diffusivity. The mixture model allows for setting the slip velocity to zero, which then makes it the homogenous multiphase model. The volume fraction equation for the secondary phase in the mixture model is then defined as follows:

$$\frac{\partial\left(\alpha_p\rho_p\right)}{\partial t} + \nabla\left(\alpha_p\rho_p\boldsymbol{U}_p\right) = -\nabla\left(\alpha_p\rho_p\boldsymbol{U}_{dr,p}\right) + \sum_{q=1}^{n}\left(\dot{m}_{qp} - \dot{m}_{pq}\right) \tag{9.35}$$

A significant advantage of the mixture model is that it can simulate non-Newtonian flows.

Eulerian wall film

The Lagrangian wall film model is a very simple film tracking method based on the discrete phase approach to determining flows of fluid particles (see Section 9.3.3). In order to calculate the structure of the film and the interaction with the gas phase more precisely, the liquid film should rather be treated as a continuous medium, as modeling thin layers of liquid flow on the surface by Lagrange method does not allow to take into account many important interactions between the liquid and the wall that affect the structure of the flowing liquid. At the same time, standard continuous-continuous models, such as full Eulerian or VOF models, often require too much computational power. To meet these demands, the Eulerian Wall Film (EWF) model was developed, which is a middle ground between the methods, as it allows for modeling of finite thickness films on the surfaces of boundary conditions.

The thickness limit of EWF is 500 microns, and the additional assumption is that properties of the liquid are constant as a function of the liquid film thickness. The thickness of the liquid should also be significantly lower than the radius of curvature of the surface on the liquid adheres to, in order to maintain the even thickness of the film. The EWF model has two important limitations: firstly, computational domains can only be three-dimensional, and secondly, the simulations performed must run in transient state.

The formation of a liquid film on the surface may be the result of the collision of DPM droplets, stretching of the flowing liquid from the Volume Of Fluid model, or vapor condensation. The liquid film may also be initiated as an additional domain boundary condition. In the case of film formation from discrete phase droplets, as in the LWF model, the droplets may stick to the surface and merge into a dripping film, but they may also rebound, spread, and splash when the droplet reflects from the surface, fragmenting into smaller droplets. The liquid forming the film and simulated with the EWF model can detach from the surface, turning into discrete phase. This can happen in two cases:
- at the edges of surfaces. The detachment of the film may be conditioned by exceeding the critical values of the Weber number or the surface bend angle. Three models can be used to determine the mass of the stream and the diameters of the tearing droplets:
  - the O'Rourke model, in which the diameter of the droplets is equal to the thickness of the film and the mass stream of the droplets is equal to the stream of the film [29];
  - the Foucart model, in which the droplet diameter is a function of the surface edge length and the mass flux of the droplets equals the mass flux of the particles [30];
  - the Friedrich model, in which the droplet diameter is equal to the height of the liquid film and the mass flux of the droplets is an empirical correlation of the Weber number [31].

- Droplets can also be torn from the liquid film in the event of high tangential stresses between the film and the flowing gas. However, this model requires a large number of dropout breakpoints that can only be determined experimentally.

For a formed two-dimensional film in the three-dimensional domain, the fluid flow is described by the continuity equation:

$$\frac{\partial h}{\partial t} + \nabla(h\boldsymbol{U_l}) = \frac{\dot{m}_s}{\rho_l} \tag{9.36}$$

where $\boldsymbol{U_l}$ is film average velocity and $\dot{m}_s$ is mass source per area unit due to droplet collection, separation or phase change. The modified momentum equation for EWF is as follows:

$$\frac{\partial h U_l}{\partial t} + \nabla(h\boldsymbol{U_l U_l}) = \underbrace{\frac{-h\nabla_s p_L}{\rho_l}}_{pressure\ gradient} + \underbrace{\boldsymbol{g} h}_{gravity\ force} + \underbrace{\frac{3}{2\rho_l}\tau}_{wall\ shearforce}$$

$$\underbrace{-\frac{3U_l}{h}v}_{viscosity\ force} + \underbrace{\frac{\dot{q}_l}{\rho_l}}_{additional\ source\ ptterm} \tag{9.37}$$

The pressure acting on the film is the sum of the static gas pressure and the hydrostatic pressure resulting from the thickness of the liquid film.

The EWF model can be coupled with the VOF model. The film model can turn into full two-phase flow model in the event that the assumption of a thin film is no longer fulfilled (e.g., due to growing liquid layer). In case of simulations in which the fluid flow affects the hydrodynamics of the gas phase flow or the phenomena occurring on the film surface are of importance, it is recommended to use the VOF model. However, such a solution will significantly extend the computation time. Due to the lack of influence of the film on the dynamics of the gas phase (in simulations without mass and heat transfer), it is recommended to calculate the movement of the liquid film based on the unfolded and convergent flow.

Figure 9.4 shows comparison between gravitational wall flow simulated with Eulerian Wall Film and Lagrangian Wall Film models. A clear difference between continuous form of flow of EWF and discrete form of LWF can be seen. The interactions between fluid parcels in LWF are greatly simplified; therefore, this model cannot simulate quantities such as surface tension, which may generate significant errors in the simulated film structure. This is why LWF is more suitable for problems where smaller amounts of separate liquid droplets are tracked on the wall boundary. On the other hand, EWF is fully capable of including internal stresses of the phases and surface tension, which makes it much more suitable for cases where continuous films form on the wall boundary.

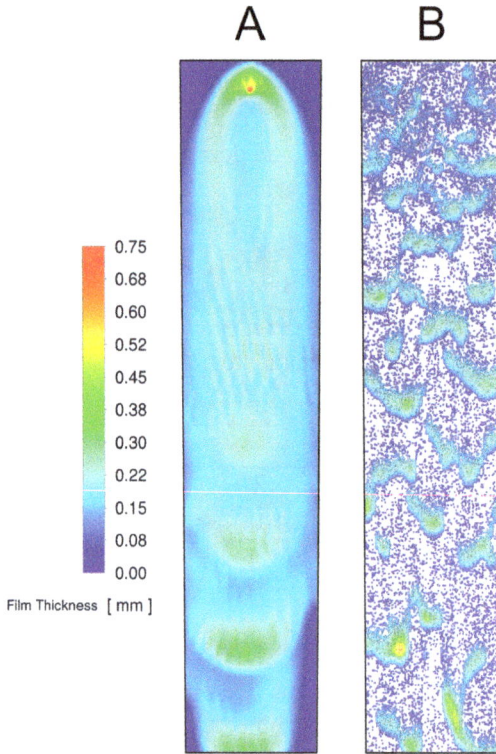

Fig. 9.4: Comparison between Eulerian wall film (A) and Lagrangian wall film (B) simulation of free gravitational wall flow. Colors correspond to film thickness.

9.4 Tools for RPB simulations

9.4.1 Rotating domains

By default, CFD solvers assume inertial reference frame, that is when the domain is considered stationary. However, many simulated processes use systems which involve either moving elements or whole moving domains. With one approach such movement can occur in an inertial reference frame, where the domain or subdomain is observed in motion. Another approach is to consider the process in a noninertial reference frame, where the "observer" moves along with the domain, and adequate force components are added to the momentum equation. Moving reference frames can be rotational as well as translational [32], but due to the scope of this chapter, only the former will be focused here.

When rotational movement is considered, the origin of rotation (point for two-dimensional cases, axis for three-dimensional cases) needs to be specified. At any radius r from that point, additional rotational velocity component is added.

$$U_{rot} = \boldsymbol{\omega_r} \times \boldsymbol{r} \tag{9.38}$$

where ω_r is rotational speed.

The moving reference frame velocity is connected to inertial velocity with the term of reference velocity:

$$\boldsymbol{U_r} = \boldsymbol{U} - \boldsymbol{U_{rot}} \tag{9.39}$$

where U is the absolute velocity in the inertial frame. With this definition, the continuity equation is expressed as follows:

$$\frac{d\rho}{dt} + \nabla(\rho \boldsymbol{U_r}) = 0 \tag{9.40}$$

When it comes to the momentum equations, they may be expressed as functions of relative or absolute velocity, depending on the solver. Equation (9.41) describes momentum equations for relative velocity formulation, with terms accounting for Coriolis forces, centripetal forces, and additional forces resulting from unsteady rotational velocity.

$$\rho\frac{dU_r}{dt} + \nabla(\rho \boldsymbol{U_r U_r}) + \rho \left(\underbrace{2\boldsymbol{\omega_r} \times \boldsymbol{U_r}}_{\substack{Coriolis \\ acceleration}} + \underbrace{\boldsymbol{\omega_r} \times \boldsymbol{\omega_r} \times \boldsymbol{r}}_{\substack{centripetal \\ acceleration}} + \underbrace{\frac{d\omega_r}{dt} \times \boldsymbol{r}}_{\substack{unsteady\,rotation \\ acceleration}} \right) = -\nabla p + \mu\nabla^2 \boldsymbol{U} + \boldsymbol{F} \tag{9.41}$$

The absolute velocity formulation momentum equation is defined as follows:

$$\rho\frac{dU}{dt} + \nabla(\rho \boldsymbol{U U_r}) + \rho \underbrace{(\boldsymbol{\omega_r} \times \boldsymbol{U})}_{\substack{rotational \\ acceleration}} = -\nabla p + \mu\nabla^2 \boldsymbol{U} + \boldsymbol{F} \tag{9.42}$$

where, due to inertial reference frame, Coriolis and centripetal forces are expressed with the use of a single term.

In a simplified approach to RPB process simulations, the internal may be the only element of interest, which has been the case for many studies (see Section 9.6). However, to properly simulate the process, inner and outer cavity zones need to be considered to account for initial liquid distribution as well as splashing and mass transfer in the zone between the internal and the casing. In such cases, a single non-inertial reference frame is not possible, as the internal subdomain rotates around the apparatus' axis, while the cavity zones are not in motion. In such cases the flow domain needs to be split into moving and stationary zones, limited by interface boundaries.

The simplest approach to such problems is the Multiple Reference Frame (MRF) model, sometimes referred to as frozen rotor approach [32]. As the alternative name suggests, it models the flow through all the zones in the domain as if it was frozen in time for the spectator to observe the flow fields. In this method, the domain is separated into rotating and stationary zones, but the whole mesh remains static. The only difference between the sections is that cells which belong to the rotating zones are modified to account for the movement, while the stationary cells solve standard momentum equations. This approach is reliable only in cases where interactions between the rotating and stationary sections are uniform. In other cases, the mixing plane model is suggested, as it turns interfaces between zones into boundary conditions which average the flow field data through the interface to enforce uniformity.

The main advantage of MRF and mixing plane model is that they are relatively simple and computationally cheap. However, such simplification often leads to incorrect simulation results. More importantly, due to "frozen in time" nature of the models, they are applicable only to steady-state approximations, and therefore they cannot be applied to any transient problems. For non-steady simulations in multiple reference frames, CFD software offers the sliding mesh approach.

Sliding meshes are a special type of dynamic meshes, in which sections of cells move, that is, "slide" along a common, rigidly specified interface. The mesh itself does not change its geometry, but the spatial conformation between zones changes with time. This approach therefore not only allows for transient simulations, but explicitly enforces the non-steady mode. The biggest downside to this method is that while one section of the mesh moves in relation to another, the original nodes between neighboring cells at the interface split from each other, leading to non-conformal faces, which often lead to high instabilities in calculations [33]. Figure 9.5 presents an example of a rotational sliding mesh. It can be seen that at the boundary between the stationary and the rotating region the nodes are not conformal due to rotation.

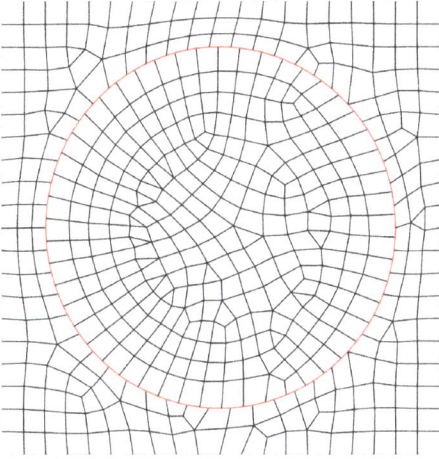

Fig. 9.5: Rotational sliding mesh with marked interface between stationary and rotating regions.

9.4.2 Flows through porous domains

In a traditional approach to CFD, the domain of the flow covers exclusively the zone where only fluids are present. Wherever the flow meets a solid, a wall boundary condition is set, creating an impassable barrier for fluid in the simulation. An exception from that rule would be dispersed granular flows, such as suspensions or dusts, in which solids are treated as a pseudo-fluid phase. Another exception would be situations in which fluids encounter permeable solid structures through which they can flow due to the solid's porosity. With the standard approach, the solid structure would be excluded from the domain, and the porous network would be accordingly meshed. Unfortunately, this method is often tedious, difficult to design, requires extremely fine meshes which slow down the simulations significantly, and often results in cells of poor quality. This issue is very common in rotating packed bed simulation tasks, as most RPBs utilize porous internals, be it metal foams, woven knit mesh, or random packings.

To overcome this problem, CFD software offers a special approach to porous domains. They exist as regular sections of the mesh, but the momentum equations applied to them include additional pressure drop, which is a quadratic function of flow velocity [34]:

$$\Delta p = C_1 \boldsymbol{U} + \frac{1}{2}\rho C_2 \boldsymbol{U}^2. \tag{9.43}$$

However, this element is not added directly to the pressure term in the momentum equation, but rather translates to an additional momentum source term:

$$\frac{\partial \rho U}{\partial t} + \nabla(\rho UU) = -\nabla p + \nabla\left(\mu\left((\nabla U) + (\nabla U)^T\right)\right) - \nabla\left(\frac{2}{3}\mu(\nabla U)\right) + \rho g + S_{por} \qquad (9.44)$$

where

$$S_{por,i} = -\frac{1}{\delta_i}\left(C_1 u_i + \frac{\rho}{2} C_2 |U| u_i\right). \qquad (9.45)$$

In eq. (9.45) the i index corresponds to the direction in n-dimensional geometry. The term needs to be expressed separately for each direction as the cell thickness δ_i may vary. The constants C_1 and C_2 are specific to the type of the porous material and fluid viscosity.

As it may be noticed, the approach to porous zones in CFD at this stage is very simplified, and thus its application to complex tasks, such as RPB simulations, may prove overly limited. For example, for continuous-continuous multiphase flows it neglects a significant number of interfacial effects that occur due to micro-mixing and solid phase vicinity. It is therefore not recommended to follow porous domain approach even when dealing with non-structured, porous internals, as even those require realistic representation of flow zone structure to produce reliable results.

9.4.3 Adaptive mesh refinement

As was stated in Section 9.3.3, the major idea behind the continuous multiphase models is their ability to track interfaces between phases. Therefore, the area in the vicinity of the interface is of particular interest in such simulations. Determination of hydrodynamic parameters of the flows is one of the major reasons for performing RPB process simulations, with phase contact area being one of the most important factors; insufficient mesh resolution may lead to significantly distorted values. On the other hand, strong increment in number of cells in the domain requires additional computational resources, resulting in a tradeoff dilemma. However, many types of CFD software offer a special tool which allows for dynamic adaptation of mesh density by selective refinement or coarsening according to an imposed condition, e.g. proximity to the interface.

The adaptation process can be divided into two steps:
- identification of the cells according to the refinement/coarsening criterion,
- adaptation of the marked cells' geometries.

After each time step, as the flow progresses, the procedure starts over, as the interface subdomain changes, and new cells need to be identified for refinement or coarsening. Several methods of mesh adaptation exist, according to the type of mesh geometry.

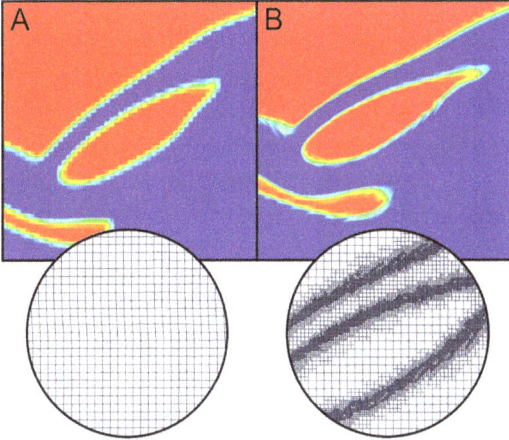

Fig. 9.6: Comparison between VOF model simulations without (A) and with (B) adaptive mesh refinement option with corresponding forms of meshes at a certain time step. Colors correspond to volumetric fraction of the primary phase.

Hanging node adaptation and PUMA

Hanging node adaptation method [35] is simpler and limited to particular types of meshes. It is appropriate for triangular and quadrilateral 2D meshes, as well as 3D meshes consisting of tetrahedral, hexahedral, wedge, or pyramid cells. It is not possible to use this type of adaptation for polyhedral meshes. This method introduces additional nodes or faces, which then split the primary (parent) cells into constitutive secondary (child) cells with predefined geometrical shapes.

Polyhedral Unstructured Mesh Adaptation (PUMA) is available only for adaptation of three-dimensional meshes. It is applicable to all types of 3D cells, including unstructured polyhedra [36]. It requires less dedicated memory for the refinement process compared to hanging node method, but it hinders some of the subsequent mesh processing methods, like dynamic remeshing or layering, or face zone fusion. Due to geometrical dependences between parent and child cells in both hanging node and PUMA method, the parent cell can be coarsened only after all its component child cells have been marked for coarsening. Additionally, only cells that had undergone refinement can be coarsened, so coarsening beyond the original mesh structure is not possible. An example of hanging node adaptation of a quadrilateral mesh is presented in Fig. 9.6.

Geometry-based adaptation

This type of mesh adaptation utilizes a slightly different approach than Hanging node adaptation and PUMA. Instead of slicing the cells to decrease the volume of cells, it focuses rather on reconstructing the polygonal face structure of neighboring cells to create smoother surfaces [37]. By projecting additional nodes onto sharp surfaces according to the desired shape condition, this method increases the number of surface elements, thus improving the approximation of the structure to the smooth reference shape.

9.5 Guidelines for RPB simulations

As can be seen in the above sections, CFD software offers a lot of different tools for simulations of multiphase flows under various conditions. In this section we present overall guidelines for simulations of RPB processes, according to characteristics of types of internals and operating conditions.

9.5.1 Rotation

Depending on the type of internal, different approaches can be taken for successful and reliable modeling of multiphase behavior. For internals which are effectively solids of revolution, that is the ones which are axisymmetric, the rotation of the domain may be simulated by putting the moving wall condition on the wall boundary. It is important to note that the near-wall layer is sufficiently fine($y^+<1$). This method can be carried out in both transient and steady states. For other types of internals, such as those made of porous materials or comprising spirally shaped canals, this approach is not possible. In such cases the sliding mesh approach is necessary with the subdomain slide interface between the stationary and rotating parts. This method does not allow for steady state simulations. Additionally, it is generally less stable, and for sufficient stability time resolution should be high enough for mesh nodes to move no further than half the distance between each other in a single time step.

9.5.2 Initial liquid distribution

Depending on the type of liquid distributor, different multiphase models can be used to simulate the flow in the eye cavity zone of the rotor. Typical full-jet nozzles can be easily simulated with the use of Volume Of Fluid model or the full Eulerian model, as the liquid flows in the form of a cohesive and uniform jet. On the other hand, when flat-fan nozzles are used, it is advised to use the discrete phase model instead, in order to limit the computational load required to track interfaces between high amounts of small fluid elements that Eulerian models would have to produce.

9.5.3 Flow patterns within the internal

The Volume Of Fluid model is the one most commonly used in simulations of hydrodynamics in rotating packed bed contactors [38]. Unfortunately, due to the fact that it calculates one common momentum equation for all phases, it tends to lose stability while simulating counter-current flows. Another problem arises with the choice of boundary conditions for the domain limit which is the gas inlet and liquid outlet at the same time. This issue can be overcome by using a hydraulic lock, for which the hydrostatic pressure is larger than the gas phase pressure drop, but it requires significantly larger computation domain. An alternative approach to this problem is defining the boundary condition as "pressure inlet," as it is the only boundary condition type that allows fluid backflow, and thus allows liquid to exit the computational domain. Unfortunately, this method requires defining the pressure value accordingly to the assumed mass flow of the phase. This, in turn, can be overcome with the use of a proper user-defined function, but it may also lead to destabilization of the simulation [39].

Full Eulerian model seems like a sensible alternative to the VOF model, as it limits its main drawbacks. Since in the Eulerian model the momentum equations are solved for each phase, the simulations, especially in counter-current, tend to be significantly more stable. Additionally, it allows the user to set "degassing" as the boundary condition, which makes it possible only for the secondary phase to escape. Unfortunately, the Eulerian model requires much more computational power than the Volume Of Fluid model, which significantly hinders its application for such complex flows. Computation speed on state-of-the-art hardware is still not high enough for practical uses of the Eulerian model in RPB applications, as the extended simulation times are not sufficiently compensated by the increased accuracy.

Alternatively, a combination of Discrete Phase Model and Eulerian Wall Film model can be utilized (Fig. 9.7). It allows the user to simulate two of three flow patterns described by Burns and Ramshaw [40]. At rotational speeds above 600 RPM, pore flow turns into separate droplet flow, which justifies the use of the coupled model of Eulerian wall film and Discrete Phase Model for the simulation of hydrodynamics within porous structures. Although EWF explicitly requires the use of three-dimensional domains, the mesh may be relatively coarse, as both models do not rely on mesh density for the calculations of dispersed phase. Unfortunately, although this approach lets the user to determine the phase contact area, it does not

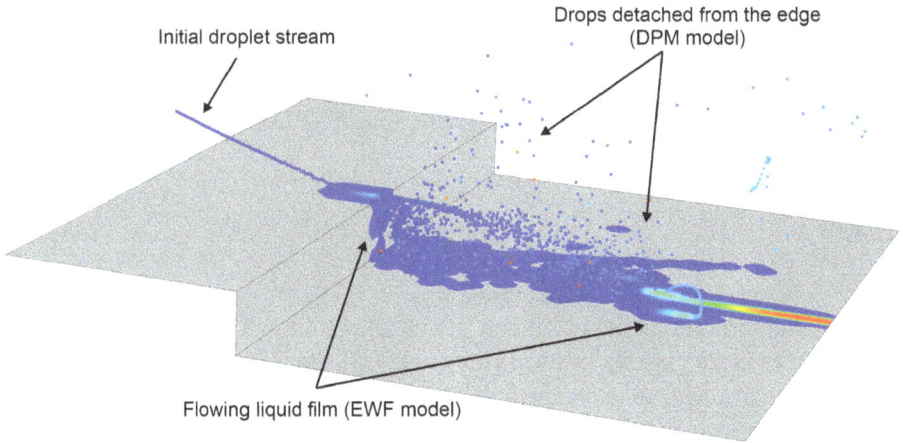

Fig. 9.7: Combination of Eulerian Wall Film and Discrete Phase Model.

allow for addition of any mass transfer modifications to the simulation, as EWF is limited to single-component films only. However, such approach may help with tentative assertions whether particular designs have potential for practical use. For processes limited by phase contact area, this method allows the simulations to give approximate results in a relatively quick time.

9.6 Literature overview

Tab. 9.1: Overview of CFD studies regarding RPB.

Geometry	Flow model	Internal	Turbulence model	Publication
2D	monophasic	woven mesh	k-epsilon	[41]
		porous	various	[42]
	VOF	woven mesh	RSM	[43, 44]
			SST k-omega	[38]
			unspecified	[45]
		porous	k-epsilon realizable	[46, 47]
		perforated plates	RSM	[48]
		coaxial mesh rings	k-epsilon realizable	[49]
	Kołodziej	woven mesh	laminar	[50]
3D	monophasic	porous	RNG k-epsilon	[51]
		woven mesh	k-epsilon realizable	[41, 52]
		random spheres	k-epsilon realizable	[53]
	VOF	woven mesh	SST k-omega	[54]

Despite being reported in literature since 2009 [51], the progress in terms of CFD simulations of RPB processes still has a long way to go. As of now, the strong majority of literature reports concern simplified, two-dimensional geometries, while only four three-dimensional CFD studies are available (see Tab. 9.1). 2D simulations work well as tentative tools providing visualizations of flow patterns, but are limited in terms of useful data for researchers.

The first reports of CFD simulations of RPB equipment were limited to single-phase flow. Since such flows are significantly easier to conduct and much cheaper computationally, three-dimensional models could be used without encountering many problems. Llerena-Chavez & Larachi [51] were the first ones to provide a study of dry pressure drops in an RPB containing a standard porous internal. Monophasic gas flow was simulated and experimentally validated for three different gas inlet setups: tangential, parallel, and perpendicular to the internal. Contributions of different RPB zones (i.e., distributor, rotor, and housing) to overall pressure drop were determined. The simulations were carried out in transient state in three-dimensional geometry and with the use of RNG k-ε turbulence model. A year later Yang et al. [41] simulated a comparative study of steady-state monophasic flows in woven knit mesh internal in full 3D as well as cross-sectional 2D geometry. The results showed that 2D simplification is well justified for monophasic flows, as it showed similar results while significantly reducing the simulation times.

The first successful attempt of liquid flow simulation (there was no gas flow, and the gas phase only filled the domain in the initial condition) was reported in 2013 by Shi et al., [43] where the geometrical model of woven mesh was simplified into 2D geometry as a set of square shaped obstacles. The Volume Of Fluid multiphase model was used with Reynolds Stress model of turbulence. The simulations were run in transient state, and the main focus of study was on determination of liquid residence times within the internal zone. The same methodology was followed by the study of Yang et al. [44], where the simulations were extended by user-defined functions of interfacial mass transfer, in order to simulate the physicochemical process of deaeration. In 2017 Xie et al. [38] published another study with analogous setup for the simulation of flow of monoethanoloamine, which is a standard absorbent of CO_2 used in the carbon capture technology. No mass transfer phenomena were investigated.

Further development of 2D simulations was provided by Ouyang et al. [46] who investigated viscous flows through porous internals using the VOF model and with the additional use of adaptive mesh refinement tools. Simulations were validated with X-ray data showing liquid holdup within the internal. Wu et al. [49] investigated a special-type internal, which consisted of several coaxial rings made of woven mesh, and intermediate liquid distributors. Liquid behavior was investigated via simulations and compared with experimental data provided by high-speed photography. However, both these studies still focused on liquid flow exclusively, and no gas flow was present.

In 2019 Li et al. [48] provided the first full 2D simulation of RPB hydrodynamics with the use of VOF multiphase model. The gas-liquid flow was in countercurrent, and the internal was in the form of coaxial perforated plates, and sliding mesh boundary was set at the limit of the stationary outer cavity zone. The simulations were compared with experimental data provided by fast camera imaging. Zhang et al. [45] used the 2D VOF method to investigate the liquid distribution in woven mesh RPBs. Liquid maldistribution index was defined and examined for a different number of liquid distributor nozzles.

Lu et al. [50] used an unusual approach, in which the liquid flow through wire mesh packing was assumed to be laminar, and the Kołodziej approach was employed to simulate two-phase flows by dividing sections of the internal into wet and dry surfaces. After the initial studies by Llerena-Chavez & Larachi and Yang et al., there was not much interest for three-dimensional simulations of rotating packed beds. In 2017 Liu et al. [52] picked up the topic for simulation of dry liquid drops in RPBs equipped with woven mesh packings and tangential gas inlet. No liquid flow was present, and the study focused on contribution of different zones to overall pressure drop. A similar study was held by Wang et al. [53], who simulated flows through a random sphere packing.

In 2019 Xie et al. [54] published the first CFD study regarding multiphase flow in 3D geometry. VOF multiphase model and k-omega SST turbulence model were employed for simulation of multiphase flow through a woven mesh packing. Due to high computational load of the simulation setup, the whole internal could not be modeled, and the researchers used a representative elementary unit. Simulations were compared with data provided with visual studies. A following study [39] expanded the approach from [54] to a simulation of a full physicochemical process, which was CO_2 absorption in solutions of monoethanoloamine by counter-current phase contact.

9.7 Take-home message

–   Computational fluid dynamics (CFD) is a numerical method of fluid flow simulation. It utilizes discrete domains consisting of finite number of polygonal cells. For each cell a separate set of mass, momentum and (optionally) energy continuity equations is solved according to their boundary conditions.
–   The no-slip wall boundary condition in CFD is satisfied with the use of wall functions, which are based on relations between dimensionless distance from the wall ($y+$) and dimensionless tangential velocity ($U+$).
–   Modeling of turbulence is one of the most complex problems in CFD due to chaotic propagation and irregularity. There are numerous approaches to modeling of turbulence, from accurate but very computationally expensive direct numerical

simulation to simplified Reynolds-averaged methods such as *k*-epsilon and *k*-omega, which are less accurate, but much cheaper in terms of computational resources.

- The Courant number is a dimensionless criterial number which sets boundary of convergence of numerical solutions. CFD tools offer preset and dynamic approaches to the Courant number.
- Multiphase flows can be approached with two main modes: Lagrangian, in which the reference point moves with particular elements of fluids, and Eulerian, where the reference frame is fixed in space.
- The Eulerian approach offers models which describe fluid elements by their independent spatial and temporal coordinates, in a way in which the domain is perceived by an outside observer. The Eulerian models can be continuous-continuous (Full Eulerian, Volume Of Fluid, Eulerian Wall Film) or continuous-dispersed (Mixture Model, special cases of Full Eulerian model).
- In the Lagrangian approach, one of the phases is considered to be continuous, while the other phases are dispersed within the continuous phase as independently tracked points in space. The dispersed elements are considered to be small enough to be treated as having no volume. Lagrangian models include the Dispersed Phase Model and Lagrangian Wall Film.
- Continuous-continuous models are used mostly in cases where the interface between the phases is of interest. When tracking the interface is not important, continuous-dispersed models, such as the Mixture Model, are recommended due to reduced computational times and easier convergence.
- Lagrangian models are usually recommended for highly dispersed systems, such as sprays, vapors, or bubble columns. In RPB processes, especially in the outer cavity zones, very fine droplets of liquid can be found, where dispersed models are particularly useful.
- In simulations of media flows within an RPB internal the interface between the contacting phases is especially important, as its surface determines the mass transfer. Thus, continuous-continuous multiphase models are usually recommended. However, to reliably determine the interfacial area, the mesh cells need to be very fine. For that, CFD software offers adaptive mesh refinement tools.
- Rotating domain simulations (such as RPB internals) in CFD software include additional forces acting on each element of fluid due to the rotational movement. Such forces may include Coriolis acceleration, centripetal acceleration and acceleration due to unsteady rotation for relative velocity formulation, or one centrifugal acceleration term for absolute velocity formulation.
- Since mass transfer in RPB units occurs not only in the rotating internal, but also in the stationary inner and outer cavity zones, relative movement between those domains needs to be taken into consideration. In CFD software this can be achieved with sliding mesh tools.

9.8 Quiz

Question 1. What are the main advantages of using CFD methods in RPB packing design?

Question 2. What does the value of the Courant number tell us about?

Question 3. What is the definition of the y^+ parameter?

Question 4. What is the main difference between the Eulerian and Lagrangian phases?

Question 5. What are the four main types of particle interactions in DPM with a wall boundary condition?

Question 6. What is the main advantage of the Multiphase Volume Of Fluid (VOF) model?

Question 7. What is the maximum permissible thickness of the liquid layer that allows the use of the Euler liquid film model?

Question 8. What are the local adaptive computational mesh refinements used for?

Question 9. What is the major limitation factor of using the Polyhedral Unstructured Mesh Adaptation (PUMA) method?

Question 10. Is it possible to perform CFD simulations with the use of several multiphase flow models simultaneously?

Question 11. Can experiments be replaced by CFD simulations?

Question 12. Which of the Eulerian models uses shared momentum equations for all phases?

Question 13. Why is y+ an important parameter?

9.9 Answers

Answer 1. Possibility to optimize the shape of the packing without the need for costly production for testing purposes.

Answer 2. It tells us how many computational grid cells have moved the fluid front in a given time step.

Answer 3. It is the non-dimensional wall distance for a wall-bounded fluid flow.

Answer 4. A computational mesh is needed to calculate the Euler phase. The Lagrangian phase can be represented in the form of material points.

Answer 5. Escape, trap, reflect, and liquid film.

Answer 6. The possibility of recreating the interphase surface.

Answer 7. The liquid layer should not exceed 500 microns.

Answer 8. In order to smoothen the results in areas of occurrence of large value gradients, calculated, for example, on the surface of interfacial contact.

Answer 9. It can only be used for 3D meshes.

Answer 10. Yes. The DPM, VOF, and EWF models can work together to describe the different nature of multiphase flow.

Answer 11. No, CFD simulations must be validated experimentally.

Answer 12. VOF.

Answer 13. By knowing the value of the y+ parameter it is possible to determine the way in which the wall layer is modeled (directly or with a function), which significantly affects the quality of the results obtained.

Nomenclature

Note: bold characters represent n-dimensional vector or matrix variables

Latin letters

a	acceleration vector	$[m/s^2]$
A	area	$[m^2]$
C	constant	[var.]
Co	Courant number	[-]
c	mass fraction	[-]
D	interfacial drag	[N]
g	gravitational acceleration	$[m/s^2]$
h	Film thickness	[m]
F	force	[N]
F_1	blending function	[-]
F_g	gravitational force	[N]
F_b	body force vector	[N]
F_v	viscosity force vector	[N]
f_D	drag coefficient	[-]

I	identity matrix	[-]
k	turbulent kinetic energy	$[m^2/s^2]$
m	mass	[kg]
$\dot{m}$	mass flux	[kg/s]
$\hat{m}$	unit vector in direction of liquid film flow velocity	[-]
$\hat{n}$	normal component unit vector	[-]
P	production	[var.]
p	pressure	[Pa]
S	sink term	[var.]
t	time	[s]
U	velocity	[m/s]
u	velocity component in x direction	[m/s]
V	volume	$[m^3]$
v	velocity component in y direction	[m/s]
w	velocity component in z direction	[m/s]
x	distance	[m]

Greek letters

α	volumetric phase fraction	[-]
β, σ	closure coefficient	[var.]
δ	cell thickness	[m]
δ_L	boundary layer thickness	[m]
ε	turbulence dissipation rate	$[m^2/s^3]$
η	turbulent diffusivity coefficient	[-]
μ	dynamic viscosity	[Pa s]
v	kinematic viscosity	$[m^2/s]$
ρ	density	$[kg/m^3]$
τ	stress tensor	[Pa]
τ_p	particle relaxation time	[s]
ω	specific turbulence dissipation rate	[1/s]
ω_r	rotational speed	[rad/s]

Indexes

f	face
q	primary phase
p	secondary phase

References

[1]   Eckert M. The dawn of fluid dynamics: a discipline between science and technology. John Wiley & Sons; 2007.

[2]   Dixon SL, Hall C. Fluid mechanics and thermodynamics of turbomachinery. Butterworth-Heinemann; 2013.

[3]   Tu J, Yeoh GH, Liu C. Computational fluid dynamics: A practical approach. Butterworth-Heinemann; 2018.

[4]   Milne-Thomson LM. Theoretical aerodynamics. Courier Corporation; 1973.

[5]   Tryggvason G. Computational fluid dynamics. Evolution. 2017;1:2.

[6]   Raase S, Nordström T. On the use of a many-core processor for computational fluid dynamics simulations. Proc. Comput. Sci. 2015;51:1403–12.

[7]   Liu H-S, Lin -C-C, Wu S-C, Hsu H-W. Characteristics of a rotating packed bed. Ind. Eng. Chem. Res. 1996;35:3590–96.

[8]   Szafran RG, Kmiec A. CFD modeling of heat and mass transfer in a spouted bed dryer. Ind. Eng. Chem. Res. 2004;43:1113–24.

[9]   Perry RH, Green DW, Maloney JO. Perry's chemical engineers' handbook (ed.), Seventh, International Edition; 1997.

[10]   Constantin P, Foias C. Navier-stokes equations. University of Chicago Press; 2020.

[11]   Richardson S. On the no-slip boundary condition. J. Fluid. Mech. 1973;59:707–19.

[12]   Moser RD, Kim J, Mansour NN. Direct numerical simulation of turbulent channel flow up to Re τ= 590. Phys. Fluid. 1999;11:943–45.

[13]   Launder BE, Spalding DB. The Numerical Computation of Turbulent Flows. In: Numerical prediction of flow, heat transfer, turbulence and combustion. Elsevier; 1983, 96–116.

[14]   Moin P, Mahesh K. Direct numerical simulation: A tool in turbulence research. Annu. Rev. Fluid Mech. 1998;30:539–78.

[15]   Alfonsi G. Reynolds-averaged Navier–Stokes equations for turbulence modeling. Appl. Mech. Rev. 2009;62.

[16]   Schmitt FG. About Boussinesq's turbulent viscosity hypothesis: Historical remarks and a direct evaluation of its validity. CR Mécanique. 2007;335:617–27.

[17]   Launder BE, Spalding DB, Lectures in mathematical models of turbulence, 1972.

[18]   Bernard PS. Limitations of the near-wall k-epsilon turbulence model. AIAA J. 1986;24:619–22.

[19]   Menter FR. Influence of freestream values on k-omega turbulence model predictions. AIAA J. 1992;30:1657–59.

[20]   Ferziger JH. Large Eddy Simulation. In: Simulation and modeling of turbulent flows. Oxford University Press; 1996.

[21]   Spalart PR. Detached-eddy simulation. Annu. Rev. Fluid Mech. 2009;41:181–202.

[22]   Courant R, Friedrichs K, Lewy H. Über die partiellen Differenzengleichungen der mathematischen Physik. Math. Ann. 1928;100:32–74.

[23]   O'Rourke PJ. Collective drop effects on vaporizing liquid sprays. NM (USA): Los Alamos National Lab; 1981.

[24]   Wakeman T, Tabakoff W, Measured particle rebound characteristics useful for erosion prediction, in: Turbo Expo: Power for Land, Sea, and Air, American Society of Mechanical Engineers, 1982, V003T05A005.

[25]   Stanton DW, Rutland CJ. Modeling fuel film formation and wall interaction in diesel engines. SAE Trans. 1996:808–24.

[26]   Kuhnke D. Spray/wall interaction modelling by dimensionless data analysis. Shaker; 2004.

[27]   Hirt CW, Nichols BD. Volume of fluid (VOF) method for the dynamics of free boundaries. J. Comput. Phys. 1981;39:201–25.

[28] Youngs DL. Time-dependent multi-material flow with large fluid distortion. Numer. Methods Fluids Dyn. 1982.

[29] O'rourke PJ, Amsden AA. A particle numerical model for wall film dynamics in port-injected engines. SAE Trans. 1996:2000–13.

[30] Fox RO. Computational models for turbulent reacting flows. Cambridge university press; 2003.

[31] Friedrich MA, Lan H, Wegener JL, Drallmeier JA, Armaly BF. A separation criterion with experimental validation for shear-driven films in separated flows. J. Fluids Eng. 2008;130.

[32] Ansys I. ANSYS fluent theory guide. Canonsburg, PA, USA: Ansys, Inc; 2020.

[33] Uriegas AC, CFD investigation on sliding mesh as a method to model wheel rotation-implementation and analysis on different rims, 2018.

[34] Feichtner A, Mackay E, Tabor G, Thies PR, Johanning L, Ning D. Using a porous-media approach for CFD modelling of wave interaction with thin perforated structures. J. Ocean Eng. Mar. Energy. 2021;7:1–23.

[35] Zhang S, Liu J, Chen Y-S, Wang T-S. Adaptation for Hybrid Unstructured Grid with Hanging Node Method. In: 15th AIAA computational fluid dynamics conference; 2001, 2657.

[36] Jasak H, Tukovic Z. Automatic mesh motion for the unstructured finite volume method. Trans. FAMENA. 2006;30:1–20.

[37] Kamkar SJ. Mesh adaption strategies for vortex-dominated flows. Stanford University; 2011.

[38] Xie P, Lu X, Yang X, Ingham D, Ma L, Pourkashanian M. Characteristics of liquid flow in a rotating packed bed for CO2 capture: A CFD analysis. Chem. Eng. Sci. 2017;172:216–29. https://doi.org/10.1016/j.ces.2017.06.040.

[39] Lu X, Xie P, Ingham DB, Ma L, Pourkashanian M. Modelling of CO2 absorption in a rotating packed bed using an Eulerian porous media approach. Chem. Eng. Sci. 2019;199:302–18. https://doi.org/10.1016/j.ces.2019.01.029.

[40] Burns JR, Ramshaw C. Process intensification: Visual study of liquid maldistribution in rotating packed beds. Chem. Eng. Sci. 1996;51:1347–52. https://doi.org/10.1016/0009-2509 (95)00367-3.

[41] Yang W, Wang Y, Chen J, Fei W. Computational fluid dynamic simulation of fluid flow in a rotating packed bed. Chem. Eng. J. 2010;156:582–87. https://doi.org/10.1016/j.cej.2009.04.013.

[42] Ouyang Y, Tang KL, Xiang Y, Zou HK, Chu GW, Agarwal RK, Chen JF. Evaluation of various turbulence models for numerical simulation of a multiphase system in a rotating packed bed. Comput. Fluids. 2019;194:104296. https://doi.org/10.1016/j.compfluid.2019.104296.

[43] Shi X, Xiang Y, Wen L-X, Chen J-F. CFD analysis of liquid phase flow in a rotating packed bed reactor. Chem. Eng. J. 2013;228:1040–49. https://doi.org/10.1016/j.cej.2013.05.081.

[44] Yang Y, Xiang Y, Chu G, Zou H, Sun B, Arowo M, Chen J-F-F. CFD modeling of gas–liquid mass transfer process in a rotating packed bed. Chem. Eng. J. 2016;294:111–21. https://doi.org/10.1016/j.cej.2016.02.054.

[45] Zhang W, Xie P, Li Y, Teng L, Zhu J. CFD analysis of the hydrodynamic characteristics in a rotating packed bed with multi-nozzles. Chem. Eng. Process. 2020;158:108107. https://doi.org/10.1016/j.cep.2020.108107.

[46] Ouyang Y, Zou H-K, Gao X-Y, Chu G-W, Xiang Y, Chen J-F. Computational fluid dynamics modeling of viscous liquid flow characteristics and end effect in rotating packed bed. Chem. Eng. Process. 2018;123:185–94. https://doi.org/10.1016/j.cep.2017.09.005.

[47] Ouyang Y, Xiang Y, Gao X-Y-Y, Zou H-K-K, Chu G-W-W, Agarwal RK, Chen J-F-F. Micromixing efficiency optimization of the premixer of a rotating packed bed by CFD. Chem. Eng. Process. 2019;142:107543. https://doi.org/10.1016/j.cep.2019.107543.

[48] Li H, Yuan Z, Liu Y, Liu H. Characteristics of liquid flow in a countercurrent rotating bed. Chem. Eng. Process. 2019;136:72–81. https://doi.org/10.1016/j.cep.2018.12.004.

[49] Wu W, Luo Y, Chu G-W-W, Su M-J-J, Cai Y, Zou H-K-K, Chen J-F-F. Liquid flow behavior in a multiliquid-inlet rotating packed bed reactor with three-dimensional printed packing. Chem. Eng. J. 2020;386:121537. https://doi.org/10.1016/j.cej.2019.04.117.

[50] Lu X, Xie P, Ingham DB, Ma L, Pourkashanian M. A porous media model for CFD simulations of gas-liquid two-phase flow in rotating packed beds. Chem. Eng. Sci. 2018;189:123–34. https://doi.org/10.1016/j.ces.2018.04.074.

[51] Llerena-Chavez H, Larachi F. Analysis of flow in rotating packed beds via CFD simulations-Dry pressure drop and gas flow maldistribution. Chem. Eng. Sci. 2009;64:2113–26. https://doi.org/10.1016/j.ces.2009.01.019.

[52] Liu Y, Luo Y, Chu G-W, Luo J-Z, Arowo M, Chen J-F. 3D numerical simulation of a rotating packed bed with structured stainless steel wire mesh packing. Chem. Eng. Sci. 2017;170:365–77. https://doi.org/10.1016/j.ces.2017.01.033.

[53] Wang J-Q-Q, Ouyang Y, Li W-L-L, Esmaeili A, Xiang Y, Chen J-F-F. CFD analysis of gas flow characteristics in a rotating packed bed with randomly arranged spherical packing. Chem. Eng. J. 2020;385:123812. https://doi.org/10.1016/j.cej.2019.123812.

[54] Xie P, Lu X, Ding H, Yang X, Ingham D, Ma L, Pourkashanian M. A mesoscale 3D CFD analysis of the liquid flow in a rotating packed bed. Chem. Eng. Sci. 2019;199:528–45. https://doi.org/10.1016/j.ces.2019.01.038.

Index

https://doi.org/10.1515/9783110724998-010